Contents

CONCLUSIONS AND SUMMARY

Page

1.	Introduction	1
2.	Changing Climate and Sea Level	3
3.	Soils	37
4.	Flora, Fauna and Landscape	53
5.	Agriculture, Horticulture and Aquaculture	67
6.	Forestry	81
7.	Water Resources	93
8.	Energy	109
9.	Minerals Extraction	121
10.	Manufacturing, Retailing and Service Industries	135
11.	Construction Industry	147
12.	Transport	159
13.	Insurance	169
14.	Financial Sector	179
15.	Health	189
16.	Recreation and Tourism	199
17.	Coastal Regions	211

References   227

ANNEX 1 Glossary   241

ANNEX 2 Acronyms   243

ANNEX 3 Membership of Climate Change Impacts Review Group   245

ANNEX 4 List of persons consulted   247

United Kingdom Climate Change Impacts Review Group

M.L. Parry *Chairman*	University College London
N. Arnell	University of Southampton
P. Bullock	Soil Survey and Land Research Centre
M.G.R. Cannell	Institute of Terrestrial Ecology (Natural Environmental Research Council)
A.F. Dlugolecki	General Accident Fire and Life Assurance Corporation p.l.c.
M. Hulme	University of East Anglia
A.J. McMichael	London School of Hygiene and Tropical Medicine
D. Malcolm	University of Edinburgh
J. Page	University of Sheffield
J. Palutikof	University of East Anglia
A. Perry	University of Wales, Swansea
C. Pollock	Institute of Grassland and Environmental Research
D. Potts	University of Nottingham
J.F. Skea	University of Sussex
K. Smith	University of Stirling
H. Thompson	National Westminster Bank p.l.c.
M. Tight	University of Leeds
K. Turner	University of East Anglia
P. Bramwell *Executive Secretary*	Department of the Environment
S.M. Cayless *Executive Secretary*	Building Research Establishment
C.J. Parry *Technical Secretary*	University College London

The authors accept liability for the content of this report but the views expressed are their own and not necessarily those of the organisations to which they belong or the Department of the Environment.

Review of the Potential Effects of Climate Change in the United Kingdom

United Kingdom
Climate Change Impacts Review Group

Second Report

Prepared at the request of the
Department of the Environment
March 1996
LONDON: HMSO

© Crown copyright 1996

Applications for reproduction should be made
to HMSO's Copyright Unit, Norwich, NR3 1BQ

ISBN 0 11 7532908 ✓

Recycled Paper

Designed by DDP Services.
Printed in Great Britain on recycled paper.
B6544. May 1996.

The front cover shows a mosaic of infrared images of the UK taken by the two Along Track Scanning Radiometer (ATSR) instruments, designed by Rutherford Appleton Laboratory and operating in orbit at height of 800km. The main purpose of both these instruments is to monitor sea surface temperature at a resolution of 1km across the globe. The second instrument is also capable of making observations of vegetation cover and state.

The mosaic includes (clockwise from top): cloud temperature variations during a storm over the North Sea, a night-time thermal image of the English Channel, vegetation patterns over western Britain during the hot summer of 1995 (including some visible radiation data), and a day-time thermal image of Ireland.

© 1995 CCLRC/ESA/BNSC/NERC

Conclusions and Summary

PURPOSE OF THIS REPORT

1. The Climate Change Impacts Review Group (CCIRG) was formed by the Department of the Environment to consider the potential impacts of climate change in the United Kingdom, and to establish what further research is needed. Its First Report **The Potential Effects of Climate Change in the United Kingdom** was published in 1991 (CCIRG, 1991).

2. This second report is, firstly, an updated assessment of the potential impacts of climate change and, secondly, a new evaluation of possible adaptive responses, both at the local and national level. It does not consider policies to reduce greenhouse gas emissions, or their environmental and socio-economic consequences.

3. As in its First Report, the Review Group considers those impacts estimated to occur under a change of climate assuming rates of increase in global greenhouse gas emissions consistent with the projections of the Intergovernmental Panel on Climate Change (IPCC).

4. This review is a synthesis of current knowledge. There are major uncertainties concerning possible greenhouse gas-induced climate change and associated impacts and responses. As more information becomes available, such as the effect of sulphur emissions, further assessments of this kind will be needed.

OVERALL CONCLUSIONS

The Changing Climate of the UK

1. There is mounting evidence that global climate is changing as a result of human activities. As stated recently by the IPCC the global warming of the last 100 years is unlikely to be due entirely to natural causes, stating that 'the balance of evidence suggests there is a discernible human influence on the global climate' (IPCC, 1996a). The cluster of warm months and seasons experienced in the UK in the last few years is consistent with this picture of a changing global climate. For example, the 1985-94 decade, both globally and for the UK, has been about 0.2°C warmer than the average of the 1961-90 period and during this decade the average global atmospheric CO_2 concentration has risen by about 5%.

This review defines a scenario [1] (termed the 1996 CCIRG scenario) of changing climate and sea level for the UK for the decades of the 2020s and 2050s assuming that no major global policies of reducing greenhouse gas emissions scenario are enacted. The effects of sulphate aerosols on future climate are not included in this scenario, (they might be expected to reduce temperature by a few tenths of a degree and modify rainfall patterns) although the report includes some examples of their possible effects.[2] Under the 1996 CCIRG scenario of the UK:

- Temperatures are expected to increase at a rate of about 0.2°C per decade, with slightly

---

[1] The term 'scenario' is used to define a description of future climate which could occur given the assumptions made about the evolution of global population, economic and technological developments, the climate system, etc.. It is not a forecast or prediction in the sense that specific probabilities can be attached to the outcome. The 1996 CCIRG scenario is consistent with the science of climate change reported in the 1996 IPCC report and is based on results from a climate change experiment completed by the UK Hadley Centre which is one of the leading centres for climate modelling in the world.

[2] Assessments of impacts under scenarios which include sulphate aerosols have not yet been made or published and remain the subject of future research.

slower rates of increase over the northwestern UK compared to the southeast and in winter compared to summer. They will be about 0.9°C warmer than the average of 1961-90 by the 2020s, and about 1.6°C warmer by the 2050s.

- Extremely warm seasons and years are expected to occur more frequently.
- Annual precipitation over the UK as a whole is expected to increase by about 5% by the 2020s and by nearly 10% by the 2050s. Winter precipitation increases everywhere, but more substantially over the southern UK. Summer precipitation decreases over the southern UK, but increases in the north.[3]
- The number of raindays and the average intensity of precipitation are expected to increase slightly.
- Average seasonal wind speeds are expected to increase over most of the country.
- Potential evapotranspiration (PE) is expected to increase over most of the UK in both winter and summer, the exception being the north of Scotland where decreases are expected to occur.
- Sea level is expected to rise at the rate of about 5cm per decade. This is likely to be exacerbated in the southern and eastern UK by sinking land and mitigated in the north by rising land.

2. Climate change may already be occurring and having a measurable effect on climate statistics. To illustrate, of the five warmest years in Central England's 337 year-old temperature record, three (1989, 1990, 1995) have occurred in the past 10 years. The summer of 1976 was the warmest ever, and that of 1995 the second warmest.

By the 2020s[4] the climate of the UK, under the 1996 CCIRG scenario, is estimated to be:

- About 1°C warmer and 5% wetter than the period 1961-90. This temperature change is equivalent to about a 200km northward shift of UK climate along a SE-NW gradient in the UK - the difference in current temperature between Oxford and Manchester.
- More geographically contrasted. The currently dry southeast will tend to become drier and the moist northwest will become wetter. Drought in the southeast and flooding in the northwest might both become more common.

By the 2050s[5] the climate of the UK, under the 1996 CCIRG scenario, is estimated to be:

- About 1.5°C warmer and 8% wetter than the period 1961-90. Average sea levels will be about 35cm higher than during the period 1961-90, and the probability of a storm surge exceeding a given threshold will have increased.
- More subject to intense precipitation events and extreme windspeeds, especially over the northern UK. General gale frequencies over the country will have increased by up to 30%.

The Likely Effects of a Changing Climate in the UK

The effects of climate change in the UK will be very mixed, with both losses and gains to the natural resource base that vary from region to region, from sector to sector and within sectors. In summary, and in very general terms, we may expect that:

---

[3] A lower degree of confidence is attached to estimates for precipitation (particularly for its regional pattern) compared with those for temperature (see page xi).

[4] 2020s = 2020-29

[5] 2050s = 2050-59

- Beneficial effects will be experienced in forestry, in some forms of agriculture (particularly pastoral farming in the northwestern half of the UK), in tourism and recreation. More adverse effects are likely on soils, wildlife, and water resources, on arable agriculture in the southern half of the UK, the insurance sector and on human health.

- Mixed effects, neither strongly net beneficial or adverse, are expected to characterise the following sectors: energy, minerals extraction, construction, transport, manufacturing and finance.

- More specifically, the following may be the most pronounced effects:

  - An increase in soil droughtiness, soil erosion and the shrinkage of clayey soils.

  - A northward shift of natural habitats, wildlife species and farming zones by about 200-300km per °C of warming or 50-80km per decade.

  - An increase in animal (especially insect) species as a result of northward migration from the Continent, and a small decrease in the number of plant species, due to loss of northern and montane types.

  - A decrease in crop yields in the southeast of the UK, with increased opportunities for both annual and perennial crops in the north and west.

  - An increase in timber yields (up to 25% by the 2050s), especially in the north of the UK (with perhaps some decrease in the south).

  - An increase in river flow in the winter, and a decrease in the summer, particularly in the south.

  - An increase in public and agricultural demand for water.

  - Enhanced potential for tourism and recreation as a result of increased temperatures and reduced precipitation in summer, especially in the southern UK.

  - Increased damage effects of increased storminess, flooding and erosion on natural and human resource assets in coastal zones.

  - Increased incidence of certain infectious diseases in humans and of the health effects of episodes of extreme temperatures.

The Capacity to Respond

All systems and sectors in the UK embody, to a varying degree, mechanisms that serve to reduce perturbations stemming from the variability of weather, and these generally can provide some resilience to possible changes of climate. Such resilience varies considerably, however, from sector to sector. It is generally high in secondary and tertiary sectors such as industry, transport and finance, but lower in primary ones such as forestry, agriculture and natural ecosystems.

It is possible to distinguish between two types of response to climate, in line with distinctions drawn by the IPCC:

- **Autonomous adjustments** are those inherent to natural or social systems which serve to respond to changing conditions. For example, migration is an autonomous adjustment by organisms to change and markets allow adjustment in economic systems. In virtually all managed systems, operating procedures are changed as new information or conditions appear: climate change is simply one extra factor, albeit one with a different degree of uncertainty.

- **Adaptation** is a deliberate and explicit response to climate change, and occurs only in managed and social systems. It may be policy-driven. An example is the revision of building design codes which are currently based on experience of past weather but which may not reflect altered frequencies and magnitudes of damaging weather events such as storms and floods which may occur as a result of climate change. Others are coastal defences and river flood protection schemes (such as the Thames Barrier).

What Would be a Significant Climate Change for the UK?

It follows from the above that the degree of impact and response implied by a given change of climate depends on the resilience of the exposed system. Our imperfect knowledge of that resilience allows only a preliminary statement of what might constitute a significant climate change in terms of potential impacts in the UK.

The following summarises information in this Report for different increments of average annual temperature rise under the CCIRG scenario:

- **Under scenario of a 0.5°C temperature increase (which is projected to occur in the 2000s):** Summer and winter precipitation increases in the NW UK by 2-3%; summer precipitation decreases in the SE UK by 2-3%. This implies that annual runoff in southern UK decreases by 5%; the frequency of 1995-type summer increases from (currently) 1 year in 90, to 1 year in 25; disappearance from the British Isles of a few 'niche' species (e.g. alpine-woodsia fern; tufted saxifrage; hart's tongue fern; oak fern); in-migration of some continental species (e.g. bearded fescue; American duckweed) and expansion of some species (e.g. red admiral and painted lady butterflies; Dartford warbler); increase in overall UK timber productivity of 3%; increase in demand for irrigation water by 21% over the increase without climate change, and in domestic demand by an additional 2%; decrease in space heating demand by 6%.

- **Under scenario of a 1.0°C temperature increase (which is projected to occur in the 2020s):** Summer and winter precipitation increases in the NW UK by about 4%; summer precipitation decreases in the SE UK by about 5%. This implies that annual runoff in southern UK decreases by 10%; frequency of 1995-type summer increases from (currently) 1 year in 90, to 1 year in 10; disappearance from the British Isles of some species (e.g. ptarmigan; mountain hare; alpine sawflies); expansion of range of most butterflies and moths and of birds such as pintail, goldeneye and redwing; increase in overall UK timber productivity of 7%; increase in demand for irrigation water by 42% over the increase without climate change, and in domestic demand by an additional 5%; decrease in space heating demand by 11%.

- **Under scenario of a 1.5°C temperature increase (which is projected to occur in the 2050s):** Summer and winter precipitation increases in the NW UK by about 7%; summer precipitation decreases in the SE UK by 7-8%. This implies that annual runoff in southern UK decreases by 15%; frequency of 1995-type summer increases from (currently) 1 year in 90, to 1 year in 3; further disappearance from the British Isles of several species; in-migration of several species; increase in overall UK timber productivity of 15%; increase in demand for irrigation water by 63%, over the increase without climate change, and in domestic demand by an additional 7%; decrease in space heating demand by 16%.

Assessment of the net impact of these various effects on different sectors in the UK would involve a complicated evaluation of a mix of natural productivity, economic and social welfare changes. Such an assessment is both beyond the scope of this report and the available published information.

Climate Change and Sustainable Development

The Review Group concludes that, under the 1996 CCIRG scenario, changes in climate may lead to significant impacts in the UK and may require explicit adaptation policies and measures in certain sectors and areas. However, in most instances other 'drivers' of change, such as developments in technology and patterns of demand are and will continue to be more powerful instruments of change. Both climatic and non-climatic drivers have significant global dimensions which are experienced in the UK through trading relationships and through participation in climate-sensitive activities overseas, for example, insurance operations. Policies for climate change, to be effective, need to take account of this wider context and° be integrated into related policy decisions including those concerned with sustainable development.

RECOMMENDATIONS

A. Implications for Policy Development

1. General Implications

i) Although the impacts of climate change in the UK are only partly understood, their associated risk could be reduced by i) formulating adaptive strategies to minimise potential negative effects and take advantage of positive ones, and ii) developing mitigative strategies to abate greenhouse gas emissions. Policies of sustainable development require a balance between such adaptive and mitigative strategies, their timing and their cost.

ii) The development of policies to adapt to climate change, both at the international, national and regional level, requires improved means of communicating current knowledge to stakeholders (such as government departments, non-governmental organisations, local authorities, the corporate sector and the consumer). This knowledge includes both up-to-date information on likely magnitudes and rates of climate change, both in the UK and worldwide, on their implied effects and on the array of potentially appropriate adaptive responses.

iii) Information on the potential impacts from and adaptations to climate change requires improved access to data. While the availability of climate change data has been improved (for example, via the Department of the Environment's Climate Change LINK project), current meteorological and other environmental data remain relatively inaccessible and are often in forms inappropriate for impact assessment. This latter situation has not improved substantially since the 1991 CCIRG Report.

iv) Careful consideration should be given to the selection of the baseline climate assumed in estimating the availability of and demand for utilities such as water, gas and electricity. Climate assumptions have implications for pricing regimes and service standards set by regulators. The setting of prices and service standards is of critical importance to consumers and supply companies.

2. Specific Implications

i) The projected adverse impacts of climate change on soils will affect future land use and biodiversity in the UK and this will need to be considered in forward planning by government departments, non-government organisations and others with responsibility for land. Overall attention will need to be given to the development of a range of soil management techniques capable of minimising these impacts. This will involve reducing the emissions from soils, an important source of greenhouse gases.

ii) Current conservation strategy in the UK is to protect areas of conservation value, often in isolated sites, against human interference. The challenge for the future is to develop a conservation strategy which recognizes that species distribution will change as climate changes, that many existing nature reserves may become less valuable as species die out, and that there may be a requirement to assist species spread by relocating them or by providing habitat corridors.

iii) The agricultural sector currently faces a reduction in the level of subsidy and a re-allocation towards the support of environmental goods. The likely impact of climate change on agricultural systems, natural and semi-natural ecosystems and other elements of the rural economy should be considered when changes in support systems are considered. For example, support payments could assist the development of wildlife corridors and, through the planting of appropriate tree, shrub and grass species could promote more effective catchment-area water management.

iv) Forests and short rotation biomass plantations play an important role in sequestering carbon, reducing reliance on fossil fuel and offsetting imports of wood products. Expansion of these land uses could provide wildlife habitat, migration routes for species dispersal, additional rural recreation and employment opportunities and make use of surplus agricultural land. The long time scale (50 years) required to achieve goals in forestry, means that policy decisions have to be made now to meet the impacts of projected 2050 regional climates. These decisions include selection of adapted tree species and populations together with the appropriate incentives for changes in land use.

v) Water resource planning and investment, over the medium and long term (more than 10 years), should consider explicitly the potential effects of climate change, alongside all the other factors already considered. Water management schemes should be capable of coping with credible future climates, or could be readily altered to do so.

vi) Present practices for the design, inspection and monitoring of buildings, minerals extraction, water and transport structures (which have a long lifetime) are based upon an understanding of current levels of risk related to weather and coastal flooding. These may need to be reviewed in the context of climate change, in particular, design codes and standards may need to be revised to account for the effect of climate change in altering levels of future risk.

vii) In certain sectors, where government fulfils an important role in guiding the development of infrastructure and where planning horizons are lengthy, adaptation to climate change may require action by government (both national and local). This could, for example, be needed in the provision of guidance on future land-use and structure planning for housing and water supply, in the transport sector, for the minerals extraction industry and, more generally, for all assets requiring coastal protection.

viii) The insurance industry should continue to co-operate with other stakeholders in the property market to reduce society's vulnerability to extreme events, by assembling and providing information on damage and best practice to assist adaptive measures such as improved construction and land development. Property owners should be encouraged to mitigate their risks through appropriate pricing, deductibles, and other insurance policy conditions. Insurers should regularly review their capabilities to handle extreme events and promote the adoption of relevant disaster plans by clients and suppliers. Legislation to allow the creation of tax-allowable catastrophe reserves is likely to be concluded soon, and insurers should integrate these into their overall plans to manage their exposure to climate risks.

ix) All parts of the financial sector should co-operate in the development of accurate and timely information on the climate risks associated with activities conducted by its clients. Reflecting climate risks in the pricing of capital will generate economic signals which would facilitate adaptation to climate change. Closer cooperation with the insurance industry would help bring this about.

x) In order to reduce the potentially negative effects of climate change on the health of the UK population, a number of health-protective measures should be introduced or strengthened (e.g. weather warnings, air pollution abatement, assurance of drinking water quality and food hygiene, appropriate vaccination, and disease surveillance). Appropriate health-care facilities and professional training would be required.

xi) The potential increase in overall tourist activity within the UK due to climate change needs careful planning and management to ensure maximum economic and social benefits. Greater tourist activity, especially in rural areas, will cause pressures on infrastructure and on the environment which policymakers need to address.

xii) Coastal zone planning and management should be increasingly orientated towards sustainable utilization of coastal resources. The effects of climate change and sea level rise should be considered alongside the impacts of human activities in the zone. Changes in natural coastal processes and systems are increasingly interrelated with socio-economic changes. UK shoreline management planning should be further integrated into coastal zone strategic planning based on the safety-first principle, with due regard for economic practicability, and the involvement of all interested parties, e.g. insurers.

B. Recommendations for Research

1. With regard to the prediction of future climate and sea level, the most fundamental need is for improved models of the climate system. Results from improved global climate models need to be linked with models which simulate regional or local ocean circulation and storm surges to enable improved predictions of changes in the local near-shore environment. With regard to climate change impact assessments, the primary need is for i) more regionally specific information about climate change both globally and in the UK and, ii) improved knowledge of altered probability distributions of weather events, especially of rainfall, windspeed and storm-surges.

2. There is a need to understand better the impacts of climate change, including doubled CO_2 levels in the atmosphere, on the size and properties of the soil carbon pool. Better models are needed to simulate changes in soil processes and the opportunities for plant species adaptation and migration. More emphasis should be given to quantifying the climate change impacts on soils rather than relying on qualitative judgements. The current lack of uniform soil data sets for the four countries of the UK inhibits the UK-wide modelling of the impacts on soils and in turn the effect on agriculture and natural habitats.

3. Our ability to forecast the effects of climate change on species survival, abundance and migration needs to be improved by:

(i) developing and testing spatially georeferenced, process-based models of species and community dynamics, and

(ii) conducting ecosystem-scale long-term experiments in the field of effects of elevated temperature and CO_2 on plant and animal responses including the effects of water-use efficiency.

There is also a need to maintain the current environmental monitoring and data collection network to provide an early warning of change and its causes.

4. Robust predictive crop and livestock production models are needed as part of the move towards computerised decision-support systems. These need to be extended from main food crops into a wider range of production systems and to enable estimations of effects of change in CO_2 and weather. The models must generate estimates of field rather than optimal performance. To do this, they must be responsive to estimates of non-optimal or deleterious factors (nutrient and water deficits, incidence of pests, weeds and diseases etc.).

5. More research is needed on the response of key tree species to increased CO_2, and to changes in temperature, precipitation and windspeed (particularly extreme occurrences of these). Knowledge of this needs then to be scaled to regional levels to match improved climatic models. Performance of existing experiments and trials should be analysed to better relate yield

to environmental variables and to detect climatically-driven changes in growth patterns. Impact of climate change on the biology of present and potential pests and diseases needs more work to better estimate the responses of forest ecosystems.

6. Improved models are required to simulate the effects of climate change on particular aspects of the water sector, specifically flood frequency and aquatic ecosystems. There is also a need for studies into impacts on specific water resource systems, expressing impacts in monetary and reliability terms.

7. The adoption of air-conditioning systems, the consequent impact on equipment markets and electricity demand and the sensitivity of both of these to climate change require investigation. This work should cover both systems for large commercial buildings and smaller room-size units for homes, shops and small offices. Such work would be helpful in identifying:

(i) the extent to which non-climatic factors are driving the greater use of air conditioning;

(ii) the extent to which recent warm summers have stimulated demand; and

(iii) any climatic thresholds which appear to trigger the adoption of new systems.

8. Interdependencies between the energy sector, the water industry, agriculture and other sectors of the economy should be explored. Such work could use techniques, such as input-output analysis, to explore the ripple effects of changed prices and availabilities throughout the economy. The capacity of energy and water users to reduce levels of consumption in response to climate-induced changes in price and availability needs to be explored.

9. In the absence of information on the potential impact of climate change on industry information on the climate-change sensitivity of other sectors on which industry depends (e.g. agriculture, forestry, energy, minerals extraction etc.) would be valuable.

10. Computer simulation of the impacts of weather, using quantitative time series of hourly observed data, is playing an increasingly important role in the assessment of the probable future performance of buildings at the design stage. Climatic data tapes are needed that will quantitatively describe climatic change at the level of detail needed to assess various aspects of future building performance properly.

11. Further research is needed to examine how individuals and corporate users of transport will adapt their behaviour to accommodate the effects of climate change. More information is also required on the likely frequency of weather events which have the potential to severely disrupt transportation services.

12. For the insurance sector more detailed information is required on the likely pattern of extreme events in the UK and abroad in areas with a significant exposure to property damage for UK insurers. Improved means of translating the impact of such events into economic terms are required.

13. In relation to potential health impacts, research is needed to:

(i) clarify the range of potential impacts, and their likely temporal and spatial distributions;

(ii) describe existing geographic gradients in climatic and health-related indices, within both the UK and the European region, (this would assist the forecasting of climate-related shifts in health risk within the UK) and

(iii) assist the development of integrated predictive modelling (especially of infectious disease outcomes).

14. The links between weather, climate and the demand for particular forms of tourism and tourism activities is not well understood. Better techniques need to be developed for portraying holiday climatic data in a way meaningful to tourists attempting to choose between competing holiday destinations.

15. For coastal regions of the UK, global climate models require further refinement to enable them to produce more localised predictions concerning the regional effects of changes in sea level and storm frequency, duration, magnitude and direction. The output from the scientific models then needs to be coupled with socio-economic model data, covering the relevant social costs and benefits of the process of 'environmental' change and policy response options in the more vulnerable sectors of the UK coastal zone. In the absence of local climate impact prediction data there is a need for further impact studies based on scenarios of possible future changes and conditions, incorporating both scientific and socio-economic analysis.

16. The impact of sea level rise on coastal and estuary sites where there are facilities for processing, refining, storing and extraction of minerals should be studied. A study of the availability of materials required and the associated costs for improving sea defences should be implemented. Movements of likely changes in the water table should be investigated. Research into the performance of floating platforms/shuttle tankers for extraction of deep level offshore oilfields in severe weather conditions needs to be continued.

17. Integrated assessments of climate change at the regional and national level are needed to understand the interdependencies between the natural, economic and social spheres.

SPECIFIC CONCLUSIONS

Changing Climate and Sea Level

1. There is mounting evidence that global climate is changing as a result of human activities. As stated recently by the Intergovernmental Panel on Climate Change (IPCC), the global warming of the last 100 years is unlikely to be due entirely to natural causes. The cluster of warm months and seasons experienced in the UK in the last few years is consistent with this picture of a changing global climate. For example, the 1985-94 decade, both globally and for the UK, has been about 0.2°C warmer than the average of the 1961-90 period and during this decade the average global atmospheric CO_2 concentration has risen by about 5%.

This review defines a scenario (termed the 1996 CCIRG scenario) of changing climate and sea level for the UK for the decades of the 2020s and 2050s assuming that no major global policies of reducing greenhouse gas emissions are enacted. It is based upon recent results from the Hadley Centre of the UK Meteorological Office which is recognised worldwide as a leading centre for climate prediction. The effects of sulphate aerosols on future climate are not included in the scenario, although some results from experiments that do consider them are mentioned. Comparisons are made with results from 11 other climate change experiments and using other values for the climate sensitivity[6] to indicate how representative this scenario is of a wider range of possible changes in climate.

The scenario implies the following:

Global Climate
- Global mean surface air temperature by the decade of the 2020s will be about 0.9°C warmer than the average of the 1961-90 period and, by the 2050s, about 1.6°C warmer. The rate of warming is about 0.2°C per decade. The range of global warming values for the 2050s is between 1.1°C and 2.4°C. The effect of sulphate aerosols on the climate system would be to reduce the global warming by the 2050s by about 0.3°C.

- The atmospheric CO_2 concentration by the 2020s will be about 430 ppmv (±10ppmv) and, by the 2050s, about 525ppmv (± 25ppmv). These concentrations represent increases of 29% and 57% respectively

---

[6] The term 'climate sensitivity' defines the eventual change in global-averaged surface air temperature which would occur if the atmospheric concentration of greenhouse gas was instantaneously doubled. In this scenario the climate sensitivity used is 2.5°C.

compared to the average 1961-90 value. The equivalent CO_2 concentration of all greenhouse gases by the 2050s is almost double the 1961-90 average.

- Global mean sea level by the 2020s will be about 19cm higher than the average of the 1961-90 period and, by the 2050s, about 37cm higher. This rate of sea level rise is about 5cm per decade. The range of sea level rise values for the 2050s is from about 26cm to 50cm. The effect of sulphate aerosols would be to reduce the sea level rise by the 2050s by about 14cm.

UK Climate

- Temperatures over the UK are expected to rise at a generally similar rate to the global mean, about 0.2°C per decade, with slightly slower rates of rise over northwestern UK compared to the southeast and in winter compared to summer. These differences in pattern and seasonality arise largely due to the influence of ocean circulation changes in the North Atlantic. The regional differences in temperature change over the UK are more pronounced than those reported in the 1991 CCIRG scenario,

- Extremely warm seasons and years are expected to occur more frequently. A hot summer such as 1995 (currently about a 1-in-90 year event) is calculated to occur about three times during the 2050s decade (i.e. a 1-in-3 year event), while a very mild winter such as 1988/89 (currently a 1-in-30 year event) is calculated to occur about twice during the 2050s decade.

- Annual precipitation over the UK as a whole increases by about 5% by the 2020s and by nearly 10% by the 2050s. Winter precipitation increases everywhere, but more substantially over southern UK. Summer precipitation decreases over southern UK, but increases in the north.

- The number of raindays increases slightly over most of the country in winter, but only over northern UK in summer. Fewer raindays occur over southern UK in summer.

- The average intensity of precipitation increases modestly in all seasons and for all regions. Return periods of heavy daily rainfall events shorten, especially over northern UK in summer.

- Average seasonal wind speeds increase over most of the country in both summer and winter. Return periods of the strongest daily mean wind speeds shorten over northern UK, but with little change in the south. The frequency of gales over the UK is calculated to increase by the 2050s by up to 30%.

- Potential evapotranspiration (PE) is estimated to increase over most of the UK in both winter and summer, the exception being the north of Scotland where decreases will occur. Summer PE in southern UK by the 2050s is about 25% higher than the average of 1961-90.

- The global rise in sea level, about 5cm per decade, is likely to be exacerbated in the southern and eastern UK by sinking land and mitigated to some extent over the northern UK by rising land. Although nothing quantitative can currently be said about changes in the frequency or magnitude of storm surges around the coasts of the UK, the probability of a storm surge exceeding a given threshold is likely to increase.

2. Although the 1996 CCIRG scenario currently represents our 'best' estimate of UK climate in the future, it describes only one possible future evolution of climate and sea level in the UK and implies a linear response of the climate system to increasing greenhouse gas concentrations in the atmosphere. Other responses are also possible and some may involve more abrupt changes in climate related to, for example, changes in the ocean circulation of the North Atlantic. It is impossible to attribute a probability to any one scenario although, since the 1996 CCIRG scenario is based on results from state-of-the-art

climate models, it is a reasonable reflection of our current knowledge of how the climate system works. This knowledge base would suggest the following hierarchy of decreasing confidence in the changes in the different climate variables:

certain: increasing CO_2 concentrations;

confident: increasing global mean temperature and sea level;

likely: geographic patterns of regional temperature change;

possible: geographic patterns of regional potential evapotranspiration, precipitation, gales, return period estimates and frequencies of extreme events.

Soils

1. The soil is important in a climate change context not only because it underpins virtually all terrestrial ecosystems and agricultural production but also because it is a significant source of and sink for greenhouse gases.

2. Soil processes and properties, particularly organic matter and soil waterbalance, will be sensitive to increasing temperature and changing amounts and distribution of rainfall, affecting the ability of the soil to sustain current agricultural crops and natural habitats.

3. The soils of southern Britain are likely to be most at risk from climate change particularly in view of the predicted increase in summer droughtiness. This will affect the functioning of terrestrial ecosystems, influence groundwater recharge and flow and give rise to increased need for irrigation for agriculture. The combination of higher temperatures and increased rainfall may be beneficial to the soils of northern Britain and allow a wider range of crops to be grown.

4. Wetland areas of southern Britain will be at risk from drying out and peat soils will be lost at increasing rates due to drying and wind erosion.

5. Most of the clay soils of southern Britain will be subject to more intense shrinkage in summer, with potentially severe implications for building foundations.

6. It is predicted that soil erosion will increase throughout Britain. The predicted drier summers in southern Britain would leave the lighter soils more prone to wind erosion and wetter winters would lead to more water erosion. Higher rainfall in northern Britain would cause increased erosion by water.

7. Over 50% of Grade 1 agricultural land in England and Wales lies below the 5m contour and is thus located where it might be affected by any rise of sea level. Much may become saline and therefore of limited use for agriculture unless protection measures such as sea defences are established and maintained.

8. Changes in the way land is managed and used will need to be made to offset many of the impacts and these could be combined with measures to increase the sink for greenhouse gases in the soil and decrease their emission from it.

Flora, Fauna and Landscape

1. The natural biota of the UK has been most profoundly altered by human activities in the past and, over the next 50 years, continued effects of land-use change are likely to have a greater impact than predicted changes of climate.

2. An increase in mean annual temperature of 0.5°C that is sustained for a decade or so will bring about noticeable changes in plant and animal distributions and abundance. A 1°C increase in temperature may significantly alter the species composition in about half of the statutory protected areas in the UK.

3. There will be significant movement of species northwards, particularly insects and ephemeral weeds, in response to rising temperatures. A mean temperature increase of 1.5°C by 2050 is equivalent to a potential northward shift

of 50-80km per decade or an altitude shift of 40-55m per decade. Conservation strategies need to be developed that recognise that the natural ranges of species will not be static and that many species may need to migrate to survive.

4. The direct effects of a 30% increase in atmospheric CO_2 concentration over the next 50 years may be small relative to the effects of changes in temperature and rainfall, but there is at present no consensus on the magnitude of CO_2 effects on plant growth and water relationships in field conditions.

5. Overall, the number of animal species (especially insect) in most parts of the UK is likely to increase due to immigration and expansion of species ranges. By contrast, the number of plant species may decrease, and a substantial number of the 506 currently endangered species may be lost, because species-rich native communities may be invaded by competitive species and some wet, montane and coastal communities will be lost.

6. There is some threat of invasion by alien weeds, pests, pathogens and viruses, but the majority of alien species that could spread do already occur in the UK.

7. A 20-30cm increase in sea level will adversely affect mudflats and some salt marshes, including nature reserves that are important for birds.

8. Climate change will occur too rapidly for species to adapt in an evolutionary sense. Mitigating measures that can be taken include the translocation and rescue of species, the provision of habitat corridors, fire control, and control of eutrophication.

Agriculture, Horticulture and Aquaculture

1. Agriculture in the UK is generally sensitive to climatic factors which contribute significantly to year-on-year variability in output. Changes in the intensity, annuality and distribution of precipitation, together with prevailing temperatures and the incidence of extreme weather events would have the greatest effects upon production.

2. Alterations of rainfall, temperature, and the incidence of extreme weather events will impact upon all components of the sector. The north-south gradients of climate variables (in particular summer temperatures and rainfall) will also be of major significance in terms of impact. Elevated atmospheric CO_2 will stimulate plant productivity but by a variable amount depending upon species, location and management.

3. Spatial variation in inputs will affect specific production systems in different ways. Limitations caused by reduced water availability in the south and east UK, coupled with higher temperatures and increased evapotranspiration may shift potential production of arable and other field crops northwards and westwards as well as placing extra pressures on water for irrigation.

4. Grassland productivity in the wetter north and west would be sustained by warmer winter temperatures, and the boundaries of forage maize cultivation may continue to move northwards. Stock damage to wetter pastures could reduce the advantages of increased grassland productivity.

5. Trout farming is likely to show the greatest sensitivity within aquaculture systems. This would be most marked in the south due to warmer temperatures and to the increased likelihood of drought restricting river flows.

6. Adverse effects on soils and increased incidence of pests, weeds and diseases could reduce or negate any yield increases attributable to climate change.

7. Warmer and drier summers will increase cultivation opportunities for novel crops including industrial and perennial biomass crops (e.g. *Miscanthus*).

8. The sensitivity to climate change of the whole sector, and the industries which it supports, is likely to be less than that of specific production systems. Altered patterns of land use will help to sustain overall performance. The effects of climate change on global production will, however, impact at the sectoral level in the UK in terms of changes in market price and in the UK's altered competitive position with respect to other food-exporting regions of the world.

Forestry

1. Expansion of UK forestry earlier this century was intended to reduce reliance on imports. Future changes in forest area will also largely be policy-driven, reflecting shifts in the balance of agricultural support, social policies and conservation values, rather than affected by responses to changes of climate.

2. The timber trade in the UK, supplying only 15% of domestic wood product demand, is dependent on conditions in exporting countries (Scandinavia, N. America) and thus relatively insensitive to change in British supply. Changes in overseas supply would alter production economics here, thus affecting expansion and species choice.

3. UK forests, uniquely in Europe, are based on introduced species, which are more productive and in greater demand by wood-using industry than native species. Most forests are young, have short rotations and thus could allow introduction of species (or origins) better adapted to the altered climate. More rapid early growth of commercial species, already occurring, may be attributable to increased ambient CO_2, nitrogen deposition and higher temperatures. Further increases due to future changes of climate would generally improve the financial return on afforestation, encourage forest expansion and reduce reliance on imports.

4. Yields of the main commercial species in central and northern UK may increase by about 25% (equivalent roughly to 1.0 $Mm^3 yr^{-1}$) by 2050 in response to increased mean and accumulated temperatures (>5.6°C) wherever water availability is not limiting.

5. In southern UK decreased precipitation and increased evapotranspiration would reduce general productivity and drive sensitive species (e.g. beech) from marginal sites affecting wood supply, amenity, recreation and conservation values. Urban trees may be particularly stressed and substitute species should be sought now.

6. Areas subject to more frequent drought might expect increased insect pest damage, fire hazard and poorer wildlife habitat.

7. Adaptation to climate change will require positive management intervention. Existing species populations cannot migrate and would therefore require deliberate replacement with better adapted material while management strategies would need to alter to accommodate changes in growth patterns and in the frequency of damaging impacts.

Water Resources

1. The high degree of sensitivity of the water industry and water users in the UK to climate variability has been illustrated by the droughts of 1988-92 and 1995, and by the floods of 1993, 1994 and 1995.

2. Under the 1996 CCIRG scenario there would be an increase in river flow in winter and a decrease, especially in the south, during summer. This would adversely affect abstractions of water from rivers in summer (for water supply, irrigation and cooling water) as well as instream uses such as navigation, recreation and ecosystem maintenance. The effects of changes in the volume and timing of flow on managed water systems will depend significantly on the size and number of reservoirs, and the degree to which different sources are linked.

3. Increased winter rainfall and wetter catchment conditions are likely to increase the frequency of riverine flooding.

4. It is not yet established whether groundwater recharge would increase in future wetter winters, because the recharge season would be shortened by increased autumn and spring evaporation. It is therefore quite possible that recharge would reduce.

5. Higher water temperatures will speed up self-purification processes and improve water quality, but could also increase the risk of algal blooms in rivers and lakes. Furthermore, lower flows in summer would reduce the dilution of effluents and pollutants. The effects on water quality of changes in temperature and flow may be less important than the effects of changes in land use and agricultural practices.

6. Sea level rise would have an easily avoidable impact on both coastal aquifers and freshwater intakes located close to the tidal limit.

7. Climate change might add an additional 5% onto the 12% increase in demand for public water supplies expected in southern England between 1990 and 2021, largely due to increased usage in gardens. The increase in peak demands could be much greater. This has important implications for the reliability of supply to domestic consumers and may require the redesign of parts of the distribution networks. Demands for spray irrigation are predicted to rise by 69% by 2021 without climate change, and could rise by 115% if temperatures were to rise.

8. Changes in the frequency of drought and restrictions on water users depend not only on changes in water availability and demand, but also on catchment geological conditions and the volume of storage available.

9. Aquatic habitats will be affected by the increase in water temperature and changes in river flows. Most fish species in Britain are unlikely to be significantly affected, but exceptions include cold-water lake fish and, most importantly, the native brown trout, which are both likely to be adversely affected by a rise in temperature. Changes in flow regimes might also affect salmon migration patterns.

10. Water management has always been an inherently adaptive process, responding to and anticipating many changes. Possible climate change is one extra pressure, with a longer time horizon than most other changes. Current management techniques can be used to adapt to change, but this may involve economic, social or environmental cost, and some systems are better able to adapt than others.

Energy

1. The impacts of climate change on the energy sector are diverse and few in themselves will be of major significance. The more important climate impacts relate to energy demand and the availability of natural resources. The technological potential for adaptation is high because the lifetime of most assets in the energy sector are shorter than the timescales associated with projected climate change. Rapid climate change and unusual weather conditions such as those experienced in recent years could however lead to a degree of mismatch between investments and climate conditions for which they were not designed.

2. Globally, the energy sector is the largest source of anthropogenic emissions of CO_2, the most important greenhouse gas. The direct impacts of climate change on the energy sector will be much lower than those resulting from market developments, technological change or mitigation policies designed to constrain greenhouse gas emissions. There have, for example, been major changes in the electricity sector, in terms of both organisation and technological choice, even since the First CCIRG Report (1991).

3. Climate change could reduce energy demand for space heating in the UK by about 5% below the level which it would otherwise have been by the 2050s, and by about 3% by the 2020s. This would benefit householders who, on average, currently spend 4.8% of their budget on home heating. Recent warmer winters have reduced space heating demand below expected levels.

Climate change would increase the use of existing air conditioning systems and could stimulate the wider adoption of new systems. Energy use for air conditioning would increase most rapidly in southern parts of the UK which is already warmer and where a greater temperature rise is projected.

4. Climate change could affect the availability of renewable energy resources. In the UK hydro-electricity is currently the most important renewable energy source but the use of wind energy and biomass could increase in the future. Photovoltaic energy may become important at the global level but is less sensitive to climate. Hydro-electric potential would be affected by both the timing and volume of run-off. Wind energy resources could increase. Higher CO_2 concentrations could lead to higher biomass availability as long as water availability is adequate.

5. All of the UK's petroleum refineries, all of its operating nuclear power stations and 70% of its conventional fossil-fuel fired power generation capacity are located on coasts or estuaries. Requirements for further protection against sea level rise are highly site-specific but are unlikely to cause any major difficulties.

6. Energy sector operations may be affected in a number of minor ways by climate change. Increased storminess would affect offshore oil/gas exploration and production operations. The availability of cooling water could affect power generation. Overhead electricity cables and various renewable energy systems are vulnerable to storms and high winds.

Minerals Extraction

1. The UK minerals extraction industry is accustomed to continuously and quickly adapting to changes in production locations, strong environmental protection pressures for onshore and offshore operations, the implications of sustainable development, privatisations and changes in market conditions over the last 30 years. Significant changes have included the rapid development of the offshore oil and gas industry and the reduction in the coal industry.

2. The UK is in a strong position for the production of energy, construction materials and certain industrial minerals but is heavily dependent on imports for base metals, iron ore, coking coal for steel production and precious metals.

3. The historical record of successful adaptation of the UK minerals extraction industry suggests that the projected changes in climate in relation to temperature and rainfall do not create any significant technical problems for the oil, gas, construction materials, industrial minerals and coal industries.

4. More importance will be attached to information about increases in the frequency and intensity of storms and the rise in sea level. These factors could cause inland and coastal flooding, coastal erosion, tidal surges, hazards for shipping and increased salinity. The impact of these changes would be to raise the costs of extraction and, possibly, would result in the relocation of production, transport and processing facilities associated with minerals extraction. Although technical solutions are likely to be developed, there is no guarantee that the producing activity would be economically viable.

5. Current estimates about the rate of increase in the sea level suggest that coastal investment associated with minerals extraction should allow adequate time for current capital investments to be recovered. An increased frequency and intensity of storms could reduce the time periods before action is required in relation to decisions about improving sea defences or relocating operational facilities.

6. Decisions to improve sea and river defences would require increased use of rock armour and large sizes of hard rock and sand from marine dredging. It is likely that this would require imports of rock armour to supplement UK production and to limit overland transport of this material

Manufacturing, Retailing and Service Industries

1. Fewer severe winters would be beneficial, reducing disruption at all stages from the supply of raw materials through processing to markets for the finished goods. A parallel increase in hot, dry summers, which has been shown to disrupt sectors with a large water requirement, is unlikely to have a substantial impact over the long-term. The water supply industry is expected to adjust appropriately, although costs may increase. Industry has demonstrated the ability to economize in both water and electricity use.

2. Overall sales volumes will be dictated by the economic environment, but buying patterns would be influenced in part by climate change, affecting the type of goods required from the manufacturing sector. For example, in the clothing industry, there may be a shift in demand from heavy woollen material towards light cotton clothing. Manufacturing and retail units which can predict and exploit relevant aspects of climate change will have a clear advantage.

3. Impacts on agriculture and forestry, both in the UK and overseas, will have knock-on effects for industries which derive their raw materials from these sectors, affecting costs and possibly the continuity of supply. Increased transport costs at any stage in the movement of goods, from the factory supplier to the consumer, caused by strategies to limit emissions, may lead to an increase in retail prices.

4. Changes in shopping habits may be expected, requiring changes in the design of shopping malls and precincts, and in opening hours. Technological advances allowing, for example, extensive adoption of electronic shopping, would make the shopping environment relatively invulnerable to the effects of climate change.

5. Industry operates in a global market place. Whether or not climate change over the UK, in relation to the impacts of climate change elsewhere, would place the national industrial base at an advantage to international competitors, remains to be seen.

6. Adaptation by industry is expected to be autonomous. The exception is the case of rising sea level. If the frequency and severity of storm surges increases, as the result of a combination of a rise in mean sea level and more frequent storm occurrence, industry located at the coast, along with other assets, will require protection. Re-location is not expected to be necessary over the time scales considered in this report.

Construction

1. The most important Industry climate change impacts on buildings and other types of construction are likely to arise from higher summer temperatures, increased winter rainfall, and, in the north of the UK, increased extreme winds. If sea defences are breached, the damage to construction in the affected areas will be very considerable.

2. Soil moisture movements on shrinkable clays will increase, so more careful attention to foundation design will be needed in vulnerable areas.

3. Thermal conditions in winter are likely to improve. The energy needed for space heating in fully heated buildings is likely to decline by around 20%. The lower mean winter indoor-outdoor temperature differential will favour greater use of passive solar energy in buildings.

4. Summer conditions in large towns and cities, in the southeast of the UK especially, will become less acceptable as there will be more hot days, and the urban heat island will make conditions worse. Pressure to expand the use of air conditioning will mount. This pressure can to some extent be resisted through climatically sensitive design using natural cooling techniques. However, urban traffic noise and pollution will continue to make it difficult to find acceptable solutions in town centres.

5. The life of buildings now being erected will extend beyond 2059, the end year of the climate change scenario period in this Report. New construction should be designed with the probable climate of the decade 2050/59 in mind. However, clients and their designers do not presently have the appropriate information to approach the implicit problems objectively in terms of costs and benefits.

6. The construction industry can respond to the expected changes by providing adaptation at the rate required. However, the actual outcome will depend on the attitudes of building owners to adaptation. They will need to be persuaded that the expected long term benefits are greater than the costs of adaptation.

7. The periodic major refurbishments of existing buildings will provide suitable opportunities to respond appropriately to climate change needs. Change will only happen if building owners understand the need for appropriate action. Such decision making needs to lie within a structure of cost benefit analysis concerning climate change and construction. This structure does not yet exist. The approach will need to include consideration of future insurance costs for construction.

Transport

1. The extent to which the predicted levels of climate change will be important for transport is very much dependent upon the extent of other predicted changes in the transport system, such as the growing need to incorporate changes which are environmentally sustainable. It is expected that changes resulting from these other factors could be significantly more important than those arising from climate change.

2. There are a few areas where transport is particularly sensitive to changes in climate and where policy measures will need to be devised in order to enable appropriate adaptations to take place. The aspects of transportation likely to be more significantly affected by climate changes include: transport links close to areas of coastline and estuaries which may be subject to increased likelihood of flooding as a result of sea level rise; an increased likelihood of disruption of all modes of transport resulting from an increase in the frequency of strong winds; and the possibility of greater levels of damage to infrastructure as a result of higher summer temperatures.

3. There are likely to be some positive effects of climate change for transportation, for example, a reduction in the numbers of days with frost and lying snow could lead to less disruption of air, road and rail transport modes.

4. The lifetimes of most transport artifacts are quite short and many, such as motor vehicles, can be expected to be replaced several times over the next half century, hence giving a high level of adaptability to potential changes. Some more long-lived pieces of transport infrastructure, such as roads and bridges, are less easily adaptable in the short term. While the potential for adaptation is good, it is dependent upon a willingness to change and upon the projected climate changes occurring sufficiently slowly.

5. Transport is a derived demand and hence is highly dependent upon changes in other sectors documented elsewhere in this report. The linkages between transport and other sectors of the economy are not clearly known. Changes in these sectors as a result of climate change (or otherwise) will have considerable secondary impacts on transport. These may exceed any direct impacts of climate change on transport itself. For example, changes in patterns of tourist activity in the UK resulting from climate change could generate considerable changes in the overall levels of demand for transport or could alter the existing balance of demand between various areas of the country.

Insurance Sector

1. Property insurance would be immediately affected by a shift in the risk of extreme weather

events. It is anticipated that changes in their frequency and/or severity will occur, which would necessitate pricing or product changes, particularly for the most vulnerable areas or activities covered. At present a detailed description of future weather is not available; thus it is not possible to make specific planned responses.

2. Since the risk of coastal flooding will increase due to sea level rise, this will expose the property insurance industry to the greatest potential losses in the UK (e.g. the insured value of the property protected by the Thames Barrier is estimated to be £10-20 billion).

3. Because of the international nature of UK insurance institutions, the impacts from climate change or sea level rise abroad are important. UK insurers incurred substantial claims in Hurricanes Gilbert (1988), Hugo (1989), Andrew (1992) and Opal (1995) directly through foreign subsidiaries, and also through reinsurance activities in London, including Lloyds. Hurricanes and other tropical storms may increase as a result of global climate change.

4. Property insurers have begun to adapt to the run of increased UK weather losses from storm and drought-induced subsidence. More efficient methods of handling claims have been introduced. Building insurance rating now recognises susceptibility to subsidence. Major exercises are or have been conducted individually and collectively to assess the approximate pattern of risk from coastal flooding, and the potential impact of storms. New taxation arrangements have been negotiated with government to facilitate the creation of funds to meet the irregular but high cost of catastrophes.

5. UK insurers are actively involved in exercises to reduce the vulnerability of buildings to wind damage. The insurance rating of weather hazards is continually reviewed together with assessing the financial impact from potential extreme events in order to control exposure.

6. Apart from property insurance, there is little literature on the impact of weather. There will, however, be impacts on human health and longevity in the UK, on the transport sector (including international marine business) and on energy exploitation. The insurers of such activities have not yet addressed the issue of climate change specifically.

Financial Sector

1. The greatest implications for the sector are primarily indirect, being felt through pressures upon customers, both commercial and personal. If the risk of flooding increases, the sector will face indirect pressures as business customers lose income through disruption, while homeowners and businesses face direct financial pressures through higher insurance premiums.

2. The availability of insurance coverage has reduced the indirect effects upon the sector; as a result little information is available from which to project figures. The price of capital to different sectors/geographic areas may increase if lenders/investors are faced with higher risks as a result of the withdrawal of adequate insurance coverage.

3. The financial sector is of great importance to the UK economy and in view of the international nature of much of the sector's business, the consequences of climate change outside the UK are likely to be of major significance. The financial sector tends to focus on the short term and discount the longer term, but can react relatively quickly to changes in its market, thereby offsetting some of the longer term issues experienced by other sectors which have more fixed investments.

4. Climate change could affect the geographical distribution and balance between business sectors within the UK, impacting differently upon individual institutions as a result of their own particular portfolios. Financial operations are vulnerable to short term disruption due to failure of communications or denial of physical access, such as resulted in October 1987 in the London area.

5. If climate change seems likely to have an effect on socio-economic activities, this will affect the appraisal of existing and new debt and equity investment opportunities, particularly in agriculture and for coastal regions.

Health

1. Climate change would have both direct health effects (e.g. heatwave-induced deaths) and indirect health effects (e.g. altered rates of infectious diseases and altered exposure to air pollutants). While some health benefits would result from climate change, the net impact upon health is likely to be adverse. In the longer term, the indirect health effects (especially from infectious diseases, allergic disorders due to pollens and spores, reduced supplies of freshwater) may become dominant.

2. Most of the data that allow quantitative forecasting of health impacts refer to the direct effects of thermal stress and extreme weather events. US studies suggest that an increase in heatwaves would cause several thousand extra deaths annually, especially in large urban settings. UK research indicates that this excess may be largely offset by reductions in winter-related deaths.

3. Despite recent observations in some tropical regions, and modelled predictions that suggest a wider geographical spread of vector-borne infectious diseases (such as malaria, dengue and leishmaniasis), the UK should not be much affected. Public health defences appear to be sufficient to cope with climate-related increases in potential transmission 'pressures' from these diseases.

4. Other food-borne and water-borne infective agents, as causes of diarrhoeal and dysenteric infections, are likely to spread more readily in warmer and wetter conditions.

5. The existing problems of urban air pollution would be exacerbated by climate change, by enhancing the production of photochemical pollutants (e.g. ozone) and, perhaps, by amplifying the biological impacts of certain pollutants. The geographical distribution and the seasonality of aeroallergic disorders (hayfever, asthma, etc.) would also change.

6. The overall burden of additional illness or premature mortality cannot yet be quantified, in particular because of the complexities of forecasting the various indirect impacts upon human health. Simplistic calculations based solely on, for example, heatwaves and weather disasters would therefore be incomplete and misleading.

Recreation and Tourism

1. Climate forms an important part of the environmental context for recreation and tourism. Climate change is likely to provide new opportunities for the industry if a trend towards warmer, drier and sunnier summers stimulates an overall increase in tourism in the UK.

2. Any climate-led changes will take place on both a domestic and an international scale but may be over-ridden by other factors, such as the availability of leisure time and the amount of disposable incomes.

3. There is likely to be an enhanced uptake of many outdoor pursuits ranging from gardening to more strenuous sports, particularly those which are water-based.

4. Any increase in either the intensity of use of leisure facilities, or the length of the operating seasons, is likely to have a positive economic impact through better employment opportunities and the enhanced viability of commercial enterprises.

5. Given an overall increase in tourist activity, there will be a need for better management of visitor pressures at peak periods in order to maintain quality. Greater problems of environmental protection could arise in key settings including crowded municipal parks, eroding beaches and moors and heaths threatened with an increased fire risk.

6. In the north of the UK, any significant increases in rainfall, windspeed or cloud cover are likely to offset the more general advantages associated with higher temperatures. The viability of the Scottish ski industry will decline if snow confidence becomes less secure than at present.

7. Potential rises in sea level will adversely affect fixed waterfront facilities, such as marinas and piers. For beaches backed by sea walls, increased erosion could lead to a loss of beach area. Other recreational locations along the coast, such as sand dunes, shingle banks, marshlands and soft earth cliffs, may also be affected.

Coastal Regions

1. A number of uncertainties still surround the prediction of future sea levels and storm intensity/frequency at the regional level. According to the 1996 CCIRG climate change scenario global mean sea level may be +19cm higher than the average of the 1961-90 period by the 2020s and +37cm higher by the 2050s. The global rise in sea level is likely to be exacerbated in southern and eastern UK by sinking land. Coastal areas could also face a significant risk of increased flooding, inundation and erosion with consequent assets damage if the frequency or severity of storm surge events also increased. Currently it is not possible to quantify this risk. Coastal lowlands around the Wash, stretches of the Norfolk and Suffolk coasts and, to a lesser extent, areas on Teeside and in southwest Lancashire seem particularly vulnerable.

2. Assets in the UK coastal zone are important environmentally, economically and politically. A significant proportion of the UK's social and economic welfare depends directly or indirectly on the availability of environmental goods and services provided by coastal and marine systems.

3. Many coastal zone resources are sensitive to change, including human intervention, and are having to accommodate a relatively high rate of change and subsequent environmental pressure.

4. Several stretches of the UK's coastline and related coastal zones are vulnerable to a combination of the effects of natural variability, climate change and human interventions. Protecting the capital (human and natural) assets, or in some cases relocating assets, will be costly if the effects of climate change are sudden rather than gradual. Any increase in the frequency with which sea and coastal defences are over-topped or breached would have major financial implications; for example, damages from the extreme flood event at Towyn in North Wales in 1990 totalled £35 million.

5. Once information on the physical effects of potential or actual climate change is available, social and economic systems will begin a complex process of adaptation, which then introduces feedback that may alter the future state and change the potential damage.

6. The limited number of economic studies undertaken for UK coastal areas suggest that the protect-response strategy is generally economical. However, there may be local areas where a strategy of either 'managed retreat' or 'do nothing' is best. An approach which emphasises human safety encompassed within an evolving integrated management strategy, moving beyond shoreline issues to a coastal zone-wide basis, has much to recommend it.

7. The potential impacts of climate change and sea level rise are likely to be most severely felt, if defences fail, in terms of local direct and indirect human health effects and economic asset damage. There may also be losses to tourism and recreation and natural systems such as wetlands and mudflats and related flora and fauna. The national effects on other sectors such as forestry, energy, transport, industrial production, agriculture, water resources, soil resources and financial services are likely to be small.

1. Introduction

1.1 BACKGROUND

The Climate Change Impacts Review Group (CCIRG) was formed by the Department of the Environment in 1990 to review the potential effects of climate change in the UK, to identify research that is needed and the implications for policy. It also enables the UK to meet, in part, its obligations under Article III of the UN Framework Convention on Climate Change (UNFCCC) to take action to reduce climate change impacts.

The First CCIRG Report, published in 1991 (CCIRG, 1991), indicated the likely impacts of climate changes on the environment and economy, suggested measures that might be needed to reduce these impacts and identified gaps in knowledge where more research was required. It assigned some sense of priority in terms of economic impact and timescales. The report built upon previous reviews (Parry and Read, 1988; Department of the Environment, 1988).

Since 1991 the quantity of published research on the effects of climate change has approximately doubled. Several large research projects on impacts in the UK have been completed, and the Intergovernmental Panel on Climate Change (IPCC) has produced its supplementary and its second full assessment reports (1992 and 1996). Although there remains substantial uncertainty both about the nature of future climate change, and the nature of its impacts, a re-assessment of this situation five years on is needed.

1.2 SCOPE OF THE REPORT

The Review Group is an independent body of experts (see Annex 3). The members of the Group are able to call on the expertise of a number of other experienced individuals (see Annex 4).

The remit of the Review Group is to:

- Assess the impacts that would be likely to affect the UK, assuming changes of climate that are currently estimated to occur, unless there are major changes in greenhouse-gas emissions as a result of international agreements.

- Consider the array of adaptations that might serve to reduce these impacts. (This was not considered in as much detail in the First CCIRG Report).

- Consider the implications of climate change for policy.

- Make recommendations for future research.

This report does not consider strategies for reducing greenhouse-gas emissions, nor the impacts that may stem from their implementation. It is very largely concerned with the potential effects of climate change likely to occur over the geographic area of the UK, but it has also considered some climate changes that might occur elsewhere in the world where the effects of these might impact on activities in the UK.

The review is a synthesis of current knowledge. As more information becomes available, such as the effect of sulphur emissions, further assessments will be needed. Effects of changes in UV due to stratospheric ozone depletion are the subject of assessment by a separate Review Group.

The areas covered include: the changing climate and sea level; soils; flora, fauna and landscape; agriculture, horticulture and aquaculture; forestry; water; energy; minerals extraction; construction; transport; manufacturing industry; insurance; finance; human health; recreation and tourism.

The report estimates impacts of and evaluates adaptive responses to a scenario of increasing greenhouse-gas emissions which is compatible with that adopted by the IPCC. This approach assumes a best-estimate trend in population, technology and affluence, with an equivalent doubling of the pre-industrial atmospheric CO_2 level consequently expected to occur in about 2025.[1] The climate scenario differs from that used in the First Report in a number of respects. It also describes the climate for two future periods of time: the decades of the 2020s and 2050s. Warming projections in the 1996 scenario are about 0.5°C less for the 2020s and the 2050s than they were in the 1991 scenario. Winters in particular do not warm by so great an amount; 0.8°C to 2.0°C by the 2050s in the 1996 scenario compared to 2.3°C to 3.5°C in the 1991 scenario. The gradient of warming over the UK is also different, with greater warming in the north and less in the south in CCIRG 1991 being inverted in the 1996 scenario so that the southeast of the UK warms by more than the northwest.

---

[1] Equivalent doubling of pre-industrial atmospheric CO_2 (estimated to be about 280ppmv) occurs when the net radiative forcing of all greenhouse gases combined reaches the equivalent of the forcing due to an atmospheric CO_2 concentration of 560ppmv. This obviously occurs sooner than the doubling of pre-industrial CO_2 concentration alone, which in the 1996 CCIRG scenario would not occur until the 2060s.

2. Changing Climate and Sea Level

SUMMARY

- There is mounting evidence that global climate is changing as a result of human activities. As stated recently by the Intergovernmental Panel on Climate Change (IPCC) the global warming of the last 100 years is unlikely to be due entirely to natural causes. For example, the cluster of warm months and seasons experienced in the UK in the last few years is consistent with this picture of a changing global climate. The 1985-94 decade, both globally and for the UK, has been about 0.2°C warmer than the average of the 1961-90 period and during this decade the average global atmospheric CO_2 concentration has risen by about 5%.

- This review defines a scenario (termed the 1996 CCIRG scenario) of changing climate and sea level for the UK for the decades of the 2020s and 2050s assuming that no major global policies of reducing greenhouse gas emissions are enacted. It is based upon results from the Hadley Centre of the UK Meteorological Office which is recognised worldwide as a leading centre for climate prediction. The effects of sulphate aerosols on future climate are not included in the scenario, although some results from experiments that do consider them are mentioned. Comparisons are made with results from 11 other climate change experiments and using other values for the climate sensitivity to indicate how representative this scenario is of a wider range of possible changes in climate.

Global Climate

- Global mean surface air temperature by the decade of the 2020s will be about 0.9°C warmer than the average of the 1961-90 period and, by the 2050s, about 1.6°C warmer. The rate of warming is about 0.2°C per decade. The range of global warming values for the 2050s is between 1.1°C and 2.4°C. The effect of sulphate aerosols on the climate system would be to reduce the global warming by the 2050s by about 0.3°C.

- The atmospheric CO_2 concentration by the 2020s will be about 430ppmv (±10ppmv) and, by the 2050s, about 525ppmv (±25ppmv). These concentrations represent increases of 29% and 57% respectively compared to the average 1961-90 value. The equivalent CO_2 concentration of all greenhouse gases by the 2050s is almost double the 1961-90 average.

- Global mean sea level by the 2020s will be about 19cm higher than the average of the 1961-90 period and, by the 2050s, about 37cm higher. This rate of sea level rise is about 5cm per decade. The range of sea level rise values for the 2050s is from about 26cm to 50cm. The effect of sulphate aerosols would be to reduce the sea level rise by the 2050s by about 14cm.

UK Climate

- Temperatures over the UK are expected to rise at a generally similar rate to the global mean, about 0.2°C per decade, with slightly slower rates of rise over northwestern UK compared to the southeast, and in winter compared to summer. These differences in pattern and seasonality arise largely due to the influence of ocean circulation changes in the North Atlantic. The regional differences in temperature change over the UK are more pronounced than those reported in the 1991 CCIRG scenario.

- Extremely warm seasons and years are expected to occur more frequently. A hot summer such as 1995 (currently about a 1-in-90 year event) is calculated to occur about three times during the 2050s decade (i.e. a 1-in-3 year event), while a very mild winter such as 1988/89 (currently a 1-in-30 year event) is calculated to occur about twice during the 2050s decade.

- Annual precipitation over the UK as a whole increases by about 5% by the 2020s and by nearly 10% by the 2050s decade. There are, however, substantial regional and seasonal departures from these average changes. Winter precipitation increases everywhere, but more substantially over southern UK. Summer precipitation decreases over southern UK, but increases in the north.

- The number of raindays increases slightly over most of the country in winter, but only over northern UK in summer. Fewer raindays occur over southern UK in summer.

- The average intensity of precipitation increases modestly in all seasons and for all regions. Return periods of heavy daily rainfall events shorten, especially over northern UK in summer.

- Average seasonal wind speeds increase over most of the country in both summer and winter. Return periods of the strongest daily mean wind speeds shorten over northern UK, with little change in the south. The frequency of gales over the UK increases by the 2050s by up to 30%.

- Potential evapotranspiration (PE) is calculated to increase over most of the UK in both winter and summer, the exception being the north of Scotland where decreases occur. Summer PE in southern UK by the 2050s decade is about 25% higher than the average of 1961-90.

- The global rise in sea level, about 5cm per decade, is likely to be exacerbated in southern and eastern UK by sinking land, and mitigated to some extent over the northern UK by rising land. Although nothing quantitative can be said about changes in the frequency or magnitude of storm surges around the coasts of the UK, the probability of a storm surge exceeding a given threshold is likely to increase.

Levels of Confidence

- Although the 1996 CCIRG scenario currently represents our 'best' estimate of UK climate in the future, it describes only *one* possible future evolution of climate and sea level in the UK and implies a linear response of the climate system to increasing greenhouse gas concentrations in the atmosphere. Other responses are also possible and some may involve more abrupt changes in climate related to, for example, changes in the circulation of the North Atlantic. It is impossible to attribute a probability to any one scenario although, since the 1996 CCIRG scenario is based on results from state-of-the-art climate models, it is a reasonable reflection of our current knowledge of how the climate system works. This knowledge base would suggest the following hierarchy of decreasing confidence in the changes in the different climate variables:

  certain: increasing CO_2 concentrations;

  confident: increasing global mean temperature and sea level;

  likely: geographic patterns of regional temperature change;

  possible: geographic patterns of regional potential evapotranspiration, precipitation, gales, return period estimates and frequencies of extreme events.

2.1 INTRODUCTION

This chapter describes the 1996 CCIRG scenario of changing climate and sea level for the UK which this review group has used in their assessment of climate change impacts and adaptations in the UK. This scenario differs from the 1991 CCIRG scenario (CCIRG, 1991) in a number of important ways (see Box 2.1). Both the magnitude and pattern of global and regional warming have been revised since 1991, reflecting developments in our basic understanding of the physical processes which control climate and in the sophistication of the models which simulate it. These developments have been summarised in the scientific reports published in recent years by the Intergovernmental Panel on Climate Change (IPCC, 1990; 1992a; 1995). Current climate models now operate at a higher spatial resolution and with a more advanced representation of the oceans than the first generation equilibrium climate change experiments of the late 1980s. The main global climate model results we have used in the 1996 CCIRG scenario derive from a transient climate change experiment completed by the Hadley Centre in 1992 (Murphy, 1995; Murphy and Mitchell, 1995). Compared to the 1991 CCIRG scenario, more detail about the spatial pattern of climate change over the UK is included, a wider range of climate variables are analysed and changes on both seasonal and daily timescales are examined.

Although the 1996 CCIRG scenario provides a more complete description than five years ago of the changing climate and sea level of the UK, it does not necessarily provide a more accurate description. Climate change prediction, in the sense of being able to attach specific probabilities to the outcomes of climate model experiments, is not yet possible. It may yet be many years before both model development and computing power reach a stage when climate prediction in this sense is achievable and, even then, the dependence of future climate on a range of inherently unpredictable social, economic and technological trends means that definitive predictions of future climate will remain elusive. It is for these reasons that the term 'climate scenario' is usually attached to the sort of portrait of changing climate described in this chapter.

Climate change scenarios present coherent, systematic and physically plausible descriptions of changing climates which may be used as input into climate change impact assessments (Carter

---

Box 2.1: Differences Between 1991 and 1996 CCIRG Scenarios

The anticipated rate of global warming due to greenhouse gases alone is less than it was five years ago. Warming projections in the 1996 CCIRG scenario are about 0.5°C less for the 2020s and the 2050s than they were in the 1991 scenario. This rate of warming would be reduced still further were the effects of sulphate aerosols considered. Future estimates of global CO_2 concentration and sea-level rise are broadly comparable between the two scenarios. Winters in particular do not warm by so great an amount; 0.8° to 2.0°C by the 2050s in the 1996 scenario compared to 2.3°C to 3.5°C in the 1991 scenario. The gradient of warming over the UK is also different, with greater warming in the north and less in the south in CCIRG 1991 being inverted in the 1996 scenario so that the southeast of the UK warms by more than the northwest. This is due to changes in the rates of deep water formation in the North Atlantic. Estimated changes in winter precipitation are much as they were five years ago, but summer precipitation in the 1996 scenario shows a strong difference between the north (wetting) and the south (drying). These precipitation changes remain highly uncertain, however, as they did in the 1991 CCIRG scenario.

et al., 1995). There are a variety of ways in which such scenarios can be constructed, although the most widely favoured methods use results from climate change experiments undertaken with global climate models (GCMs). Such results can be used either in isolation or in conjunction with observed climate data to provide more refined and detailed scenarios (e.g. through down-scaling techniques, the use of weather generators, etc.). Recently, a number of High Resolution Limited Area Models (HRLAMs) have also been used to conduct climate change experiments (e.g. Jones *et al.*, in press). Such models can be used to produce climate change scenarios at finer spatial resolution, although these models are still driven by the results from the coarse resolution global climate models. Results from global climate model experiments can also be linked to the output from simpler, one-dimensional climate models to enable a wider range of assumptions about future greenhouse gas emissions, climate feedbacks and climate sensitivities to be represented in the scenarios. This latter approach of linking results from a GCM and a simple climate model has been used here and is described in the following section.

2.2 APPROACH

There are five key stages in the construction of the 1996 CCIRG scenario: i) selecting the greenhouse gas emissions scenario to be used; ii) calculating the associated projection of global warming; iii) using results from a global climate model experiment to define the geographic patterns of change; iv) defining the reference period; v) and using a statistical model which calculates changes in the frequencies of daily temperature events for specific sites. Wherever possible the insights and recommendations of the IPCC have been adopted and consistency ensured between the 1996 CCIRG scenario and climate scenarios constructed by the Climate Impacts Link Project, funded by the UK Department of the Environment (Viner and Hulme, 1994). The five stages in the scenario construction are briefly summarised in turn.

i) In 1992, the IPCC defined a set of six greenhouse gas emissions scenarios for the world (named IS92a to f; Leggett *et al.*, 1992) which continue to be used by many climate change analysts (Alcamo *et al.*, 1995). These scenarios define possible pathways of greenhouse gas emissions in the absence of new policies to reduce them and are therefore regarded as 'non-intervention' scenarios. For the 1995 IPCC scientific assessment (IPCC, 1996a) four of these six scenarios were modified to account for the more rapid phasing out of CFCs and HCFCs than was envisaged at the start of the 1990s (the emissions scenarios IS92d and IS92e had already made this assumption). These modified scenarios are now referred to as the augmented IS92 scenarios. The 1996 CCIRG climate change scenario uses the augmented IS92a emissions scenario (see Box 2.2), the global warming from which falls roughly in the middle of the six IS92 scenarios (Figure 2.1).

ii) To convert this emissions scenario into a projection of global warming a simple one-dimensional upwelling-diffusion energy balance model has been used (see Wigley and Raper, 1987; 1992; 1993; Raper *et al.* (1996). This model has been, and still is being, widely used in IPCC assessments and has evolved to take into account the major scientific developments reported by the IPCC over the last six years. The version of the model used for the 1996 CCIRG scenario is the same as that used in the 1995 IPCC report (Kattenburg *et al.*, 1996). A climate sensitivity of 2.5°C (the 'best guess' IPCC value) has been used in the calculations shown here. The climate sensitivity is a measure of how sensitive global climate is to a specified change in the geochemical composition of the atmosphere. As used here it is defined as the eventual warming of global climate which follows a doubling of CO_2 concentration in the atmosphere.

> **Box 2.2: The IS92a Emissions Scenario**
>
> The IS92a scenario adopts intermediate assumptions compared to the other IPCC scenarios, assuming the intermediate World Bank global population projection of 11.3 billion by 2100, an average annual global economic growth rate of 2.3% between 1990 and 2100 and a mix of conventional and renewable energy sources in the future. It assumes that the price of solar power will drop and that the price of nuclear energy will increase until they are at about the same level. Net CO_2 emissions from deforestation remain roughly at 1Gt/yr., but then decline after 2025. Diminishing land clearing results in net storage of 0.1 Gt/yr. by 2100. While there is a large range of other possible futures with regard to greenhouse gas emissions, the IS92a scenario remains a sensible one to use at the present time given that the emissions fall roughly midway in the range of scenarios recently reviewed by the IPCC (Alcamo *et al.*, 1995).

Table 2.1 shows the changes for a number of future decades in three important indicators of the global climate system derived from this model - global mean CO_2 concentration, global mean temperature and global mean sea level. The global warming by the 2020s decade is about 0.9°C, rising to about 1.6°C by the 2050s, with a sea level rise of about 19cm by the 2020s, rising to 37cm in the 2050s. Some of the uncertainties regarding these calculations are discussed at the end of this chapter. Climate will not stop changing after the 2050s decade. Although this decade is used as the time horizon of this report, Table 2.1 shows that under the IS92a scenario the world continues to warm at a rate of slightly more than 0.2°C per decade. Furthermore, if emissions of greenhouse gases should be stabilised at some future time, global concentrations, climate and sea level would not stabilise at the same time. There is a time lag between the stabilisation of the various components of the system: emissions, concentrations, global temperature and global sea level. For example, Wigley (1995) has shown that global temperature would continue to rise for several decades following a stabilisation of greenhouse gas concentrations and that global sea level would continue to rise for several centuries following

Figure 2.1: The calculated rise in global mean temperature (with respect to 1961-90) for three different greenhouse gas emissions scenarios (augmented IS92a, IS92c and IS92e) using a climate sensitivity of 2.5°C. The vertical bars show the decades 1985-94, the 2020s and the 2050s as used in Table 2.1. Historic global mean temperature anomalies from 1854-1994 (with respect to the 1961-90 average) are also shown. No effects due to sulphate aerosols are included. Results derive from the simple climate model used in Kattenburg *et al.*, (1996) and developed by Wigley and Raper (Raper *et al.*, 1996).

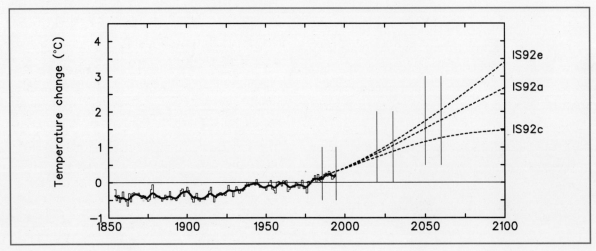

Table 2.1: Observed and modelled values of temperature and sea level change, and CO_2 concentration, for two historic and three future periods. All changes are shown with respect to the 1961-90 average. Modelled results assume the augmented IS92a emissions scenario and IPCC 'best guess' model parameters (Kattenburg et al., 1996). No effects due to sulphate aerosols are included (see Table 2.6 for calculations including these effects).

	1961-90	1985-94	2020s	2050s	2090s
Observed Central England ΔT (°C)	0.0	+0.19			
Observed global-mean ΔT (°C)	0.0	+0.19			
Observed CO_2 concentration (ppmv)	334	352			
Modelled global-mean ΔT (°C)	0.0	+0.23	+0.92	+1.63	+2.55
Modelled CO_2 concentration (ppmv)	334	355	433	525	677
Modelled global-mean ΔSL (cm)	0.0	+4.0	+19.2	+36.9	+63.1

such stabilisation. Under the scenario used here, if greenhouse gas concentrations were held at 2050s levels an additional 0.4°C of global warming would still occur after the 2050s decade. This change is referred to as the climate change commitment of a given date and emissions scenario.

iii) A large number of climate change experiments using global climate models have been performed over the last ten years. Many of these have been equilibrium experiments in which the sensitivity of the global climate system to an instantaneous doubling of greenhouse gas concentrations is explored. Results from five such experiments were used in the 1991 CCIRG scenario. Since then, a number of transient climate change experiments have been completed in which the more realistic, slowly evolving nature of climate change has been better simulated by global climate models (Gates et al., 1992; Kattenburg et al., 1996). Results from one such transient experiment were used for the 1996 CCIRG scenario. This experiment was completed by the Hadley Centre during 1991/92 using their coupled ocean-atmosphere climate model and is referred to as UKTR. The full details of this experiment are reported in Murphy (1995) and Murphy and Mitchell (1995). The UKTR experiment produces a reasonable simulation of current climatic patterns, at least compared to other global climate models (Hulme, 1994). The implications of choosing just one model experiment to define the CCIRG scenario are discussed in Section 2.7.

The geographic patterns of climate change calculated by the UKTR model during the last simulated decade of the experiment (model years 66-75) were used to define the scenario. By this time the UKTR model world had warmed by about 1.8°C with respect to the modelled control climate. These patterns of change were scaled according to the IS92a global warming projection to generate scenarios for the decades of the 2020s and 2050s (i.e., the changes from the final decade of the UKTR experiment were multiplied by 0.9/1.8 for the 2020s decade and by 1.6/1.8 for the 2050s decade). This approach assumes a simple linear evolution of the pattern of climate change over time; this may be a poor assumption. The spatial resolution of the UKTR experiment, although good for a global climate model, is still coarse. The land area of the UK, for example, is represented by only four gridboxes. Many of the smaller-scale influences on UK climate (such as the configuration of the coastline and local topography) are not simulated in the model. The spatial patterns of climate change shown in subsequent maps in this chapter therefore largely reflect the large scale response of the climate system to greenhouse gas forcing and not the response of local climatic influences.

iv) The baseline period has been defined as 1961-90. This enables the baseline climatology constructed for the UK by Barrow *et al.* (1993) to be used and represents the most recent 30-year period for which full climate data are available. Table 2.1 shows the average 1961-90 values for a number of relevant climate variables and also the observed and modelled changes from these reference values for the most recent 10-year decade, 1985-94. Both the Central England Temperature (Parker *et al.*, 1992) and global temperature (Jones, 1994) records reveal that this decade has been nearly 0.2°C warmer than the average of 1961-90. This change is quite well simulated by the simple climate model used here, the observed value of 0.2°C falling about mid-way between the modelled values of 0.15°C and 0.23°C representing, respectively, global warming with and without the effects of sulphate aerosols included (cf. Tables 2.1 and 2.6). Confident attribution of this warming to greenhouse gas and sulphate aerosol-induced climate change is nevertheless still difficult (see Box 2.3).

v) For calculations of changes in the frequency of extreme temperature events for specific sites in the UK, and in accumulated degree days above and below certain thresholds, a statistical model has been used. This model, called SPECTRE (see Barrow and Hulme, 1996), was developed during 1994 for the UK Ministry of Agriculture, Fisheries and Food. The results from SPECTRE shown in this chapter are fully consistent with the 1996 CCIRG scenario.

The five steps described above are used to construct the 1996 CCIRG scenario of climate and sea level change for the UK. This scenario emphasises changes for the decades of the 2020s and the 2050s. These decades are, respectively, those in which people currently in their mid-30s and the current pre-school population would reach retiring age. Box 2.1 summarises the differences between the 1991 and 1996 CCIRG scenarios.

2.3 CHANGES IN TEMPERATURE

2.3.1 Changes in mean seasonal temperature

Changes in mean winter and summer surface air temperatures over the UK for the decades of the 2020s and 2050s are shown in Figure 2.2. A gradient of decreasing warming, from southeast to northwest, is evident in both seasons, winter warming by the 2050s ranging from about 2°C

Box 2.3: Detecting Human-induced Climate Change

Considerable efforts have been made in recent years, both to detect statistically significant climate change in a variety of instrumental records and to attribute any such change to human pollution of the atmosphere (Santer *et al.*, 1996). Detection studies have often examined global mean datasets (such as the record of surface air temperature shown in Figure 2.1) while attribution studies have more commonly used datasets which contain information about the spatial pattern of climate. The 1995 IPCC report has concluded that the recorded change in global mean temperature over the last century is unlikely to be due entirely to natural variability of the climate system. Exactly how important human-induced changes in atmospheric composition have been for global climate change remains impossible to quantify at present. Since the UK forms less than 1% of the surface area of the Earth, changes in UK climate are not necessarily very significant indicators of global scale climate change. Nevertheless, the warming of the 1985-94 decade with respect to 1961-90, both globally and for the UK, is at least consistent with the theoretical expectation of human-induced climate change which forms the basis of the CCIRG scenario.

Figure 2.2: Change in mean winter (a; December, January and February) and summer (b; June, July and August) temperature (°C) for the 2020s (left) and 2050s (right) decades with respect to the average of 1961-90.

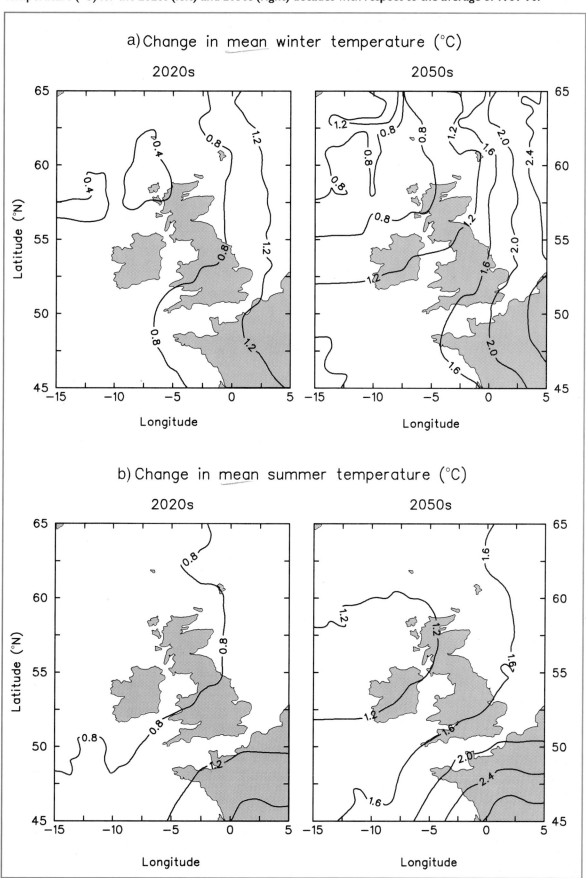

Figure 2.3: Change in mean winter (a) and summer (b) diurnality (°C) for the 2020s (left) and 2050s (right) decades with respect to the average of 1961-90.

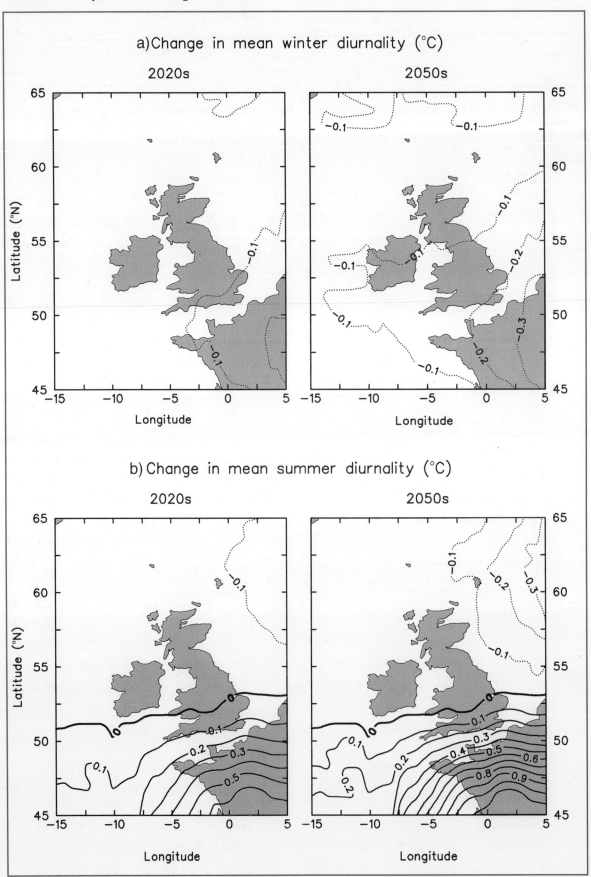

to about 0.8°C and summer warming from about 1.8°C to about 1.2°C. This pattern of warming is different from that reported in the 1991 CCIRG scenario, where higher latitudes over the UK warmed more rapidly than more southerly ones. The difference between the 1991 and 1996 CCIRG scenarios is a consequence of using results from a climate change simulation in which the dynamics of the ocean are explicitly modelled. The sequestration of heat into the deep ocean in the area of North Atlantic Deep Water formation is simulated by the UKTR experiment and this sequestration reduces the surface air temperature warming in the vicinity of the North Atlantic Basin. This phenomenon was not simulated by the earlier equilibrium experiments used in the 1991 scenario. Also in contrast to the 1991 scenario, the warming rate between winter and summer is similar over most of the UK. This difference from the 1991 scenario also results from the stronger influence of the oceans in the UKTR experiment. In the earlier equilibrium experiments the ice-albedo feedback mechanism dominated the winter temperature response to greenhouse gas forcing even at these mid latitudes.

Figure 2.3 shows a further feature of this warming over the UK, namely changes in the mean diurnal temperature cycle. A slight decrease in diurnality occurs in winter, with minimum temperatures rising more rapidly than maxima, whereas in summer there is an increase in diurnality over southern Britain and little change further north. The marked increase in summer diurnality over the more southerly latitudes of the country results from reductions in summer cloudiness and precipitation (as will be shown later).

There are also likely to be important secondary effects of this general warming tendency which will have health (Chapter 15) and tourism (Chapter 16) implications. Higher air temperatures (particularly in summer) combined with increased local emissions of NO_x and hydrocarbons (particularly from the transport sector) will lead to changes in lower tropospheric ozone concentrations, air quality and the formation of photochemical smog in urban areas. Some of these issues have been, or are being, considered in the Department of Environment's Review Groups on the Quality of Urban Air and on Photochemical Oxidants.

2.3.2 Changes in return periods of extreme seasonal anomalies

In addition to the above changes in mean seasonal temperatures, it is also useful to have some idea about how the likelihood of certain seasonal mean temperature extremes may change in a warming climate. The 1991 CCIRG Report made the simplifying assumption to estimate such changes that only mean temperature would change and that temperature variability would remain the same. The summer and winter temperatures from the 336-year Central England Temperature record compiled by the late Gordon Manley and now updated by the Hadley Centre (Parker et al., 1992) were described statistically assuming a normal distribution and the occurrence probability of a number of extreme seasons calculated. These distributions were then shifted according to the mean warming in the scenario (the variability was held constant) and the occurrence probabilities recalculated for conditions of climate change.

The same methodology was followed here and the result is illustrated in Figure 2.4 for summer temperatures in Central England. Table 2.2 shows the changing return periods for some recent extreme seasonal temperature anomalies over the UK. The two sets of results in Table 2.2 refer to two different choices of baseline period. Since summers during 1961-90 have been slightly warmer than the full 336-year record, the estimated return period for the 1995 summer temperature using 1961-90 as the baseline period is only about 90 years compared to 230 years if the baseline period is 1659-1994. In either case, it is estimated from the 1996 CCIRG

Figure 2.4: Distribution of summer mean temperatures for Central England: (a) based on the 1961-90 period average and (b) assuming the 1996 CCIRG scenarios for the 2020s and 2050s decades. Shaded areas denote changes in approximate exceedance probabilities for the hot summer of 1995.

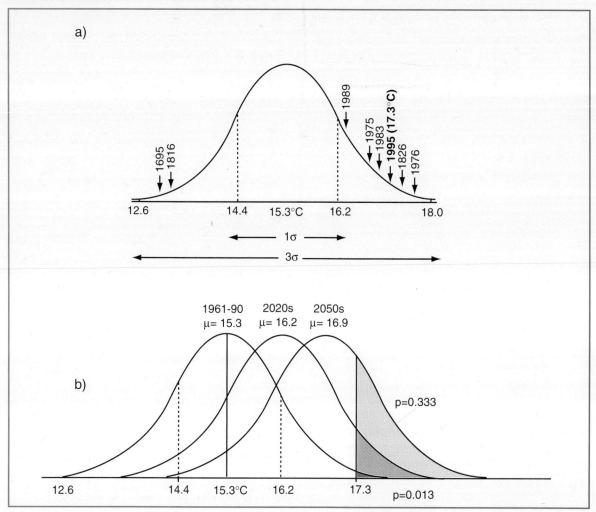

Table 2.2: Extreme annual and seasonal temperature anomalies of recent years and their approximate estimated return periods (in years) under 'current' climate and under the 1996 CCIRG scenario for the 2050s decade. The two sets of results refer to different definitions of what constitutes the baseline period: 1659-1994 or 1961-1990. Estimates derive from the Central England Temperature record (Parker et al., 1992; and updated).

		with respect to 1659-1994			with respect to 1961-1990		
	temp. (°C)	anom. (°C)	current (years)	2050s (years)	anom. (°C)	current (years)	2050s (years)
Annual 1990	10.6	+1.4	85	1.8	+1.1	65	1.6
Summer 1976	17.7	+2.5	1110	8	+2.4	310	5.5
Summer 1995	17.3	+2.1	230	4	+2.0	90	3
Winter 1988/89	6.5	+2.8	70	6	+2.5	30	4
Winter 1962/63	0.4	-3.3	205	∞	-3.6	230	∞

scenario that the 2050s decade alone would experience on average about three such summers. Assuming these seasonal extremes are independent events, the probability of two successive 1995-like summers occurring during the decade is about 0.9 (i.e., very likely). Such an event would have severe implications for agriculture (see Chapter 5) and water resources (see Chapter 7) in the UK. The very mild winter of 1988/89 would recur on average between every four and six years (i.e., twice during the 2050s) compared to between one and four times a century at present. By the 2050s decade, the chance of a very cold winter such as 1962/63 is extremely small.

This analysis makes certain assumptions which cannot be fully evaluated at present: first, that the variability of seasonal temperatures remains unchanged and, second, that the normal distribution is a reasonable description of the frequency distribution of these seasonal temperatures and that this distribution will continue to describe future temperatures. There is also the problem hinted at above. The UK climate is currently not stationary and therefore it is difficult to know which period should form the baseline for calculating the distribution parameters. Despite these limitations, the type of analysis described above does give a rough indication about how the frequency of seasonal temperature extremes may change in the future.

2.3.3 Changes in frequencies of daily temperature extremes

Changes is the frequency of *daily* temperature extremes are also likely to accompany a climatic warming and such changes will have important consequences for species (Chapter 4), agriculture (Chapter 5) and human health (Chapter 15). Using the SPECTRE model (Barrow and Hulme, 1996), Table 2.3 presents estimates of the changes in mean annual frequencies of very warm days and of air frosts for six locations representing a range of UK climates. The method uses a variety of statistical distributions to describe the daily temperature regimes and then perturbs these distributions (both the mean and variance) in accordance with the results from the UKTR climate change experiment.

By the 2050s, frost frequencies fall by about 50% at all six locations, with the largest reduction at Plymouth where on average only eight air frosts are estimated to occur by the 2050s compared to 21 at present. The average frequency of very warm summer days increases throughout the country (except at Fortrose where even in the 2050s the climate is not warm enough to generate days warmer than 25°C), with the mean frequency at Oxford nearly doubling from 12 at present to 20 by the 2050s.

2.3.4 Changes in degree days

A further consequence of climatic warming in the UK will be changes in the degree days above and below certain temperature thresholds. De-

Table 2.3: Changes in mean annual frequencies of daily temperature extremes for six locations around the UK. Results obtained from the SPECTRE model (Barrow and Hulme, 1996). Santon Downham is in Norfolk, Hillsborough is near Belfast and Fortrose near Inverness.

	Latitude (°N)	1961-90	2020s	2050s
		Mean annual frequency of days with $T_{min}<0°C$		
Plymouth	50.4	21	10	8
Oxford	51.8	42	26	18
Santon Downham	52.4	79	60	44
Hillsborough	54.6	41	27	19
Durham	54.8	56	39	28
Fortrose	57.6	31	22	15
		Mean annual frequency of days with $T_{max}>25°C$		
Plymouth	50.4	3	4	8
Oxford	51.8	12	15	20
Santon Downham	52.4	11	17	26
Hillsborough	54.6	1	2	3
Durham	54.8	3	5	7
Fortrose	57.6	0	0	0

gree days are the number of degrees above (or below) a specific threshold temperature, accumulated over all days in the year or season on which the temperature is above (or below) the same threshold value. These thermal indices are used in a variety of impact sectors including agriculture and forestry (growing degree days; Chapters 5 and 6) and energy and the built environment (heating and cooling degree days; Chapters 8 and 11). Table 2.4 shows scenario values for a number of such indices for the same six locations in the UK used above, again derived from the SPECTRE model. Growing degree days (GDDs; accumulated mean temperatures above 5.5°C) increase by between 24% and 34% by the 2050s decade, with Durham, for example, recording as many GDDs by the 2050s as Plymouth at the present time. Heating degree days (HDDs; accumulated mean temperatures below 15.5°C) reduce by a smaller percentage (between 16% and 23%) and the latitudinal shift of zones is also not as great as for GDDs with Durham still recording more HDDs by the 2050s than does Plymouth under current climate. Cooling degree days (CDDs; accumulated mean temperatures above 18°C) increase everywhere, although owing to the small baseline totals the percentage changes are not always meaningful. Nevertheless, CDDs nearly double by the 2050s for locations such as Oxford and Santon Downham in the south of England.

Table 2.4: Changes in three thermal indices by the 2020s and 2050s decades (growing degree days, cooling degree days and heating degree days) for six locations around the UK. Results obtained from the SPECTRE model (Barrow and Hulme, 1996). Percentage changes from 1961-90 to the 2050s are shown in parentheses in the final column.

	1961-90	2020s	2050s
Growing degree days (T_{mean} >5.5°C)			
Plymouth	1956	2234	2454 (+25%)
Oxford	1909	2171	2372 (+24%)
Santon Downham	1689	1927	2113 (+25%)
Hillsborough	1415	1668	1870 (+32%)
Durham	1479	1724	1921 (+30%)
Fortrose	1343	1601	1805 (+34%)
Heating degree days (T_{mean} <15.5°C)			
Plymouth	1992	1723	1528 (-23%)
Oxford	2306	2041	1849 (-20%)
Santon Downham	2751	2487	2299 (-16%)
Hillsborough	2625	2341	2137 (-19%)
Durham	2712	2435	2237 (-18%)
Fortrose	2616	2318	2106 (-20%)
Cooling degree days (T_{mean} >18.0°C)			
Plymouth	42	83	123 (+193%)
Oxford	136	191	236 (+74%)
Santon Downham	142	199	247 (+74%)
Hillsborough	2	20	44 (very high)
Durham	33	71	105 (+218%)
Fortrose	0	1	7 (n/a)

2.4 CHANGES IN PRECIPITATION

2.4.1 Changes in mean seasonal precipitation

Mean winter and summer precipitation for the 2020s and 2050s are shown in Figure 2.5, expressed as percentage changes with respect to 1961-90. Winters are wetter throughout the UK in the 1996 CCIRG scenario, up to 10% in south eastern England by the 2050s. The pattern of precipitation change in summer shows a contrast between northern and southern UK - the north getting wetter and the south getting drier with the dividing line through the north Midlands and north Wales. Summer precipitation declines by up to 9% in the extreme south of England by the 2050s.

The 1991 CCIRG scenario did not contain changes in precipitation *patterns* over the UK. The two reasons given were that climate models were not capable of simulating the observed patterns of precipitation at regional scales and that different climate models generated different patterns and magnitudes of change when forced by the same increased greenhouse gas concentrations. The former problem is perhaps not as great as a few years ago. The newer gen-

Figure 2.5: Percentage change in mean winter (a) and summer (b) precipitation for the 2020s (left) and 2050s (right) decades with respect to the average of 1961-90.

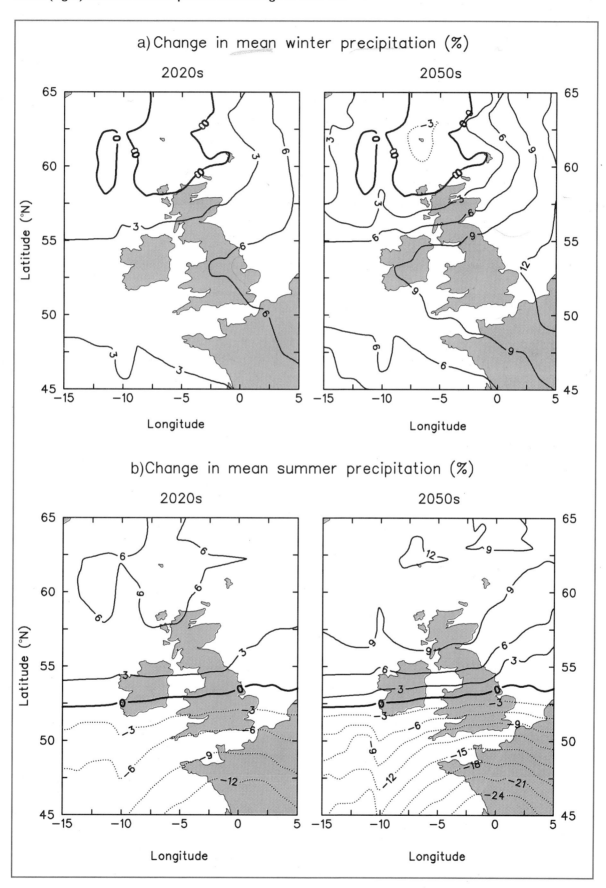

eration of high resolution climate models (of which UKTR is one) are performing markedly better than the older coarse resolution models at reproducing the mean regional patterns of precipitation, at least for some regions including Europe (Hulme, 1994). The latter problem persists, however, and Section 2.7 illustrates the range of precipitation changes simulated for the UK by a set of 11 global climate model experiments.

The development of transient climate change experiments has, however, highlighted a new difficulty about defining the patterns of climate change which are due to greenhouse gas forcing, a difficulty which is present to some extent for all of the variables described in this scenario, but which is particularly acute for precipitation. Precipitation is a climate variable which has a high degree of natural interannual and interdecadal variability, certainly higher than for temperature and probably higher than for most other surface climate variables. Since transient experiments simulate the gradual evolution of climate as the atmosphere is subject to gradually increasing concentrations of greenhouse gases, the natural variability of precipitation obscures for at least several decades any changes which may be due to greenhouse gas forcing. The precipitation changes which have been defined from the final decade of the UKTR experiment (model years 66-75) are therefore a mixture of natural variability and greenhouse gas induced changes. Separating out these two contributions to the changes shown in Figure 2.5 is very difficult.

The problem is illustrated in Figure 2.6 which shows the year-by-year values of summer and winter precipitation from the 75 years of the UKTR experiment for regions in both northern and southern UK. The values are expressed as percentage changes from the model control climate. The smooth curves are retrospective 10-year moving averages of the yearly values and one can see that there is considerable variation from decade to decade in the magnitude, and even the direction, of precipitation change simulated by the model. For example, if the years 35-44 were used to define the precipitation scenario then over northern UK in summer a substantial (20%) increase in precipitation would result, whereas if the following decade were used (years 45-54) then a small (5%) decrease in precipitation would occur. Nevertheless, there is some visual indication from Figure 2.6 of a fairly consistent increase in winter precipitation over northern UK and a slightly less consistent decrease in summer precipitation over southern UK.

2.4.2 Changes in daily rainfall regimes

Obtaining a scenario of changes in rainday frequencies from a climate model is not straightforward, since the nature of a model 'rainday' is substantially different from one in the real world. An observed rainday is usually defined as a day with at least 0.1mm of precipitation falling at a specific site. Most climate models generate more raindays than exist in reality. The main reason for this is that climate models are simulating rainfall over very large areas (10,000s km^2) rather than at individual sites. In some months and at some latitudes virtually every model day is a rainday. For this reason, raindays were defined here as occurring only if the model generated more than 2mm rainfall within a gridbox. Comparison with an observed climatology has shown that this threshold produces somewhat similar mean rainday frequencies between the climate model and reality (Hulme *et al.*, 1994).

Figure 2.7 shows the changes in mean winter and summer frequency of raindays, thus defined, for the 2020s and 2050s. The patterns shown here are broadly similar to the changes in mean seasonal precipitation, with increases in winter raindays over most of the country and changes in summer raindays depending on latitude (increases over northern UK and decreases over southern UK). All these changes are rela-

Figure 2.6: Year-by-year values of winter and summer precipitation change from the UKTR experiment expressed as percentage change from the model control climate for areas representative of northern (a) and southern (b) UK.

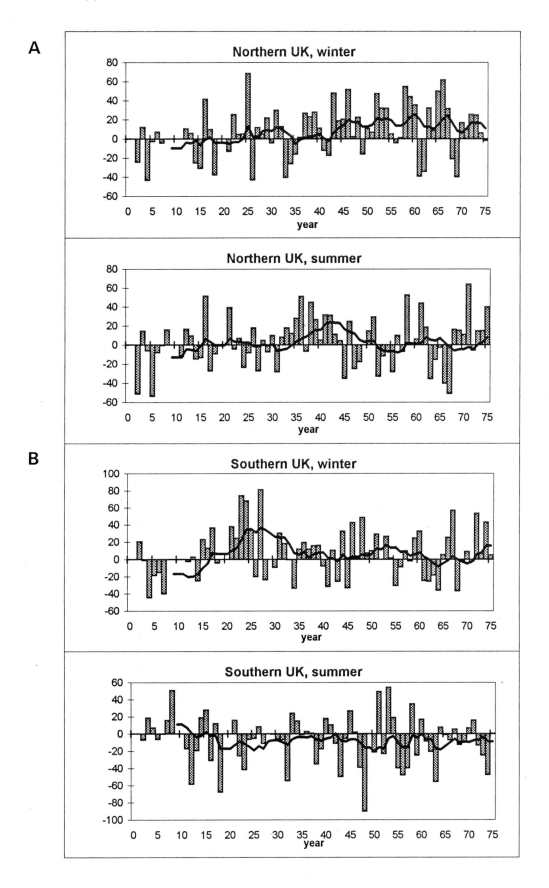

Figure 2.7: Change in mean winter (a) and summer (b) rainday frequency for the 2020s (left) and 2050s (right) decades with respect to the average of 1961-90. See text for discussion of rainday definition.

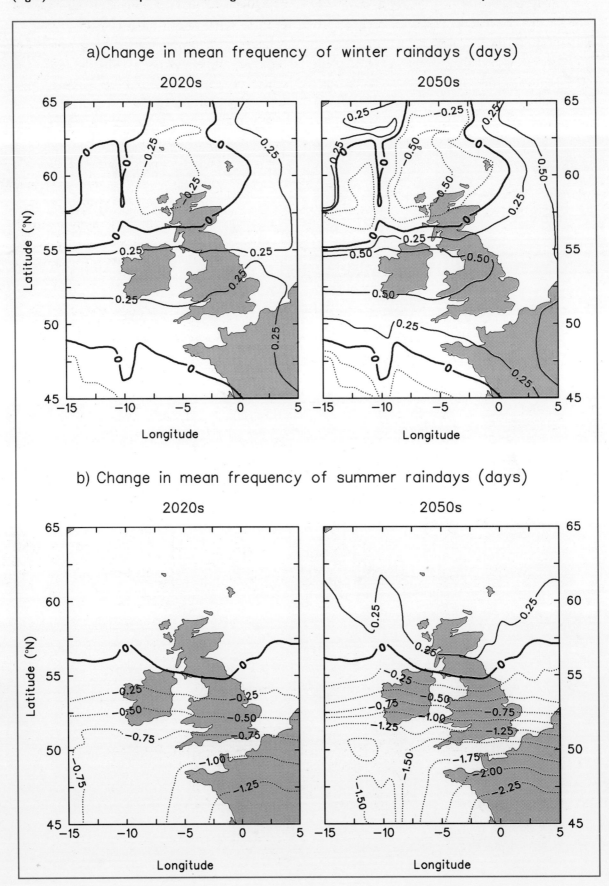

tively modest, however, amounting to less than two days per season. The most significant changes are those occurring over the extreme south of the country in summer, where a reduction of one or two raindays is equivalent to about a 5% decline in rainday frequency. The proportional changes in rainday frequencies, however, are generally less than those in precipitation, implying a general increase in precipitation amounts on raindays.

For many applications it is changes in the frequency of the heaviest rainfall events that will be of most significance, for example, for soil erosion (Chapter 3), agriculture (Chapter 5) and flood estimation (Chapter 7). Although only 10 years of simulated daily rainfall data are available from UKTR, it is possible to examine these data to identify the direction and relative magnitudes of the change in return period of the heaviest rainfall events (Figure 2.8). It is important to remember that these curves do not accurately portray the *absolute* magnitudes of daily rainfalls above certain thresholds: the spatial resolution of these models means that they are not suitable for this purpose. Neither can changes in rainfall intensities on shorter time steps, for example hours or minutes, be calculated since such model data are rarely archived. Summer daily rainfall intensities are estimated to increase over northern UK. For example, a

Figure 2.8: Estimated return periods for summer and winter daily rainfalls for northern (a) and southern (b) UK under current 1961-90 climate conditions ('control') and for the 2050s decade. Both control and 2050s return periods are calculated from the daily rainfall distributions from the last 10 years of the UKTR experiment.

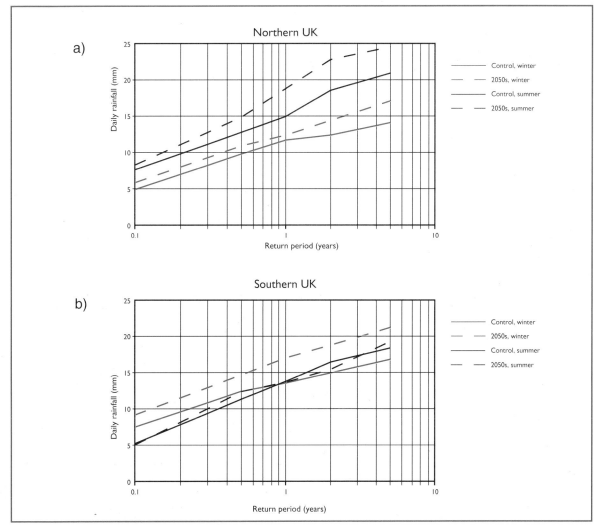

20mm daily total occurs almost once a year by the 2050s compared to once every four years under the model control climate (see Figure 2.8 caption for explanation). There is little change over southern UK. In winter, rainfall intensities increase in both regions, although more substantially over southern UK where, for example, a 15mm daily rainfall total occurs twice a year by the 2050s compared to once every two years under the control climate.

It is prudent to assume that the same caveat about the extent to which these changes are greenhouse gas-induced change as opposed to being caused by natural variability applies as much to the above discussion about rainfall frequency and intensity change as it does to changes in mean seasonal precipitation.

2.5 CHANGES IN OTHER VARIABLES

Changes in other climate variables apart from temperature and precipitation are also important for a range of climate change impacts on the UK. Thus changes in snowfall will be important for transport (Chapter 12) or recreation (Chapter 16), in wind speed for forestry (Chapter 6) and construction (Chapter 11), in radiation for agriculture (Chapter 5) and construction (Chapter 11) and in potential evapotranspiration for agriculture (Chapter 5) and water resources (Chapter 7).

2.5.1 Snowfall

It is not possible to obtain a model-derived scenario of change in either snowfall or snow cover duration since the UKTR experiment greatly underestimates the occurrence of both of these variables over the UK. All that can be said in a qualitative sense is that there is likely to be a reduction in snow cover (for example, fewer days with snow lying) whereas reductions in the number of snowfalls, while likely, may not be so great or so certain. Changes in the synoptic origins of the winter precipitation may also be important in determining its effect on snowfall (Harrison, 1993). It would be possible to employ a dedicated snow cover model, using scenario temperature and precipitation changes as inputs, to calculate the affect of the 1996 CCIRG scenario on UK snow cover, but such an exercise has not been attempted.

2.5.2 Relative humidity

Vapour pressure increases everywhere over the UK, with the largest absolute increases of about 1.4hPa occurring over northern UK in summer (not shown). Owing to rising temperatures, however, relative humidities do not increase in the same way and over southern parts of the country relative humidity actually declines throughout the year, in summer by up to 6% (Figure 2.9). These changes in relative humidity have important implications for changes in potential evapotranspiration (see Section 2.5.5).

2.5.3 Solar radiation

Changes in summer incoming solar radiation (direct and diffuse) broadly follow the pattern of precipitation change shown earlier (Figure 2.5). The strongly latitudinal pattern of radiation change in summer (Figure 2.10) closely follows changes in cloud cover (not shown) and precipitation. Summer radiation changes are in the range ±5%, but incoming solar radiation shows little absolute change in winter.

2.5.4 Wind speeds

Changes in mean seasonal wind speeds are shown in Figure 2.11 and these increase throughout the country and in both seasons. Mean winter winds over southern England are up to 7% higher than the average of the 1961-90 period by the 2050s. Using daily wind speed output from UKTR, a similar return period analysis was performed on wind speeds as for daily rainfall intensities (Figure 2.12). The same caveat applies as before - this analysis does not produce reliable values of the *absolute*

Figure 2.9: Change in mean winter (a) and summer (b) relative humidity (percentage units) for the 2020s (left) and 2050s (right) decades with respect to the average of 1961-90.

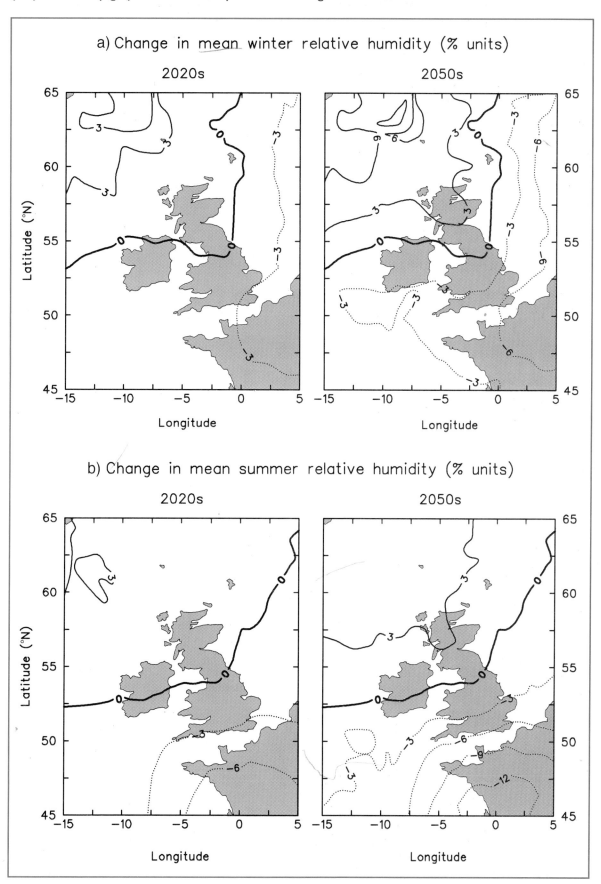

Figure 2.10: Change in mean winter (a) and summer (b) incoming solar radiation (Wm^{-2}) for the 2020s (left) and 2050s (right) decades with respect to the average of 1961-90.

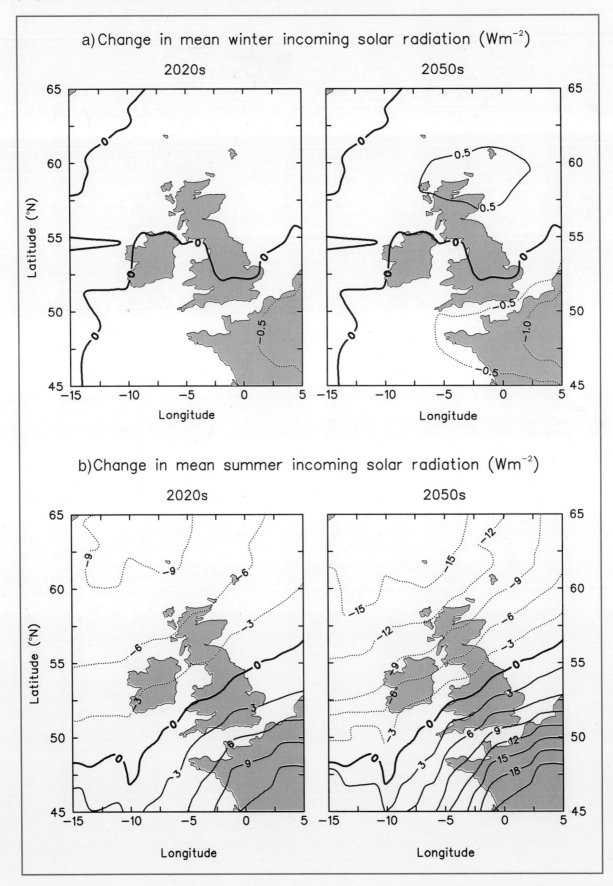

Figure 2.11: Percentage change in mean winter (a) and summer (b) wind speed for the 2020s (left) and 2050s (right) decades with respect to the average of 1961-90.

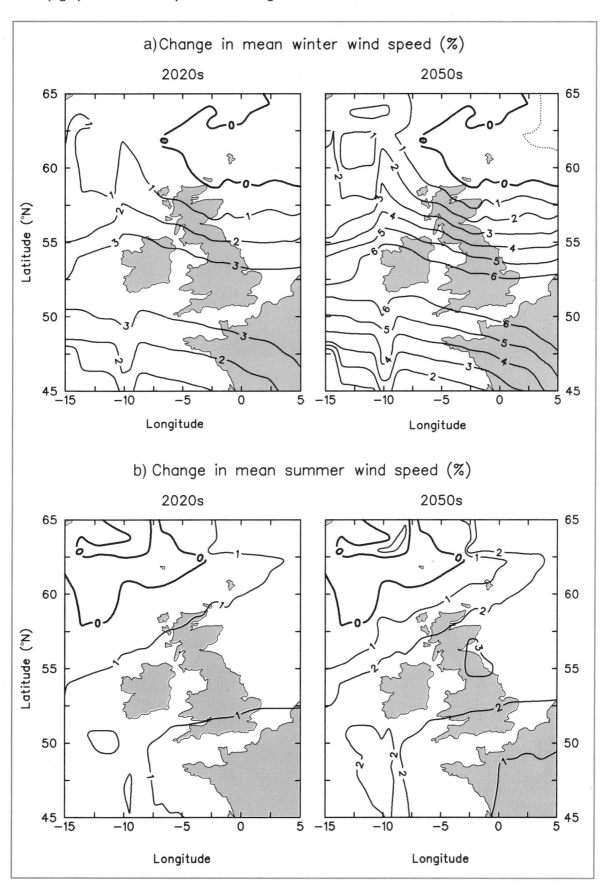

Figure 2.12: Estimated return periods for summer and winter daily windspeeds for northern (a) and southern (b) UK under current 1961-90 climate conditions ('control') and for the 2050s decade. Both control and 2050s return periods are calculated from the daily windspeed distributions from the last 10 years of the UKTR experiment.

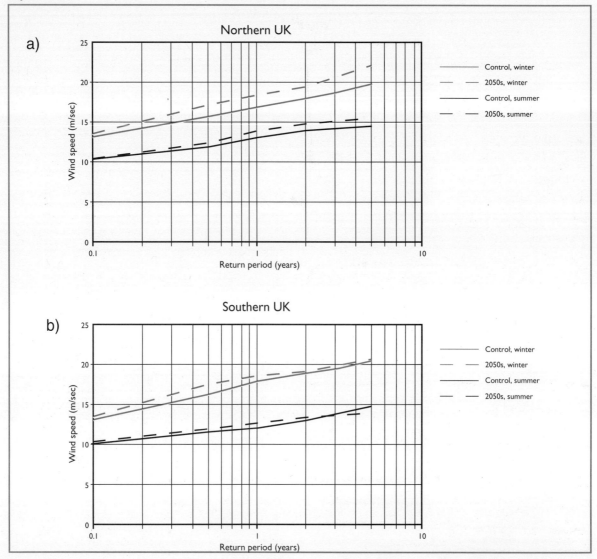

magnitudes of wind speeds at certain return periods, but the changes do give an indication of the sign and relative magnitude of the change which might be expected by the 2050s. Also, these changes are for mean daily wind speeds and not for maximum instantaneous gusts. Return periods generally shorten over northern UK in winter, where a daily mean wind speed of 18 m sec^{-1} changes from a one-in-two year event to an annual event. Return periods shorten by smaller amounts over northern UK in summer. Over southern UK there appears to be little change in return periods for the strongest winds in either summer or winter, despite the general increase in mean seasonal wind speed depicted in Figure 2.11.

An alternative approach to examining changes in wind speeds is to use a gale index. Carnell *et al.* (in press) have performed such an analysis for the UKTR experiment and calculate that gale frequencies over the UK increase by up to 30% for a 1.6°C global warming. This result may be model-specific, however, since two other model experiments analysed for changes in UK storminess do not show any similar increase in storm frequencies. Furthermore, the same caveat about distinguishing greenhouse gas-induced changes from those due to natural variability which was cited above in connection with the precipitation scenario also applies to these wind changes.

2.5.5 Potential evapotranspiration

Potential evapotranspiration (PE) is a calculated quantity and may be defined as the maximum quantity of water capable of being lost to the atmosphere as water vapour under a given climate by an extensive stretch of vegetation freely supplied with water. PE is not output directly by climate models, but can be calculated offline using one of the established PE formulae and the appropriate variables which are generated by the model. PE was calculated here using a standard Penman formula (Penman, 1948) with mean temperature, vapour pressure, wind speed and global radiation (direct and diffuse solar radiation) as inputs. Each of these variables has a different relative influence on the eventual PE estimate, with temperature and relative humidity having the strongest influences. Figure 2.13 shows the mean seasonal changes in PE for the 2020s and 2050s. Large relative increases in winter PE are generated, although in absolute terms these remain rather small at less than 1mm day^{-1}. In summer, PE increases by up to 30% over southern England, but further north PE increases are smaller and over northern Scotland PE decreases by up to 15%. The main reason for this summer pattern of PE change is to be found in the patterns of temperature and relative humidity change shown earlier. Temperatures rise most strongly over southern UK, whereas relative humidity increases in the north, but decreases in the south. Since PE is most strongly affected by these variables a strong latitudinal contrast in PE change is generated.

It should be noted that in calculating PE change no allowance has been made for changes in plant stomatal conductivity resulting from higher CO_2 concentrations (see Flora, Fauna and Landscape, Chapter 4). Previous sensitivity work in the UK has suggested that such water use efficiency gains could offset a substantial proportion of any climatically-induced increase in PE (Arnell and Reynard, 1993).

2.6 CHANGES IN SEA LEVEL

As the global temperature increases, global mean sea level is expected to rise due to thermal expansion of the oceans and to increased melting of mountain glaciers and the Greenland ice sheet. The most likely response of the Antarctic ice sheet to global warming is increased accumulation (increased precipitation more than offsetting any melting around the ice sheet margins) and, if so, this would contribute a slight fall in sea level.

Global mean sea level under the augmented IS92a emissions scenario will rise with respect to the average of the 1961-90 period by about 19cm by the 2020s, by about 37cm by the 2050s and by about 63cm by the 2090s decades (Table 2.1). Nearly 60% of the sea level rise by the 2050s is due to thermal expansion of the oceans and nearly 30% due to changes in the mass balance of land glaciers. These calculations make no allowance for regional changes in sea level induced by changes in ocean circulation or atmospheric pressure, although it is known that such regional differences will arise (Mikolajewicz *et al.*, 1990; Gregory, 1993). Indeed, Gregory (1993) analysed the output from the UKTR experiment to estimate global mean sea level rise due solely to thermal expansion of the oceans (climate models do not yet simulate ice melt from land glaciers). His global mean result was broadly comparable with the result shown here. Due to the low spatial resolution of UKTR it is not possible at the present time to determine differences between global and regional or local eustatic sea level changes.

It is possible, however, to estimate future crustal movements relating to isostatic rebound and since these will either exacerbate or reduce the sea level rise around the coast of the UK they are potentially important to include in the sce-

Figure 2.13: Percentage change in mean winter (a) and summer (b) potential evapotranspiration (PE) for the 2020s (left) and 2050s (right) decades with respect to the average of 1961-90. PE is calculated using the standard Penman formula.

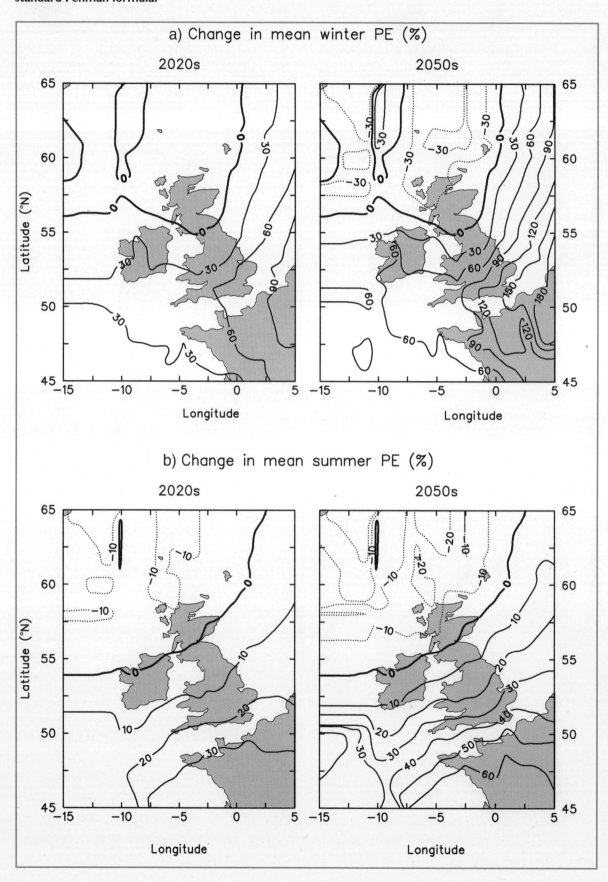

nario. Figure 2.14 shows the net change in sea level for two regions of the UK coastline (East Anglia and northwest Scotland) after allowing for these vertical land movements (Shennan, 1989). The southeast of England is likely to experience a more rapid net sea level rise due to subsidence (for example nearly 50cm by the 2050s) and the north of Scotland a lesser net rise (about 25cm by the 2050s).

Figure 2.14: Rise in sea level by the 2020s, 2050s and 2090s with respect to 1961-90 for the global mean (first bar), net rise for East Anglia (second bar) and net rise for northwest Scotland (third bar). The net changes have projected vertical land movements added to the global mean sea level change. The vertical bars show the range of projections derived from the simple climate model of Wigley and Raper, with the 1996 CCIRG scenario indicated by the horizontal lines. Vertical land movement data come from Shennan (1989).

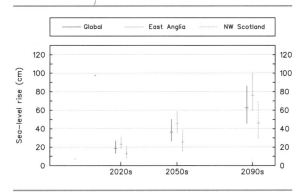

The majority of impacts of sea level rise are likely to be experienced in the coastal zone and in estuaries through changes in the frequency or severity of storm surge events. Global climate models are not capable of simulating such local storm surge episodes and no relevant scenario work has yet been completed for the UK coastline using high resolution tidal ocean models linked to model output from climate change experiments. Given an increase in mean sea level of about 37cm by the 2050s, the probability of a storm surge exceeding a given threshold is likely to increase. If one also assumes an increase in mean wind speed of a few per cent, as suggested in the 1996 CCIRG scenario, then this increase in overtopping risk seems likely to rise further, all other things remaining equal. A final assessment of the changed risk of damaging storm surges under conditions of changing climate and rising mean sea level, however, needs to consider the full interaction between changes in instantaneous wind speeds, wave heights, storm tracks and intensities, tide surges, river flooding *and* elevated sea levels. This detailed assessment has yet to be made.

2.7 UNCERTAINTIES

It has already been indicated that a range of future emissions scenarios is possible and that different emissions scenarios will have different implications for future climate and sea level (although the differences for global climate by the 2050s is not too great, cf. Figure 2.1). The 1996 CCIRG climate scenario has used the augmented IS92a emissions scenario and based subsequent calculations on this assumption. Even with greater insight into the future of the world economy and the nature of technological change and if it were therefore possible to predict future greenhouse gas emissions with confidence, current knowledge about the climate system and its response to increased atmospheric emissions would be inadequate to predict future climate with great certainty. This Section describes four of the reasons which contribute to this uncertainty and which explain why the portrait of UK climate and sea level for the next century just described can only be regarded as one out of a range of plausible scenarios.

For two of these reasons it is possible to estimate the uncertainties involved: i) an unknown climate sensitivity; and ii) systematic uncertainty in global climate models. Figure 2.15 summarises the effect of different assumptions about the climate sensitivity on projections of future global mean temperature. Projections over the next century vary considerably depending on whether the climate sensitivity is low (1.5°C) or high (4.5°C). By the 2050s, the range of projections of global warming is from 1.1°C to 2.4°C, compared to the 1996 CCIRG value of 1.6°C.

Figure 2.15: Projected rise in global mean temperature (with respect to 1961-90) for the augmented IS92a emissions scenario and for three different values of the climate sensitivity. The vertical bars show the decades 1985-94, 2020s and 2050s used in Table 2.1. Historic global mean temperatures from 1854-1994 (with respect to 1961-90) are also shown. No effects due to sulphate aerosols are included. The simple climate model of Wigley and Raper (Raper et al., 1996) is used (cf. Kattenburg et al., 1996).

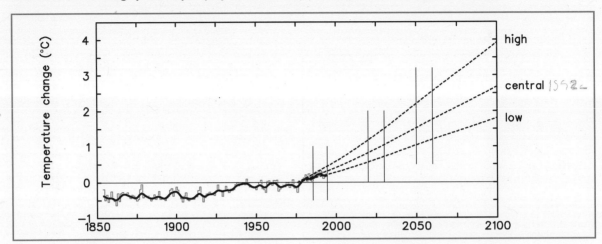

The effect of different values of the climate sensitivity on the rate of climate change in the UK is summarised in Figures 2.16 and 2.17 which show the mean seasonal changes in temperature and precipitation for the 2050s decade (cf. Figures 2.2 and 2.5), but with assumptions of a low and high climate sensitivity. Mean summer warming for central England by the 2050s is therefore estimated to lie between about 1°C and 2.3°C, with the magnitude of the summer precipitation change ranging between -4% and -8%.

The geographic pattern of future climate change for the UK in the 1996 CCIRG scenario derives from one climate model experiment (UKTR). Experiments with different climate models, but with similar forcing, yield different patterns of climate change over northwest Europe and it is important that this uncertainty is appreciated. Figure 2.18 summarises one aspect of this uncertainty by presenting results from 11 different climate model experiments for an area representing central England. The annual temperature warming over this region by the 2050s ranges from about 1.1° to 2.8°C and the annual precipitation change ranges from slight drying to a wetting of about 15%. The UKTR experiment used in the CCIRG scenario generates a relatively wet response to global warming over central England compared with other climate model experiments, and a rather subdued temperature rise. Table 2.5 shows the actual precipitation changes for this region from each of the 11 model experiments and for the two primary seasons as well as the yearly change. Precipitation change in all three seasons varies

Table 2.5: Percentage changes in mean precipitation by the 2050s with respect to 1961-90 derived from 11 global climate model experiments for a region representing central England. The acronyms refer to the respective climate modelling centres (and model versions in the case of the UK Meteorological Office); see Glossary. The effect of different model climate sensitivities has been eliminated.

GCM experiment	Annual	Winter	Summer
UKHI	14.7	24.6	6.0
UKTR	9.3	11.9	-0.5
UKLO	8.3	4.4	4.4
CSIRO	7.0	8.1	1.8
LLNL	6.0	11.1	8.5
CCC	4.9	11.1	-4.0
GFDL	4.5	10.1	-7.3
GISS	1.5	3.3	1.6
OSU	1.1	4.7	-5.2
ECHAM	0.7	4.0	-5.6
BMRC	-0.7	7.3	-5.6

30

Figure 2.16: Change in mean winter (a) and summer (b) temperature (°C) for the 2050s decade with respect to the average of 1961-90, assuming a low (1.5°C; left) and high (4.5°C; right) climate sensitivity.

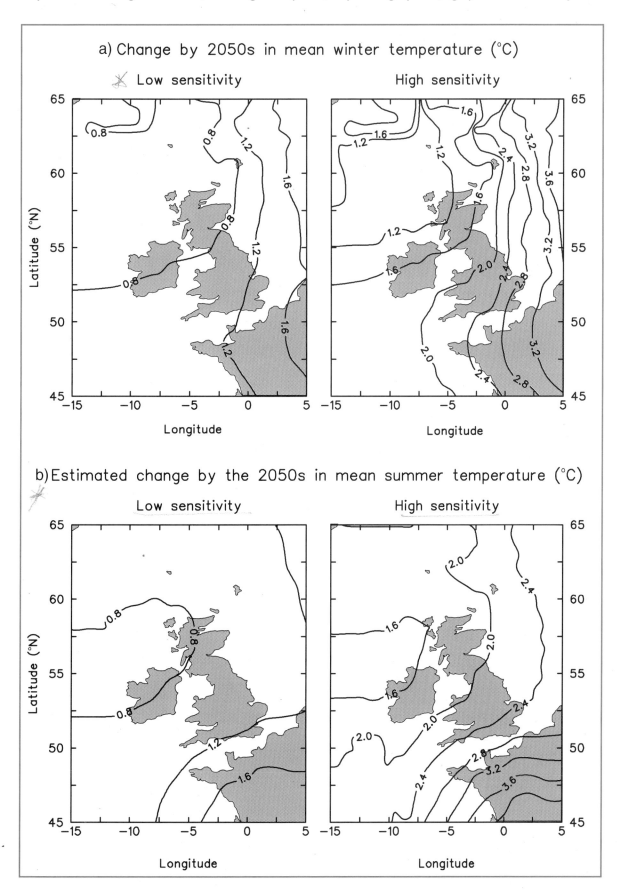

Figure 2.17: Percentage change in mean winter (a) and summer (b) precipitation for the 2050s decade with respect to the average of 1961-90, assuming a low (1.5°C; left) and high (4.5°C; right) climate sensitivity.

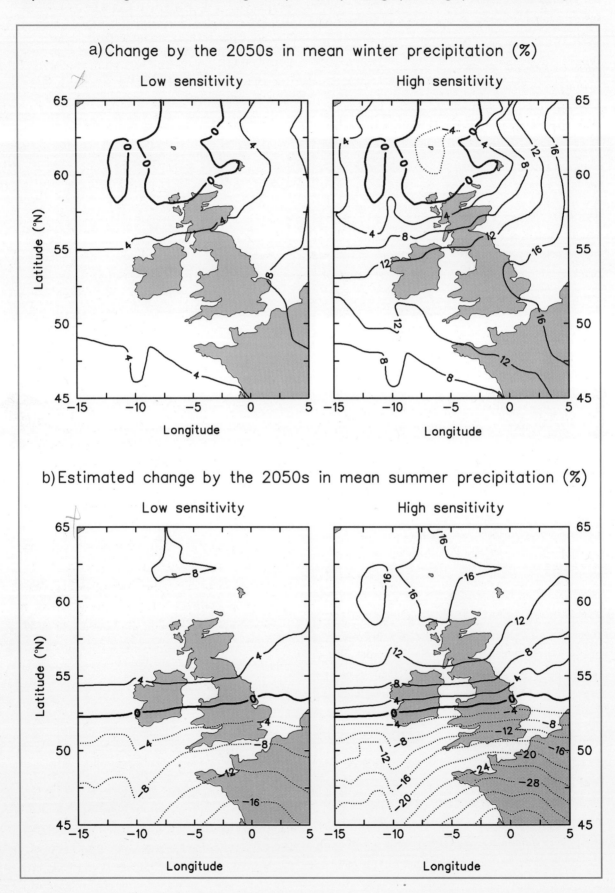

widely between the models, but for summer this range straddles zero change, whereas for winter and for annual (nearly) all the changes are positive.

Two other reasons which limit confidence in regional scenarios of climate change as predictions of future climate and which cannot be quantified adequately at present are, iii) the unknown predictability of regional climate change using climate models; and iv) the effect of sulphate aerosol forcing on UK climate.

The first of these problems is a philosophical one concerning how predictable is climate using the type of global climate models which have been developed. Figure 2.18 showed a range of climate changes for central England derived from 11 different climate model experiments. While this indicates a possible envelope of changes within which the actual future climate might lie (the UKTR result is merely one realisation of future climate, albeit one that derives from a seemingly 'good' model), this envelope cannot be assumed to define the extreme limits of what is possible. Non-linearity in the climate system which is not captured by the structure of the climate model, the role of feedback processes which have not yet been quantified in climate models, or inadequacies caused by poor spatial resolution, are all factors which make it impossible at present to quantify the degree of confidence in future regional climate changes simulated by global climate models. For example, relatively abrupt changes in the circulation of the North Atlantic Ocean which are possible, would have major implications for the climate of the UK and are not necessarily simulated by the climate model.

The second of these problems relates to the effect that other anthropogenic pollutants of the atmosphere, in particular sulphur dioxide (SO_2) emissions, have on climate. Theoretical and modelling work has now shown that the sulphate aerosols, which derive from SO_2, have a strong local/regional-scale effect on climate and

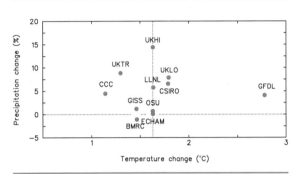

Figure 2.18: Changes in mean annual temperature and precipitation for the 2050s (with respect to 1961-90) from 11 global climate model experiments for a region representing central England. The acronyms refer to the respective climate modelling centres (and model versions in the case of the UK Meteorological Office); see Glossary. The effect of different model climate sensitivities has been eliminated.

on these spatial scales can effectively mask warming which might otherwise have occurred due to rising greenhouse gas concentrations (Taylor and Penner, 1994; Wigley, 1994). The effect of sulphate aerosols on *global* warming projections has been estimated by Wigley and Raper (1992) Mitchell *et al.* (1995) and Raper *et al.* (1996) and these effects are summarised in Table 2.6 (cf. Table 2.1). At a global scale, under the augmented IS92a scenario it would appear that by the 2050s sulphates may suppress global warming by about 0.3°C and sea level rise by about 14cm. At regional scales, however, it is the *pattern* of the aerosol-induced changes which is important and, as yet, there are only two or three climate model experiments which have addressed this issue (see Box 2.4). None of the results shown in the 1996 CCIRG scenario have therefore incorporated the effects of sulphate aerosols on climate.

2.8 RESEARCH NEEDS

With regard to the prediction of future climate and sea level, the most fundamental need is for improved models of the climate system. There are a number of aspects to such improvements. First is access to more powerful computers to

Table 2.6 Observed and modelled values of temperature and sea-level change for one historic and three future decades. All changes are shown with respect to the 1961-90 average. Modelled results assume the augmented IS92a emissions scenario, the IPCC 'best guess' parameters (Kattenburg et al., 1996) and either with or without SO_2 effects.

	1961-90	1985-94	2020s	2050s	2090s
Central England ΔT (°C)	0.0	+0.19			
Observed global-mean ΔT (°C)	0.0	+0.19			
Modelled global-mean ΔT (°C)	0.0	+0.23	+0.92	+1.63	+2.55
Modelled global-mean ΔT (°C) with SO_2	0.0	+0.15	+0.71	+1.31	+2.29
Modelled global-mean ΔSL (cm)	0.0	+4.0	+19.2	+36.9	+63.1
Modelled global-mean ΔSL (cm) with SO_2	0.0	+2.1	+12.0	+23.4	+41.9

enable longer climate simulations to be performed and at higher resolution. Second is the incorporation of additional feedback processes into such models, such as atmospheric chemistry interactions or feedbacks with surface vegetation and land cover changes. Third is the basic scientific research which leads to better understanding of the physics involved in such feedback processes. A good example here is the need to observe and understand better the role of clouds in modifying the energy budget of the atmosphere. Fourth is the investigation of the ability of, and reasons for, climate models to simulate abrupt changes in climate, as have been postulated to have occurred in the past and which may occur in the future and which alter the quasi-linear view of climate change implied in most scenarios. All of these developments should also benefit work which is directed towards the detection and attribution of climate change, assuming that commensurate improvements are made in both instrumental and pre-instrumental observational datasets, essential for detection purposes.

There are also a number of other research areas which are important specifically with regard to the construction and application of regional climate scenarios. Even if global climate models in the future are run at high resolution there will remain the need to 'downscale' the results from such models to individual sites or localities for impact studies. Downscaling methodologies are still under development and more work needs to be done in intercomparing these methodologies and quantifying the accuracy of such methods (see Wilby, 1994). Related to this is the need to link results from global climate model experiments to models which simulate regional or local ocean circulation and storm surges in order to be better able to predict changes in the local near-shore environment. Some experimental work along these lines has been started, but it is still in its infancy.

Box 2.4: Global Climate Model Experiments Including Sulphate Aerosol Effects

Until 1995, all climate change experiments simulating near-future climate using global climate models had considered only the effects of increases in greenhouse gas concentrations. The Hadley Centre have now, however, completed the first global climate model experiment which simulates human-induced climate change from 1860 to 2100 considering emissions of both greenhouse gases and sulphur dioxide, the precursor for sulphate aerosols (Mitchell et al., 1995). Results from this experiment only became available during the preparation of this CCIRG scenario and, to date, have still not been used as the basis of scenarios used in any published climate change impact work. The experiment confirms, however, that sulphate aerosols reduce the rate of greenhouse gas-induced global warming by several tenths of a °C per century and also alter the geographic pattern of climate change. For example, if results from this new experiment had been used as the basis for the 1996 CCIRG scenario, then the warming over the UK by the 2050s would have been a few tenths of a °C less than shown in this chapter and the warming contrast between the northwest and southeast of the country (see Figure 2.2) would have been reduced. This latter result arises from the fact that sulphate aerosol concentrations are higher over the continent of Europe than over the North Atlantic, thus the aerosol cooling effect (offsetting greenhouse gas-induced warming) is greater over the southeast of the country than in the north. Future SO_2 emissions are assumed to decrease throughout Europe in this IS92a emissions scenario. The effect of sulphate aerosol forcing on other climate variables has not yet been properly evaluated.

3. Soils

SUMMARY

- The soil is important in a climate change context not only because it underpins virtually all terrestrial ecosystems and agricultural production but also because it is a significant source of and sink for greenhouse gases.

- Soil processes and properties, particularly organic matter and soil water balance, will be sensitive to increasing temperature and changing amounts and distribution of rainfall, affecting the ability of the soil to sustain current agricultural crops and natural habitats.

- The soils of southern Britain are likely to be most at risk from climate change particularly in view of the predicted increase in summer droughtiness. This will affect the functioning of terrestrial ecosystems, influence groundwater recharge and flow and give rise to increased need for irrigation for agriculture. The combination of higher temperatures and increased rainfall may be beneficial to the soils of northern Britain and allow a wider range of crops to be grown.

- Wetland areas of southern Britain will be at risk from drying out and peat soils will be lost at increasing rates due to drying and wind erosion.

- Most of the clay soils of southern Britain will be subject to more intense shrinkage in summer, with potentially severe implications for building foundations.

- It is predicted that soil erosion will increase throughout Britain. The predicted drier summers in southern Britain would leave the lighter soils more prone to wind erosion and wetter winters would lead to more water erosion. Higher rainfall in northern Britain would cause increased erosion by water.

- Over 50% of Grade I agricultural land in England and Wales lies below the 5m contour and is thus located where it might be affected by any rise of sea level. Much may become saline and therefore of limited use for agriculture unless protection measures such as sea defences are established and maintained.

- Changes in the way land is managed and used will need to be made to offset many of the impacts and these could be combined with measures to increase the sink for greenhouse gases in the soil and decrease their emission from it.

3.1 INTRODUCTION AND BACKGROUND

The soil, through its multiple functions, influences a wide range of land uses and is itself, an important ecological habitat. The ways in which the soil responds to climate change could have significant impacts on the economy, particularly with respect to agriculture and forestry, and on the environment, for example with respect to the location of and change to terrestrial ecosystems.

The impacts of climate change on soils were treated in the first CCIRG Report (1991). At that time, there had been little or no research linking soils and climate change. Based on what was known about soil processes, initial estimates involving 'best scientific judgement' were made of the likely impacts of climate change on these processes. Now, five years later, it can be confirmed that the most sensitive impact issues were correctly identified in the previous report but that now, with the hindsight of research, some of the main impacts are better understood but still not quantified. These are treated in Section 3.4. There is a need for much more quantitative information to support many of the more qualitative judgements.

The soil is important in a climate change context not only because it underpins most terrestrial ecosystems but also because it is, apart from the oceans, the most important natural source of, and sink for, greenhouse gases. Thus the pool of carbon in the soil is normally 2-3 times that in the natural vegetation and standing crops which grow in it and twice that of the carbon in the atmosphere (International Society of Soil Science, 1990; Armstrong-Brown et al., 1995). Most of the soil carbon is associated with peat soils and in minerals soils with topsoil organic matter and decaying roots deeper in the soil but in calcareous soils, carbonate-C can be significant. On the one hand, during the natural processes of decomposition of organic matter, CO_2 is released to the atmosphere and on the other hand, the soil can sequester carbon from the atmosphere, particularly under grass and forest vegetation.

Globally, soils are also an important source of methane and nitrous oxide, contributing over half of the emissions of CH_4 (Batjes, 1992) and of N_2O (Seiler and Conrad, 1987; Bouwman et al., 1993). In the UK, natural wetland soils contribute methane, though this contribution is small (2% of the UK methane emissions) compared to global wetland emissions (about 25% of global emissions of CH_4 come from wetlands). Landfills are the other main soil-related contributor to methane emissions in the UK. In addition to emitting CH_4 and N_2O, the soils in the UK can also act as a small sink for both gases from the atmosphere.

The soil, therefore, is important in the fluxes of gases to and from the atmosphere and management of the soil offers opportunities for reducing the amounts of important radiative forcing gases reaching the atmosphere.

3.2 ESTIMATED EFFECTS OF CLIMATE CHANGE AND SEA LEVEL RISE

3.2.1 Sensitivity of soils to weather and climate

Climate is one of the five principal factors affecting soil formation and any change in climate will have repercussions for soil formation. Most soils take hundreds of years to form and where soils have been unable to benefit from forming in unconsolidated glacial deposits, it can take as much as 1000 years to develop one centimetre of soil from hard rock. Some soil processes, such as mineral weathering, are slow and the impact of climate change on these will be long term, i.e. hundreds of years. Other processes, e.g. changes in organic matter, many of which are critical for the plant-soil interactions, are much more sensitive to climate change and impacts can be experienced within months to decades.

Soil properties and processes are most sensitive to two main climatic parameters: i) temperature, which affects the rate at which many soil processes take place and ii) rainfall, because soil moisture is an important driving force for most soil processes and is critical in determining the ability of the soil to sustain a particular agricultural crop or natural habitat. It is not only the amount that is important but also the frequency and intensity, the latter being an important factor in soil erosion. Until now, little could be said apart from the results of climate change experiments with general circulation models (GCMs). The 1996 CCIRG scenario with respect to rainfall is important in describing regional differences in the impacts of climate change on soils.

Soils have the capacity to store large amounts of water, depending on soil type. Over a normal year this water supply is extracted for the growing plants in the spring and summer period and water lost is replenished during the winter period. Soil water thus undergoes seasonal changes and, other than in exceptionally dry years, can supply the needs of the vegetation, though in some agricultural and horticultural situations this may need to be supplemented by irrigation. Any change in soil water balance will have major effects on the soil ecosystem and on the land use that it supports. The drought years of 1975/76 and 1990/91 - 1991/92 were clear indicators of the problems that can be caused by changes in the climate. Indications of likely changes to climate (see Chapter 2) include a decrease in rainfall in summer in southern Britain. This coupled with an increase in temperature of 1.5 - 2°C could have a significant effect on the soil water balance in summer, with major implications for crops, forests and other habitats, as well as for the soil ecosystem.

The other main sensitive component is organic matter, the decomposition rate of which increases with increasing temperature. The impact of increased CO_2 in the atmosphere on organic matter has still to be clarified, and it is not yet known whether the extra above-ground growth will create sufficient extra soil organic matter to compensate for losses associated with increased temperature and in some areas decreased rainfall. Organic matter is an extremely important soil component and a decline in levels or quality will deleteriously affect soil fertility, structure, stability, water holding capacity and biodiversity, thus decreasing soil quality and increasing the difficulties of managing the soil for particular land uses.

Other soil-related impacts that are sensitive to the changes of climate described in the 1996 CCIRG scenario include soil subsidence, soil erosion and the flooding of Grade 1 soils in England and Wales by marine incursion. The combination of hotter and drier conditions in summer in southern Britain will enhance the likelihood of soil shrinkage and place building foundations under increased stress and encourage increased soil erosion. Some 57% of the best quality land in England and Wales (Grade I) is below the 5 metre contour and is at risk from the proposed sea level rise (Whittle, 1990).

Although many of the impacts, as described above, will lead to increasing soil-based problems and, in particular, endanger the ability of the soil to sustain current crops and other land uses, some areas of the UK, e.g. northern Britain, may experience an improvement in soil conditions.

3.2.2 Effects of climate change

Soil moisture balance
The soil moisture balance has a large effect on the functioning of a range of terrestrial ecosystems, influences groundwater recharge and river flows and controls irrigation water requirements. Changes in soil moisture balance will affect many of the major soil processes, e.g. leaching, oxidation-reduction, organic matter decomposition, as well as the ability of a particular soil to support a given vegetation. Changes in soil moisture deficit are complex

since they can result from direct climatic effects (precipitation inputs and temperature effects on evapotranspiration), climate-induced changes in vegetation type, different plant growth rates and cycles, different rates and depths of soil water extraction and the effect of enhanced CO_2 on plant respiration. Changes in the depth and duration of waterlogging is likely to result from changes in the amount, distribution and intensity of rainfall.

There is little doubt that, under the 1996 CCIRG scenario for the 2020s and the 2050s, there will be significant changes in the soil water balance. In southern Britain, a combination of warmer summers with increased evapotranspiration and a decrease in rainfall would be expected to lead to a reduction in the amount of soil water in summer. In other periods and regions, the outlook is less serious. Southern Britain appears to become warmer and wetter in winter by the 2050s which may provide some problems for soil cultivation and trafficability by machinery and livestock but could be more suited to crop growth. It is probably in northern Britain that the soils are likely to benefit most from climate change. Increasing temperature and rainfall are likely to be beneficial for the growth of crops and their extension into previously marginal areas.

Plant available soil water. The changes of climate over most of southern Britain imply a large increase in soil droughtiness and resulting water stress in crops. Although this may be compensated to some extent by an increase in winter rainfall, the stress on the crop during its growing season is likely to lead to a marked decline in yields unless aided by irrigation water. The winter rainfall is important for recharging soil moisture and re-filling on-farm reservoirs (Leeds-Harrison and Rounsevell, 1993).

Using the Thomasson (1979), Jones and Thomasson (1985) and Thomasson and Jones (1992) concept of soil droughtiness:

Droughtiness = AP-MD (crop adjusted) mm where AP (mm) is the soil available water and MD is the meteorological moisture deficit, it is possible to identify the impact of the 1996 CCIRG scenario of climate change on soil droughtiness with respect to different crops. In the example illustrated in Figure 3.1a,b, with an increase in temperature of 2°C and a fall in rainfall of 10%, virtually all of East Anglia, currently very important for wheat production in the UK, would become moderately to strongly droughty, making most of that area only marginally suited to wheat production (Brignall and Rounsevell, 1994, 1995; Brignall et al., 1995). Crops other than wheat would also be affected and some currently well-suited to the area could become marginally suited (see Chapters 5 and 6).

Soil wetness. The disturbance of the current soil water balance by climate change will ultimately affect a wide range of agricultural and environmental land properties. Among these is the depth and duration of soil waterlogging. The concept of wetness class (Hodgson, 1976; Robson and Thomasson, 1977) has been used to map the distribution of soils with different water regimes in England and Wales (Brignall and Rounsevell, 1994). The climate sensitivity of soil wetness can be illustrated by comparing the distribution of the six wetness classes (Hollis, 1989). The classes, which range from I (soil is not wet within 70 cm depth for more than 30 days in most years) to VI (soil is wet within 40 cm depth for more than 335 days in most years) are derived using the median field capacity period (Figure 3.2a). The effects of using the upper quartile (dry) field capacity period and the lower quartile (wet) field capacity period are shown in Figures 3.2b and 3.2c respectively. Unfortunately, equivalent data for Scotland and Northern Ireland are not yet available. Brignall and Rounsevell (1994) also compared the frequency of the six wetness classes for the wet median and dry quartile field capacity periods and used the comparison to illustrate the effects of increasing climate dryness in England and Wales. Evidently, a large shift occurs in the frequency of each wetness class. In particular, increasing climatic dryness markedly reduces the numbers in classes V and VI with those in classes II and III becoming more common. This is the direction of change likely to apply in southern England with the potential that wet soils could generally dry out, leading to a marked change in flora and fauna in natural and semi-natural habitats and to changes in agricultural cropping and management (see Chapters 4, 5 and 6). In northern Britain where both winter and summer rainfall increase in the 1996 CCIRG scenario, there may be a swing to more soils occurring in wetness classes V and VI. The overall message from these studies is that there could be large shifts in the distribution of wet soils, with major implications for moisture-loving species.

Soil structure and mechanical properties.
Soil water, and particularly the wetting and drying cycles of soils, strongly influence soil structure and the mechanical properties of soils. Generally the warmer, more moist climate that the 1996 CCIRG scenario predicts for northern Britain is associated with the formation of adequate quantities of organic matter and a large biodiversity, both conducive to good quality structure formation, a good network of pores and mechanical properties which facilitate working of the soil. By comparison, the intense drying-out of the soils that may become more typical of the summers in southern England may lead to a decline in organic matter and a decrease in biodiversity and to more difficult soils to work mechanically. More intense drying out will also lead to increased soil cracking and, particularly on soils with a loam or clay texture, a more extensive network of pores than at present (Armstrong et al., 1994).

As a result, many of the soils of southern Britain, could become more difficult to work in summer and early autumn and may become less trafficable in winter. Workability and trafficability of soils are both dependent on soil moisture re-

Figure 3.1: Winter wheat suitability calculated a) for current climate (1961-90) and b) for temperature +2°C and precipitation 90% of current climate. (After Brignall and Rounsevell, 1995).

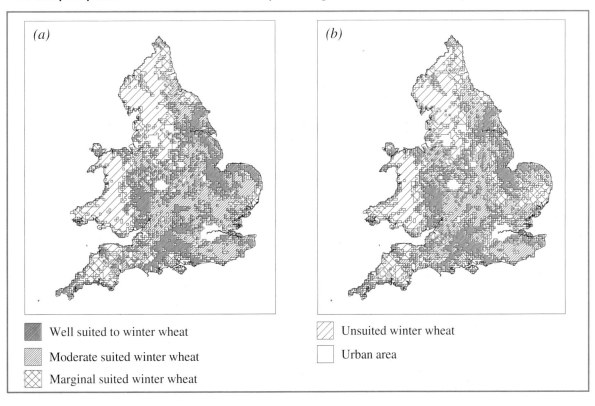

Figure 3.2: Wetness classes in England and Wales derived using a) the median field capacity period based on current climate, b) the upper quartile (dry) field capacity period based on current climate and c) the lower quartile (wet) field capacity period based on current climate. (After Brignall and Rounsevell, 1995).

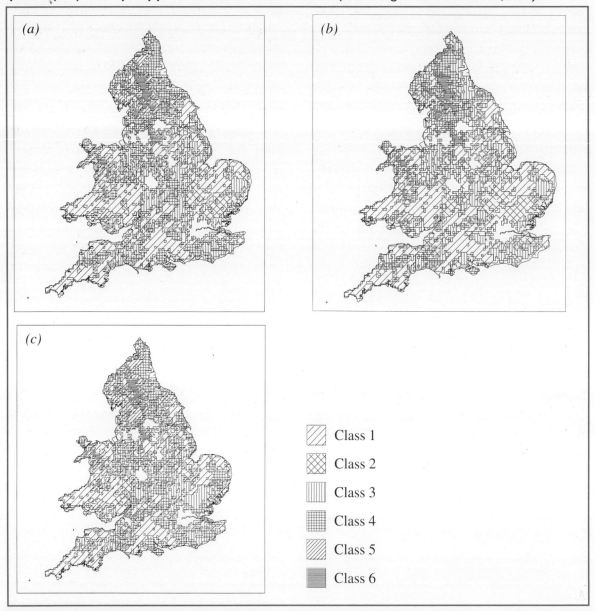

Definition of Wetness Classes (after Hollis, 1989)

I   The profile is not wet within 70 cm depth for more than 30 days (not necessarily continuous period) in most years (more than 10 out of 20 years).

II The profile is wet within 70 cm for 31 to 90 days in most years or, if there is no slowly permeable horizon within 80 cm depth, it is wet within 70 cm depth for more than 90 days but not wet within 40cm for more than 30 days, in most years.

III The profile is wet within 70 cm depth for 91 to 180 days in most years or, if there is no slowly permeable horizon within 80 cm depth, it is wet within 70 cm depth for more than 180 days, but only wet within 40 cm depth for between 31 and 90 days in most years.

IV The profile is wet within 70 cm depth for more than 180 days, but not within 40 cm depth for more than 210 days in most years or, if there is no slowly permeable horizon within 80 cm depth, it is wet within 40 cm depth for 91-210 days in most years.

V   The profile is wet within 40 cm depth for 211 to 335 days in most years.

VI The profile is wet within 40 cm depth for more than 335 days in most years.

gime (Rounsevell, 1993; Rounsevell and Jones, 1993). Dry soils such as those likely to occur in southern Britain may become baked hard and subsequent mechanised disturbance may lead to the formation of large persistent clods. Dry cloddy topsoils present a problem for seed bed preparation for autumn-sown cereals and oil-seed crops and the mechanised harvesting of maincrop potatoes (Brignall and Rounsevell, 1994). In terms of working the land, the number of potential machinery workdays will increase in most parts of the UK making soil tillage more easy (Figure 3.3a,b) except where the soil has dried out excessively. (Rounsevell and Brignall, 1994; Rounsevell et al., 1994).

Soils with high clay contents, especially those with a smectitic mineralogy, have the potential to shrink when they are dry and swell as they wet-up again. This behaviour results in the formation of large cracks and fissures. Drier climatic conditions would be expected to increase the frequency and size of crack formation in soils, especially those in temperate regions which currently do not reach their full shrinkage potential (CCIRG, 1991). One consequence of this would be the more rapid and direct movement of water and dissolved solutes from surface soil to aquifers through so-called 'by-pass' flow via cracks with the result that the soil profile would not rewet as before and the rooting zone would be deprived of valuable water (Armstrong et al., 1994). The more direct movement of water to aquifers would decrease the filtering effect of soil and increase the possibility of nutrient losses and pollution of ground and surface-waters. Where animal manures or sewage sludge are applied to land there would be a greater risk of organic contaminants and micro-organisms reaching aquifers. In the case of acid soils, increased cracking could lead to an increase in the quantity of metals (e.g. aluminium, manganese, iron, cadmium) entering aquifers. In these situations the combined effects of changes in soil chemical and biological cycles and the increased cracking need to be assessed.

Other impacts associated with soil moisture balance. In the UK the dominant movement of water is downwards through the soil rather than upwards via capillary rise. The implication of this is that basic cations (nutrients) are leached from the soil and the soils tend to become more acid. In agricultural and some forestry situations the loss of nutrients and acidification is countered by the use of fertilisers and lime. According to the 1996 CCIRG scenario most of northern Britain will experience increased rainfall which could be reflected in increased leaching of nutrients and increasing acidity. The consequence would be that the soils become more nutrient deficient and more acid unless there is compensation for this trend.

Of concern is the fact that much crop production in East Anglia will need to be sustained by the use of irrigation and water for this may not be available (see Chapters 5 and 7).

Organic matter

Organic matter is one of the most important soil components since it influences most soil functions and is a key to maintaining a healthy productive soil. It is instrumental in nutrient cycling, it enhances the soils' waterholding capacity, it increases soil stability, improves structure and provides the substrate for huge numbers of soil organisms which are required for many soil processes.

Climate change can influence organic matter contents in three ways (Howard and Howard, 1994): (i) by changing the rate at which the organic matter in the soil and entering the soil decays; (ii) by altering plant productivity rate, and hence the annual return of carbon C to the soil; and (iii) by altering plant species distributions and/or land use patterns and hence changing the type of plant material entering the soil.

Changes in land use brought about by man also have significant effects on organic matter stocks in the soil and these changes may have greater impacts than those associated with climate change. Thus a change from grassland to arable

Figure 3.3: Good autumn machinery work-days (Wda) calculated a) for current climate and b) for temperature +2°C and precipitation 90% of current climate. (After Rounsevell and Brignall, 1994).

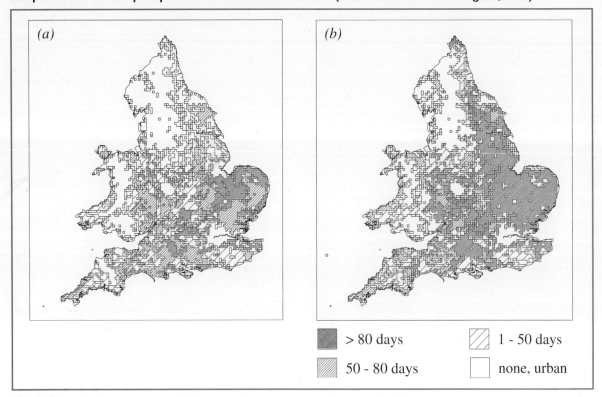

cultivation may reduce soil organic matter levels by 50% in a period of 25 years (d'Arifat and Warren, 1965; Johnston, 1973) (Figure 3.4). Organic matter levels will also be reduced when forest soils are converted to arable use. In both cases changes in the reverse direction would lead to an increase in soil organic matter but at a much slower rate than it was originally lost.

There are two main considerations with respect to the impact of climate change on soil organic matter: one relates to the fate of organic (peat) soils, the other to the maintenance of adequate levels of organic matter in mineral soils to maintain their fertility and stability.

A recent assessment of the amounts of organic carbon in UK soils (Institute of Terrestrial Ecology, 1995) has revealed that the soils contain some 11 billion tonnes of carbon. Of this, some 87% is in Scottish soils 75% in the peat soils of Scotland (Figure 3.5). Peat soils are extremely fragile, particularly with respect to erosion and oxidation following drying and could thus be particularly vulnerable under the 1996 CCIRG scenario. The release of C from oxidation of peat is also important as it is part of a large global flux.

In northern Britain, under the 1996 CCIRG scenario with increases in temperature and rainfall, most of the peat soils may be less at risk from loss by oxidation than those in southern Britain, assuming the projected increase in rainfall will compensate for the increased evapotranspiration associated with the warmer temperature. Given the warmer conditions, some of the peat soils at lower altitudes may become suitable for cultivation of crops. Such a change in land use could give rise to significant loss of peat through oxidation, shrinkage and erosion (as has occurred in the English Fens).

In southern Britain, the already endangered lowland peat soils, e.g. the Fens, will become even more at risk, the combination of warmer, drier conditions in summer causing more rapid oxidation and shrinkage (Burton and Hodgson, 1987). Most, if not all, peat soils in southern Britain will be at increased risk, with those currently supporting natural and semi-natural habitats being subject to declining water tables in summer and consequent drying out, sometimes irreversibly (see Chapter 4).

Whether, and the extent to which, organic matter levels in mineral soils in southern Britain may decrease under the CCIRG scenario even without a change in land use is still not well established. Within wide temperature limits mineralisation in response to changing temperatures may be reasonably predictable. Thus, between 15°C and 25°C, the amounts of organic C and N mineralised from a UK grassland soil were roughly doubled in accordance with the usual van't Hoff rule that an increase in temperature of 10°C doubles the rate of decomposition (Joergensen *et al.*, 1989). Although temperature increases are certain to be less than this, nevertheless there will be a temperature impact which, coupled with a decrease in rainfall, could lead to decline in the organic matter contents of mineral soils in southern Britain where already some 15% of the soils have contents below 2%. However, the situation is likely to be even more complicated. For example, if summers are warmer, biological activity and decomposition rates may be increased but if they are drier, the increased rate will only be sustained for a shorter period. At present it is impossible to resolve the effects of these potentially balancing factors on rates and duration of processes given the uncertainty of future climatic conditions. There is also the question of the positive role of increased CO_2 content on soil organic matter levels. There is evidence that gains of C by soils under enhanced CO_2 levels can exceed losses due to temperature rise in some parts of the biosphere (see Box 4.1). The extent of these gains will depend on how far the increased CO_2 is the limiting factor. It is unlikely to be the limiting factor, except perhaps in intensive agriculture. Under these conditions it is conceivable that primary productivity and the input and quality of crop residues could change (e.g. Tinker and Ineson, 1990) with implications for cycling of C

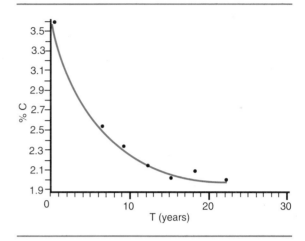

Figure 3.4: Decline in the percentage of organic carbon in the Rothamsted Highfield grass-to-arable conversion experiment (source: d'Arifat & Warren, 1965; Johnston, 1973).

and N, although this is uncertain. In practice, primary productivity in most ecosystems is limited by factors other than CO_2, e.g. drought or nutrient deficiency.

Soils of northern Britain, under the warmer temperatures and increased winter and summer rainfall of the 1996 CCIRG scenario, are unlikely to experience significant changes in organic matter contents due directly to climate change. However, these soils could experience a change in land use which, depending on use, could increase or decrease organic matter levels.

Franz (1990) considered that soil organic matter has a range of response times of changing soil conditions of the order of 10 to 100 years. On this basis, changes in soil organic matter as a result of global climate change should be measurable within the next 20 to 80 years. Soil microbial biomass measurements can provide sensitive indications of changing carbon inputs long before such changes can be detected by classical chemical analysis of total organic matter (e.g. Powlson *et al.*, 1987; Chander and Brooks, 1991). Such measurements could also, theoretically, provide early indications of changing C inputs by plants due to global warming although this hypothesis awaits evaluation (K. Goulding, personal communication).

Figure 3.5: Carbon in soils of Great Britain in kilotonnes/square kilometer (After ITE, 1995).

Soil fertility and nitrogen dynamics

The relatively small increase in temperature under the CCIRG scenario would not perceptibly alter rates of inorganic reactions, e.g. ion exchange, adsorption and desorption, in the short term but in the longer term the cumulative changes could become important. Changes in the surface properties of the clay fraction of soils may be much faster than changes in its bulk composition or crystal structure, influencing the physical and chemical soil characteristics (Brinkman, 1990). A change in soil moisture content as would occur in summers in southern Britain under the 1996 CCIRG scenario, could significantly affect rates of diffusion and thus the supply of mineral nutrients such as phosphorus (P) and potassium (K) to plants. This could alter the species composition of plants in natural systems and may require adjustments to nutrient management and fertiliser use in agriculture. A small temperature change, and the likely consequent land use changes, would have a much greater effect on the chemical and microbial transformations of organic matter, as described above. However, faster rates of mineralisation and thus supplies of nitrate, phosphate and sulphate would need to be sustained by adequate inputs of organic matter.

All microbiologically-facilitated processes are affected by moisture and temperature, but especially mineralisation/immobilisation and denitrification. A warming of soil will increase the rate of these processes. Mineralisation/immobilisation requires moist soil and denitrification requires water-saturated (anaerobic) soil, or at least, saturated microsites. Drier soils will therefore denitrify, and if very dry, mineralise/immobilise less; wetter soils may denitrify more and some, previously very dry soils, mineralise/immobilise more. As both processes produce nitrous oxide (N_2O) there could be a positive or negative feedback to climate change depending on whether the drier or wetter soils predominate. However, all these processes depend on the amount of N and C available. A much slower cycling of N will result if soils become very dry or very wet.

Using a model of the nitrogen cycle to investigate the impact of some climate change scenarios for the UK, Bradbury and Powlson (1994) indicate that should CO_2 fertilisation of crops on the soil lead to more carbon being returned to the soil, this additional input of C would, for sometime, maintain the organic nitrogen content. Eventually, assuming a continuing rise in temperature, the rate of turnover would be sufficient to cause a decline in organic N content.

Many of the 'new' crops that could be grown, particularly in southern Britain, under a warmer climate (e.g. maize, sunflowers, various legumes) are not frost-tolerant and so are spring sown. Consequently, larger areas of soil than at present would have no crop cover during winter. This is likely to increase the leaching of nitrate, and other nutrients, to aquifers unless corrective measures are taken.

Soil biota

Despite its importance and the resurgence of interest in biodiversity, there is still only a small amount of research literature which addresses the impact of climate change on soil biota, and then usually in a general way.

There could be shifts in the abundance of species within the soil microbial and faunal populations, although the direct effects from changes in soil moisture or temperature would be much smaller than those caused by changes in land use. It is difficult to predict whether there would be a change in biodiversity. The increasing concentration of atmospheric CO_2 could lead to changes in the composition of organic carbon compounds entering the soil from roots and root exudates, in addition to increasing its quantity. This may alter the species composition of the rhizosphere population which, in turn, could alter the extent to which plant roots are infected by soil-borne pathogens. Whether such changes would be beneficial or deleterious is unknown.

The temperature optima for soil organisms are usually quite broad, and a shift of one or two degrees will normally make little difference (Tinker and Ineson, 1990). The question of whether the microflora will change in line with the conditions, or will lag well behind these, has not yet been fully addressed. Soil organisms with more specific characteristics such as plant pathogens, symbiotic organisms and soil fauna, may be slower in adjusting to new conditions. For plant pathogens and symbiotic organisms there is the additional complication that they are, to varying degrees, dependent upon specific types of vegetation. A vegetation zone shift caused by temperature/precipitation changes will only have reached full equilibrium when both the vegetation and the appropriate micro-organisms have stabilised themselves together.

Powlson *et al.* (1988) compared the minimum temperatures at which the denitrifying populations functioned in soil from a temperate and a subtropical region: southeast UK and central Queensland, Australia. The soil from the cooler region had a markedly lower temperature threshold, suggesting an adaptation of the populations to the environments. It is not known what period is needed for such adaptation and, indeed, it is likely to differ between different groups of organisms. It is possible that a change in climate that is more rapid than has previously occurred might dislocate the normal adaptive mechanisms: the consequences of this are not clear.

There is some information on the rates of spread of soil-borne diseases, and the recent rapid movement of the barley yellow mosaic virus across England provides an example of the time scales involved (Hill and Walpole, 1989). In the seven years since the initial reports of the occurrence of this soil borne virus in England, the organism is now established over much of the country, and has spread at a rate which clearly exceeds those anticipated for climate change. However, there has been no real analogy since the end of the last Ice Age, when the rate of spread and change was probably (though not certainly) slower.

Soil erosion

There is general acceptance (Ministry of Agriculture, Fisheries and Food, 1993; Department of the Environment, 1994) that soil erosion has increased in the UK over the last few decades. Much of this is due to the fact that vulnerable soils (e.g. sandy and silty soils) low in organic matter are farmed intensively, a system of farming that they are unable to sustain. The changes in climate under the 1996 CCIRG scenario are likely to exacerbate the situation; in southern Britain, the climate would be warmer and drier in summer than at present and warmer and wetter in winter. The implication of this is that the soils could lose organic matter (see Section 3.2), decreasing their stability and making them more vulnerable to erosion. Intense drying out of the soil followed by re-wetting are ideal conditions under which particles disaggregate and grains become more available for transportation. In effect, the climatic conditions would be similar to those where erosion is more widespread, e.g. further south in Europe. The fact that the frequency of high intensity rainfall is likely to increase is an additional matter of concern for soil erosion. In northern Britain, much will depend on the balance between the impact of a temperature rise on the soil and the effect of increased summer and winter rainfall. Generally, an increase in the amount of rainfall, its frequency and intensity will give rise to an increase in soil erosion. Thus the whole of Britain could become more prone to soil erosion.

These conclusions are supported by the work of Boardman *et al.* (1990) and Boardman and Favis-Mortlock (1993) who point to increases in winter rainfall, summer storm frequency, the area of irrigated land and the introduction of erosion-susceptible crops such as maize, as areas of concern. Favis-Mortlock (1994) considers that for the South Downs an increase in temperature would have little direct influence on soil erosion rate but showed on the basis of equilibrium simulations that a 15% increase in rainfall would lead to a 27% increase in soil erosion.

Soil shrinkage

Despite the importance of soil shrinkage in relation to the subsidence of building foundations little new research on the problem has been undertaken since 1991 although expert systems have been developed to identify those areas most at risk (Hallett *et al.*, 1994, 1995; Doornkamp, 1993). The 1991 CCIRG report identified clearly the problems associated with soil shrinkage particularly in relation to the damage caused to foundations of buildings. The 1996 CCIRG scenario implies that soil shrinkage may be greater in southern England than was suggested in 1991. The combination of warmer temperatures and decreased rainfall in summer is particularly conducive to shrinkage of soils containing more than about 15% of clay (Institution of Structural Engineers, 1994; Dlugolecki *et al.*, 1995). Increased rainfall in winter should give rise to wetter conditions and soil swelling, particularly where the soils contain smectitic clay minerals. This enhanced shrink-swell behaviour that occurs in the top two metres would cause major stresses to the foundations of buildings (see also Chapter 11).

In northern Britain, where there are few soils with a high shrink-swell potential and where the 1996 CCIRG scenario does not show a marked season of drought, the problem for foundations from shrinkage may be a negligible one.

3.2.3 Effects of sea level rise

There has been very little research published since the 1991 CCIRG report concerning the effects of sea level rise on soils. Although the 1996 CCIRG scenario indicates a smaller rise in sea level, the implications for parts of the coastal zone are still serious. Sea level rise would lead to the flooding of areas of good quality soils currently supporting agriculture or important ecological habitats (Boorman *et al.*, 1989). One of the most serious consequences could be that over 50% of the Grade I land of England and Wales, currently lying below the 5m contour, could be affected by such a rise (Whittle, 1990). The areas that would be most affected are inland from the Wash, parts of the Norfolk and Suffolk coastal zone and coastal valleys and the North Kent Marshes. Some parts of these areas would become saline through inundation or groundwater intrusion (see Chapters 5 and 7).

The main implications of a sea level rise for UK soils are: i) loss of marsh soils in some areas, though they may be created in others; ii) a higher water table than at present, causing poor drainage and affecting current habitats and use of the land for agricultural crops; and iii) salinity, which as the floods of 1953 demonstrated, make the soils more difficult to manage effectively for agriculture.

3.3 ASSESSMENT OF POTENTIAL ADAPTATION

The following assumes that there will be a Government strategy to reduce greenhouse emissions with a view to decreasing the amount of change likely to take place. Better soil management can certainly contribute to a reduction in greenhouse gases by, for example, the use of techniques to sequester carbon, a better fertiliser use strategy, (Armstrong-Brown *et al.*, 1995).

It is important that attention is paid to managing soil water particularly in southern Britain where drier summers and wetter winters may be more likely than at present. Methods include improving irrigation efficiency and crop-water use and the development of soil conditions that minimise loss of water in the summer periods, e.g. use of appropriate cultivation techniques, addition of wastes to realise reasonable levels of organic matter, reduced tillage, and erosion conservation measures.

A major contribution both to maintaining and improving organic matter levels and to enhancing the water-holding capacity would be to increase organic matter content by careful use of various forms of waste, but great care would need to be taken to prevent inputs of metals and organic pollutants to soils. Otherwise one aspect of soil quality is damaged in an effort to enhance another. There is little doubt that with the right quality of organic waste and sufficient incentives to incorporate it into the soil, the quality of the soil would be enhanced by increased nutrients, better waterholding capacity, structure formation and decreased erosion.

The soils of northern Britain could be subjected to increased leaching, leading to loss of nutrients and increasing acidity. This would need to be mitigated by increased use of fertiliser and lime.

A whole range of conservation measures have been developed around the world to prevent soil erosion but few have been used in the UK. Measures that could be used to control erosion include contour ploughing, grass-strip waterways, contour grass strips, reduced slope length, and set-aside equivalent on critical erosion pathways.

If southern Britain becomes much drier in summer, the development of crops requiring less water will be important. The use of irrigation can, of course, offset drought but its use will depend on water availability and will need close monitoring with respect to salinisation. There may be a need for more on-farm reservoirs as advocated by Leeds-Harrison and Rounsevell (1993) to take advantage of winter rainfall.

Intensive agriculture is not a viable system on peat soils and alternative land uses should be sought, e.g. grass, forestry.

Subsidence of houses is difficult to mitigate against although in some instances it can be treated by under-pinning. However, it is increasingly possible to identify those soils which provide the most risk, thus allowing opportunities for treatment of present foundations and precautionary improvements to foundations of houses to be built in the future (see Chapter 11).

Adaptations to sea level rise are mainly technological and ultimately will depend largely on the economic incentive to make such adaptations.

3.4 UNCERTAINTIES AND UNKNOWNS

Uncertainties still remain with the pattern, frequency and intensity of rainfall and these inhibit clearer statements about some of the impacts on soil processes.

One of the major uncertainties is the influence of enhanced CO_2 concentrations on plant growth and the effect this may have on levels of organic matter, decomposition rates, nutrient balances, etc.

Although, under the 1996 CCIRG scenario it is possible to describe in some cases the effects of climate change on soil wetness, soil cultivation and nutrient turnover, it has not been possible to translate this into impacts in any detail, particularly in natural and semi-natural habitats. Similarly, no detailed analysis of the likely shifts in crops and agricultural systems as a result of climate change impacts on the soils has been achieved.

A number of the identified potential adaptations have not been tried under UK conditions and the efficacy of a range of adaptive methods will need to be explored.

3.5. PRINCIPAL IMPLICATIONS FOR OTHER SECTORS AND THE UK ECONOMY

There is a major implication for agriculture in the impacts of climate change on soils, in particular with respect to where and what crops can grown. Some soils would become more flexible and available to a wider range of crops, whereas some regions, e.g. southern Britain, may become more restricted in the crops that can be grown (see Chapter 5).

The changes to UK soils as a result of the impact of climate change are likely to cause disruption to many natural and semi-natural habitats, leading to changes within them and in some cases to migration of species to new locations (Chapters 4 and 6).

In southern Britain, because of soil moisture deficits, there is likely to be a need for more water-use for irrigation to support many crops (see Chapter 7). Soil subsidence will have impacts on the construction industry, (Chapter 11), insurance (Chapter 13) and financial sectors (Chapter 14).

Loss of land, especially Grade I land, through marine incursion has implications for the agricultural sector and possibly also for the insurance and financial sectors. (Chapters 5, 13 and 14).

3.6. RESEARCH AND POLICY ISSUES

3.6.1 Future research effort

Eight research needs can be identified:

- A much better understanding of the impacts of climate change on the populations and diversity of soil fauna and microflora.

- Through well focussed research to understand more clearly the impacts of doubled CO_2 in the atmosphere on soil organic matter quantity and quality under different climate/soil/land use scenarios.

- A better understanding of the size of the carbon pool, the chemical forms of the organic carbon in soils and the extent to which the pool of organic carbon can be manipulated with respect to, for example, nutrient turnover and adsorption-desorption properties.

- Development of soil management techniques to protect fragile ecosystems better against the more extreme events associated with climate change.

- Improved models, better equipped spatially, to link changes in soil processes and the opportunities for species adaptation, both crop and natural vegetation.

- The development of agricultural systems that are sustainable under climate change and which maintain soil quality.

- Generally, more fundamental research to examine the impact of climate change on soil processes and provide a stronger quantitative base with which to assess the scale of impact. Attention has so far been focussed more on above ground processes than on below ground processes.

- The combination of soil datasets for England, Wales, Scotland and Northern Ireland as a basis for extending current and future models to the four countries.

3.6.2 Policy issues

- A thorough assessment needs to be made of the impact of the likely changes to soils on the future land use of the UK, both agricultural and non-agricultural. Certainly, the effects of climate change on the soils of southern Britain could lead to major changes in flora, fauna, forestry and agriculture.

- Soils are a major source of and sink for greenhouse gases. Attention should be directed to forms of soil management that will minimise emissions from the soil and maximise sinks.

- More droughty summers and wetter winters in southern Britain would be likely to cause increased shrinkage and swelling of soils respectively, leading to greater subsidence risk to buildings. Directives may be needed with respect to future building construction in order that foundations may be strong enough to resist the future subsidence.

- Soil erosion has increased in the UK in the last 50 years and the climate change is likely to create soil conditions even more conducive to erosion. It would be timely to introduce soil conservation measures into the more vulnerable areas.

- The impact of sea level rise on low-lying land adjacent to the sea or likely to be influenced by marine incursions needs to be assessed and decisions made about which land should be protected.

4. Flora, Fauna and Landscape

SUMMARY

- The natural biota of the UK has been most profoundly altered by human activities in the past and, over the next 50 years, continued effects of land use change are likely to have a greater impact than predicted changes of climate.

- An increase in mean annual temperature of 0.5°C that is sustained for a decade or so will bring about noticeable changes in plant and animal distributions and abundance. A 1°C increase in temperature may significantly alter the species composition in about half of the statutory protected areas in the UK.

- There will be significant movement of species northwards, particularly insects and ephemeral weeds, in response to rising temperatures. A mean temperature increase of 1.5°C by 2050 is equivalent to a potential northward shift of 50-80 km per decade or an altitude shift of 40-55 m per decade. Conservation strategies need to be developed that recognise that the natural ranges of species will not be static and that many species may need to migrate to survive.

- The direct effects of a 30% increase in atmospheric CO_2 concentration over the next 50 years may be small relative to the effects of changes in temperature and rainfall, but there is at present no consensus on the magnitude of CO_2 effects on plant growth and water relations in field conditions.

- Overall, the number of animal (especially insect) species in most parts of the UK is likely to increase due to immigration and expansion of species ranges. By contrast, the number of plant species may decrease, and a substantial number of the 506 currently endangered species may be lost, because species-rich native communities may be invaded by competitive species and some wet, montane and coastal communities will be lost.

- There is some threat of invasion by alien weeds, pests, pathogens and viruses, but the majority of alien species that could spread do already occur in the UK.

- A 20-30 cm increase in sea level will adversely affect mudflats and some salt marshes, including nature reserves that are important for birds.

- Climate change will occur too rapidly for species to adapt in an evolutionary sense. Mitigating measures that can be taken include the translocation and rescue of species, the provision of habitat corridors, fire control, and control of eutrophication.

- Conservation strategies need to be developed that recognise that the natural ranges of species will not be static and that many species may need to migrate to survive.

4.1 INTRODUCTION AND BACKGROUND

This chapter considers the likely effect of climate change on the native fauna and on the natural and semi-natural plant communities that cover about 45% of the UK land surface. In particular, it considers impacts on the ranges of species, species abundance, the fates of rare species and impacts on communities.

Most of the general statements that were made in the 1991 CCIRG report on the potential effects of climatic change on the flora, fauna and landscape in the UK are still supported today. In order to make this report comprehensive, the important statements are repeated here. But, in addition, new evidence or opinion is presented, resulting from research conducted in the last few years.

The main sources of new material come from results emerging from the Natural Environment Research Council Terrestrial Initiative in Global Environmental Research, from the Department of Environment 'Core Modelling Programme', aimed at demonstrating climate change impacts (Parr and Eatherall, 1994), from research on how climate change might affect Britain's endangered species (Elmes and Free, 1994), birds (Austin et al., 1993; Burton, 1995) and the spread of alien species in England (Hill et al., 1994), and from observations on the impacts of the mild winters and hot summers that occurred in the UK during the period 1988-1990 (Cannell and Pitcairn, 1993).

There are three features about the biota of the UK that must be borne in mind when considering the impacts of climate change.

First, the natural landscape of the UK has been profoundly altered by man. Indeed, the most potent forces that are acting on UK vegetation at the present time arise from the direct effects of human activity, such as habitat destruction by agriculture, forestry, industry, human settlements, overgrazing, and indirect effects, such as eutrophication through groundwater and atmospheric pollutants and phytotoxicity resulting from aerial and soil contamination. The overall effect of such widespread habitat disturbance has been an increase in fast-growing plant species and a decrease in slow-growing, stress-tolerant parts of the flora typical of unimproved grassland, lowland heath and old woodland (Grime, 1996). Other important consequences of the widespread human disturbance to natural and semi-natural vegetation, combined with a mobile and affluent human population, is an increased chance of the spread and establishment of alien species, while the natural migration of native species may be hampered by the dissected landscape (Eversham and Arnold, 1992). Also, the responses of the natural flora and fauna to climate change will clearly be affected by future changes in land use driven by social and economic forces, some of which may occur in response to climate change itself.

Secondly, the UK has a maritime climate with a small seasonal amplitude in temperature over the range that is critical for life. A small increase in mean temperature greatly extends the period when the temperature is above the threshold for plant growth and animal activity. Also, the UK spans critical latitudes, over which there is 4.5°C north-south gradient in average summer temperatures and of sub-zero temperatures in winter. Because of these gradients, a large proportion of the flora and fauna have part of their northern limits in the UK, often coinciding with isotherms. For instance, about 80% of British resident butterfly species reach the northern limit of their ranges in Britain, as do 19 of the 52 UK species of wild mammals. Consequently, there is a high probability that species ranges have the potential to expand northwards following climatic warming, while those whose southern boundaries occur in Britain may withdraw northwards.

Thirdly, Britain is an island, and has in consequence an impoverished native flora and fauna, consisting largely of those species which invaded between about 10000 BP, when the land became free of ice, and 6000 BP, when the land connection with the continent of Europe was severed. Britain has fewer species in most animal and plant groups than an area of similar areal extent on the continent lying closest to southeast England. However, unlike isolated oceanic islands, Britain has few endemic species: most species that occur in Britain also occur on the continent of Europe. But, partly because of our history, a large number of species have been introduced from overseas, including the huge variety of species found in gardens.

4.2 ESTIMATED EFFECTS OF CLIMATE CHANGE AND SEA LEVEL RISE

Information and hypotheses about the impacts of climate change on the biota of the UK have been obtained by correlating the current distributions of plants and animals with current climate, by using long-term records to relate historic changes in the abundance of plants and animals to changes in climate and weather, and by using mathematical or word models based on understanding of the ecophysiology of plant species and communities.

4.2.1 Sensitivity of flora, fauna and landscape to weather and climate

There is no doubt that the natural biota are sensitive to changes in almost all aspects of climate and weather. Observations, analyses and experimentation in recent years all confirm the judgement that any sustained rise in temperature of about 1°C would have a significant effect on the UK flora and fauna - altering the abundance and ranges of many species and communities. Analyses of long-term records have shown a positive relationship between the northerliness of the Gulf Stream and both the abundance of zooplankton in Lake Windermere and the productivity of vegetation on road verges at Bibury - a teleconnection that may operate through the frequency of anticyclonic weather over western Europe (Willis *et al.*, in press).

Extreme events, such as droughts, unseasonal frosts and high windspeeds, have had the greatest impact on the flora and fauna in the past and are likely to continue to do so in the future. In recent years, the hot, dry summers of 1989, 1990 and 1995 were associated with tree dieback and loss of species of moist habitats, a high incidence of fires, increases in abundance of warmth-loving species such as American duckweed and bearded fescue, and increases in the abundance, activity and spread of many insects (Cannell and Pitcairn, 1993). Populations can recover from extreme events when they occur infrequently, but they may be permanently depressed when extreme events occur so frequently that they have little time to recover.

The next most important change resulting from a change in climate would be an increase in the magnitude and duration of soil water deficits resulting from decreased rainfall and/or increased evapotranspiration. Clearly, the wetlands would be adversely affected by droughts and species adapted to dry conditions in southeast England would be adversely affected by increased rainfall. Overall, over the next 50 years any significant changes in patterns of rainfall are likely to have a greater effect on the distribution of rare species of plants and animals in the UK than the predicted changes in temperature (Elmes and Free, 1994).

Changes in temperature per se are likely to become important once they exceed about 0.5°C and have persisted for a decade or so. The most important consequences of warming would be to reduce or eliminate critical periods of freezing or chill temperatures, to lengthen the growing season, and to improve conditions for the reproduction and dispersal of warmth-loving species. The responses of individual organisms to climatic warming depends critically on the magnitude of a set of temperature thresholds for their growth and development. These thresholds include: i) base and optimum temperatures for plant growth; ii) threshold temperature conditions for the emergence in spring of buds, insects, hibernating animals etc; iii) chilling requirements to release bud and seed dormancy (duration of temperatures -2 to 8°C); iv) threshold temperatures for flowering, pollen tube growth and seed production; and v) developmental thresholds for invertebrates for the completion of generations and winter survival.

The potential effects of the projected increase in atmospheric CO_2 levels on carbon fixation, water use and the growth of plants have been well-established (Box 4.1). However, the overall quantitative effect of increasing CO_2 levels on the UK flora over the next 50 years is still difficult to evaluate. One view is that the response will be small, because: i) increases in photosynthesis and growth in long-term experiments rarely exceed 20% even when CO_2 levels are double those at present (Poorter, 1993); ii) many species are unresponsive because the photosynthetic system down-regulates, and the responses of many native plant communities will be less than those of field crops because of limitations imposed by nutrient supply or internal sink capacity (Stitt, 1991; Luxmoore, 1991); and iii) it is difficult to detect any major effect of the 80 ppmv increase in CO_2 levels that has occurred since pre-industrial times (eg in long-term records of yields at Rothamsted or in tree rings

Box 4.1 Direct effects of CO_2 concentration on photosynthesis, water use and carbon accumulation

Almost all plants in the UK possess the C_3 phytosynthetic pathway, which evolved when atmospheric CO_2 concentrations were many times higher than at present. This pathway is based on a CO_2-reduction enzyme, Rubisco, which has a poor affinity for CO_2. The present CO_2 concentration is well below that which saturates Rubisco in the chloroplasts. Also, when CO_2 concentrations are low relative to oxygen, as at present, some of the Rubisco reacts with oxygen leading to a loss of carbon by photorespiration. Consequently, increasing atmospheric CO_2 concentrations should increase photosynthetic rates and depress rates of photorespiration. In the short-term, the photosynthetic rates of C_3 plants have been shown to increase by 25-75% in response to a CO_2 doubling, but in the long-term the response may be less owing to acclimation of a phytosynthetic process and nutrient limitations.

Increasing CO_2 concentrations also cause the stomatal pores in plant leaves to progressively close, so that the resistance to water loss is increased and less water is lost per unit of carbon gained. Total water loss by vegetation would therefore be less in elevated CO_2 if total leaf areas remained the same. But, in many cases, especially in nutrient rich or managed stands, total leaf areas may increase in response to elevated CO_2 so that total water loss may increase. It is uncertain whether stomatal responses to elevated CO_2 will offset the increase in evaporative demand at higher temperatures in the CCIRG climate scenario.

The combination in increased photosynthetic rates and increased water use efficiency is likely to increase overall plant growth in the UK, except in those areas severely affected by droughts. Increased temperatures and nitrogen deposition from the atmosphere will also generally benefit plant growth. Current research suggests that, in most areas, the increase in carbon fixation could be greater than the increase in carbon loss by an accelerated decomposition of soil organic matter, so that net ecosystem productivity will increase. Consequently, the total store of carbon on the land (especially in soil organic matter and in trees) will increase. This is the so-called CO_2 fertilization negative feedback in the carbon cycle, which is now accepted as a major sink in the global carbon budget.

(Graumlich,1991). An alternative view is that: i) the expansion of fast-growing perennial herbs in the UK may be partly due to the fact that they are more responsive than other species to elevated CO_2 levels, as well as to the increase in availability of disturbed fertile sites (Diaz et al., 1993; Grime, 1996); ii) the responses of many plants to elevated CO_2 will be enhanced as temperatures increase (Long, 1991); and iii) variation among species in the extent to which elevated CO_2 restricts water use (by decreasing stomatal conductance) may have profound effects on the performance of species, especially in dry summers (Kersteins et al., in press.

4.2.2 Effects of climate change

The main issues are the composition of existing communities, the migration and invasion of species, losses and extinctions, outbreaks of pests and diseases, productivity, and landscape changes.

Changes in the composition of existing communities

Populations of animals, particularly of invertebrates, that occur at the edges of species ranges tend to fluctuate greatly from year to year in response to changes in climate, whereas populations near the centre of species ranges fluctuate much less and perhaps more in response to population density (Thomas et al., 1994). Also, natural enemies may be absent at range boundaries and suitable habitats may be fragmented. Consequently, populations at the margins of distributions may be most affected by climatic change (increasing in size, extent and stability) in ways that depend on the conditions at those margins.

Cold-blooded vertebrates (the amphibia and reptiles) may become more active and noticeable during mild winters and earlier springs, as was the case in 1989 and 1990, but they will not necessarily become more numerous (Cannell and Pitcairn, 1993). Their populations are already limited more by the availability of habitats than by climate, and warmer and/or drier conditions could have adverse as well as beneficial effects (e.g. on habitats, hibernation and disease).

The British Trust for Ornithology analysed fluctuations, from 1973 to 1992, in the populations of rare bird species (with less than 300 breeding pairs) in the UK in relation to weather variables (Austin et al., 1993). There was a clear trend for the numbers of many species to be greater in seasons following increased rainfall, possibly as a result of increased food supplies. For instance, fifteen of the 26 breeding species that overwinter in Europe appeared to do better when seasons were wetter than average. Consequently, the projected increase in both summer and winter rainfall in northern Britain may favour species such as the pintail, goldeneye and redwing. Increased temperatures alone may benefit about one-third of our rare breeding birds (where rainfall is maintained), including the Dartford warbler, which is at the northern edge of its range. However, for most bird species, population fluctuations are due to a complex of factors concerning their habitats, food supplies and particular weather patterns or events, which makes it impossible to predict how they will respond to a climate change.

Changes in plant communities may be seen first in montane areas, in response to warming, and among species occupying damp, cool refugia (such as hart's tongue fern, oak fern and many mosses), in response to drier summers in southern Britain, as occurred in 1984, 1990 and 1995. Attempts to predict more general changes in plant communities have used the classification of species into competitive, stress-tolerant and ruderal functional types (Grime, 1974; Grime et al., 1988). This approach would suggest that, in northern Britain, where it may get warmer but remain moist, competitive species (which are the more abundant common species) will be favoured at the expense of stress-tolerant species, which include most of the rare and endangered species. (Rare species tend to be those living in conditions of high stress, resulting in niche specialisation and restricted ranges). By contrast, in southern Britain, where it may get warmer and drier, stress-tolerant species, especially those which are deep rooting, may actually do better - but only if the habitats in which they live are conserved (Hunt, 1992). Hunt and Colasanti (personal communications) have developed an expert system (TRISTAR3) which predicts the steady-state composition of vegetation in given climates and with different managements, at any location in Britain, in terms of comprehensive knowledge of plant functional types (see Parr and Etherall, 1994).

The main effect of the high temperatures per se in 1984, 1990 and 1995 was an increase in the abundance, activity and geographic spread of many insects. Some of those insects were pests, such as aphids in 1989 and wasps in 1990 and 1995. Other insects may be said to have enriched the natural environment, such as butterflies, moths, crickets and honey bees. The responses of species to single warm dry years may be different from their responses to a succession of such years. For instance, over a prolonged period of warming the populations of some insects will be stabilised by the buildup of predators. But, even so, it is to be expected that one of the most immediate and noticeable impacts of climatic warming will be on the greater abundance and expanded distributions of invertebrates (Cannell and Pitcairn, 1993).

Another noticeable and predictable impact of warmer temperatures will be that spring will arrive earlier. Plants will leaf out and flower earlier, birds will begin their breeding behaviour earlier, insects and hibernating animals will emerge earlier, and so on. Some species may benefit from this shift in phenology more than others. In 1989 and 1990, the plant species that took greatest advantage of the early springs were those that normally leaf out and flower early, ie they did so even earlier. Late-growing species that start growing earlier in response to warming may lack sufficient frost hardiness to tolerate late spring frosts (MacGillivray and Grime, in press).

Finally, high temperatures may increase the incidence of algal blooms in freshwaters which have high nutrient levels. During the hot summer of 1989 the National Rivers Authority recorded potentially toxic blooms of blue-green algae on 25% of lakes and reservoirs in England and Wales (National Rivers Authority, 1990). Botulism, which can cause large-scale mortality among aquatic birds, is also likely to become more prevalent at higher temperatures.

Migration and invasion of species

A shift of 1°C mean annual temperature in the UK is equivalent to northward shift of 200-300 km or an altitudinal shift of 150-200 m. Consequently, the current scenario of a mean temperature increase of about 1.5 °C by 2050 is equivalent to a northward shift of 50-80 km per decade or an altitudinal shift of 40-55 m per decade.

The first organisms to migrate (see glossary, Annex 1) will be those that are most mobile, particularly the warmth-loving invertebrates, some birds and mammals, where food supplies and habitats permit. Observations made in 1989-1991 showed that major changes can occur in a few years in both the abundance and ranges of invertebrates (Cannell and Pitcairn, 1993). Burton (1995) speculated that warmer, drier conditions could lead to the return to Britain of birds (as breeding species) such as the red-backed shrike and wryneck, and to colonisation by the bee-eater and roller. He also recorded the northward spread of the long-winged conehead bush-cricket from the south coast to Oxfordshire during the period 1970 to 1992.

As mentioned, a large part of the flora and fauna in the UK has a northern range boundary in the UK, and all of it has an altitudinal limit. The potential and likely spread in the UK of about 100 plant species has been modelled by defining 'climatic envelopes' that encompass the range of the species in Europe and also considering the potential availability of suitable habitat types, the time taken to reach reproductive maturity and the distance over which dispersal can take place (Parr and Eatherall, 1994). An example is shown in Figure 4.1. The plants that are predicted to migrate first are the weedy ephemeral species, with high fecundity and rapid dispersal, often species of agricultural land. The plants most resistant to movement will be those with a long lifespan and clonal plants, including trees, shrubs and many of the stress-tolerant species of unproductive heathland and grassland. In all cases, natural migration is likely to occur in spurts, following extreme events which kill vegetation and allow new species to colonise.

The threat of invasion and spread of alien and invasive plant species is, perhaps, less than might be imagined. As mentioned, the UK already has a flora that is continental in character, so that species introductions have a less damaging effect on native communities than they do on remote oceanic islands. The majority of alien species that could spread do, in fact,

Figure 4.1: The spread of *Agrostis curtisii* in Britain predicted by the Core Model Demonstrator developed as part of the Department of Environment Core Modelling Programme (Parr and Eatherall, 1994). The prediction is based on an increase of 2.5°C in mean global temperature by 2100, with a mean warming in the UK of about 0.3°C/decade. A habitat 'mask' was used based on geological and land cover information. The species was assumed to take 5 years to reach reproductive maturity and to be able to disperse its seeds 20 km. From left to right, the figure shows the distribution at present, in 2040, in 2080 and in 2100 if dispersal processes were not limiting. (Figure supplied by S. Wright and M. O. Hill)

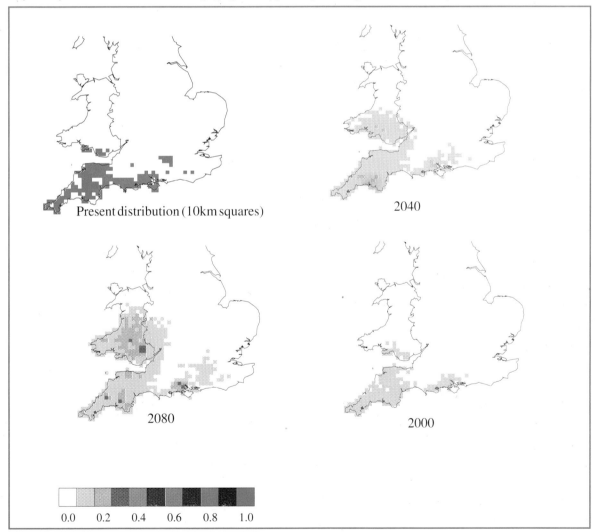

already occur in gardens and waste places throughout the UK and so have no dispersal difficulties. Hill *et al.* (1994) considered that there is little likelihood of alien species invading woods, grasslands and heaths. By contrast, cliffs, dunes and strandlines are not only strongly invadable but also have a relatively large suite of species that may colonise them. Securing shrubs on dunes and dense shrubs, such as *Cotoneaster*, on rocks are examples of such threats. A known serious threat is posed by the Hottentot fig (*Carpobrotus edulis*) which forms dense mats on some coastal cliffs and dunes, is spread by gulls, and is almost impossible to eradicate.

Losses and extinctions

The general consensus is that: i) the number of animal species in some parts of the UK may increase, because new and/or rare species of invertebrates, birds and mammals will spread and the gains are likely to exceed any losses; but ii) the number of plant species is likely to decrease, because species-rich native communities may be invaded by competitive species, some wet and montane habitats may be lost as a result of climatic change, alien species are unlikely to spread rapidly except in coastal regions, and climatic change will be too rapid to allow evolutionary adaptation.

A substantial number of the 506 species that are listed in the Red Data Book as being endangered, vulnerable or rare (313 of which are plants) could be lost as a result of climatic change. Elmes and Free (1994) used the Institute of Terrestrial Ecology Biological Records Centre data to draw 'climatic envelopes' around the current distributions of 279 Red Data Book species. They then estimated the percentage of 1 km squares in Britain which would move outside the 'climatic envelopes' with different scenarios of climatic change, to give the worst case extent of species loss. (No account was taken of spread to new areas.) About 22% of the species, which already occur in a broad range of climates, were unaffected by climatic change in temperature and rainfall. With a temperature rise alone of 2°C about 71% of species were predicted to show no loss of distribution, but 14% were reduced to half their current extent, 6% were virtually eliminated by a 2°C rise and 9% were eliminated by a rise in temperature of less than 1°C (Figure 4.2).

The current landscapes in the UK, in which most land is agricultural, will hamper the natural spread of many species, which will become stranded and vulnerable to local extinction in isolated patches of semi-natural habitat. This situation may be contrasted with that which existed when species migrated during the last glaciation.

The main species or communities that may become endangered by climate change include the following:

- Montane/alpine and northern/Arctic plant and animal species which have nowhere to go if it becomes warmer. Good plant examples are the tufted saxifrage (*Saxifraga cespitosa*) and alpine woodsia fern (*Woodsia alpina*). Good animal examples are the mountain hare, ptarmigan, snow bunting and whitefish.

- Species confined to particular locations from which they cannot readily escape, because, for instance, they occupy cool damp refugia, isolated habitats, or are dependent on particular other species for pollination, food or to complete their life cycle.

- Species of salt marshes and soft coastal communities that cannot retreat landward in the face of sea level rise.

Pest and disease outbreaks

The climatic and biotic variables leading to pest and disease outbreaks are normally complex and difficult to predict, especially when they involve coupled predator-prey and herbivore-plant interactions. Nevertheless, there is a common view that pest and disease outbreaks could become more frequent as the climate warms. The warm conditions in 1988-1990 were associated with serious outbreaks of aphids, slugs and wasps, requiring the use of large quantities of chemicals to control. Also, there were more calls for the control of pests of buildings, such as cockroaches, fleas and mites (Cannell and Pitcairn, 1993)

Plants that become stressed by drought or high temperatures are likely to become more susceptible to attack by certain insects (although they may have lower internal nitrogen concentrations and more defense compounds as CO_2 concentrations increase). In particular, trees may become

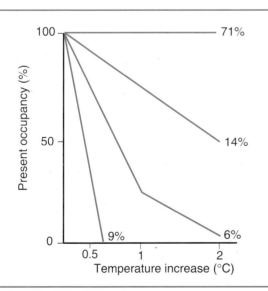

Figure 4.2: Decrease in the percentage of 1 km squares in Britain occupied by Red Data Book species with an increase in mean annual temperature of up to 2°C. (The Red Data Book lists species which are considered to be endangered.) The species are divided into four groups, which suffer either no loss (71% of the 279 species examined), a loss of about half (14% of species) or dramatic losses (6% and 9% of species). Note that potential 'gains' (by species spreading to new squares where the climate becomes favourable) are not included, so this indicates the worst case scenario. Taken from Elmes and Free (1994).

more susceptible to bark beetles, aphids and mature-foliage eaters (Jones and Coleman, 1991). Insects that feed on newly emerging foliage may be adversely affected if the synchrony between egg hatch and bud emergence is disrupted (Dewar and Watt, 1992).

Some fungal diseases of native plants and animals could become more prevalent following mild winters. The frequency and distribution of bird diseases could change, such as duck virus enteritis and avian tuberculosis, as could mammalian diseases and parasites.

Productivity

In general, any increase in temperature in the UK, combined with an increase in CO_2, is likely to increase total plant productivity, provided rainfall is maintained and mineral nutrients are available. The main factor increasing productivity in northern and upland regions will be the increase in length of growing season. Where net primary productivity is increased, there is a theoretical expectation that some invertebrate and other animal populations will also increase.

Landscape changes

The changes in plant communities, species migrations, losses and gains mentioned above will, in time, change many of the landscapes with which we are familiar.

- Montane plant communities may be lost.

- Heaths may be subject to more frequent fires as southern Britain becomes warmer and experiences dry summers, such as those in 1990 and 1995 (Chapter 2).

- Wetlands may dry out more frequently, with a change in species composition, especially if current water abstraction from aquifers were increased.

- Coastal dunes and rocks may be invaded more rapidly by alien species such as the Hottentot fig.

- Salt marshes and brackish water habitats may be lost as sea level rises.

- Some broadleaved woodland in drier areas of Britain may decline further in response to increased frequency of summer droughts, particularly in the south, where summer droughts are forecast to be more frequent and severe (Chapter 2).

Of course, major changes in the landscape may also occur as a result of changes in land use for agriculture and forestry. Some of these changes in land use may be driven or made possible by climate change itself. One of the principal controls on the distribution of natural habitats in the UK is the use of land for agriculture. If, for instance, cereal production were to shift or expand to more northerly regions than at present, this would probably result in a loss of natural habitats and would certainly alter the landscape (Parry et al., in press).

4.2.3 Effects of sea level rise

Approximately 10% of the notified nature reserves (NNRs and SSSIs) of the UK occur near sea level on the coast. The effect of sea level rise on these reserves, and on all areas of salt marshes and mud flats, will depend on whether or not they are prevented from moving inland by sea walls and other defences. A 20-30 cm increase in sea level will certainly affect the lowest mudflats and may significantly affect areas of salt marshes, particularly if landward expansion were prevented.

The invertebrate fauna of intertidal flats is likely to become poorer and less diverse. The productivity of the surviving species may be reduced by high loads of suspended sediments in the inshore waters brought about by higher tide levels. These changes would greatly reduce the numbers of many species of birds that roost, feed or breed on the UK coasts. The UK coasts are the wintering grounds for over half of Europe's waders and 60% of the UK redshank nest in salt marshes.

4.3 ASSESSMENT OF POTENTIAL ADAPTATION

'Adaptation' by the UK flora and fauna, in the sense of changed behaviour and slow genetic change in populations, will occur as each species responds to maximise its chances of survival and of leaving decendents. Also, feedbacks will occur within ecosystems to establish new equilibria between plant species, invertebrates, their predators and animals higher in the food chain. These adaptations and feedbacks will prevent some of the worst case scenarios of species loss, migration and pest outbreak occurring, but are unlikely to be sufficient to prevent major change.

'Adaptation' by man, in the sense of taking measures to mitigate or minimise some of the undesirable impacts described above, could include:

- The active or passive management of reserve areas to enable new species assemblages to develop as the habitats change character, while maintaining important existing populations.

- The translocation and 'rescue' of species which occur in isolated habitats or are otherwise unable to migrate to new locations as their existing habitats change.

- Focusing special attention on the preservation of wetlands in southern Britain, coastal marshes and montane communities.

- Strengthening measures, such as public education, public access and water supplies, to prevent and control fires, especially on heathlands.

- Strengthen measures to prevent the eutrophication of lakes.

- Use opportunities to provide habitat corridors along railways, roads and other unfarmed land to ease natural species migration.

Adaptations may also take advantage of beneficial effects of climatic warming. For instance, opportunities will arise for the expansion of species of attractive flowering plants, rare butterflies and perhaps animals, if suitable habitats are provided. Examples include the provision of flower-rich grassland for the spread of the Adonis blue butterfly, of well-grazed grassland for the large blue butterfly, and of open ground for the sand grass, *Vulpia ciliata* (Elmes and Free, 1994; Firbank *et al.*, 1995).

4.4 UNCERTAINTIES AND UNKNOWNS

Assessments of impacts will necessarily be uncertain until very detailed predictions can be made of future climates, in particular of the frequency of extreme events, and even then, uncertainty will remain about future land use change. Also, it is generally agreed that, over the next 50 years or so, any major changes in the Common Agricultural Policy, fiscal incentives for forestry, local and national initiatives to establish forests, or changes in land drainage or water abstraction, could have a greater effect on the UK flora, fauna and landscape than climate change.

Even given a detailed scenario of climate change, there would still remain some important uncertainties about the responses of the UK flora and fauna.

Information on the potential spread of species is currently based mainly on the 'climatic envelope' or correlative approach, with a limited amount of work on the constraints to spread imposed by the biology of species, available habitats and their interconnections. This work, which was begun in the Department of Environment Core Modelling Programme, has much further to go to incorporate physiological and ecological processes in order to obtain more accurate (and probably less extreme) assessments of the migration of plants and animals.

Assessments of the species that are likely to be lost and of those that could invade and become a problem have so far been based on expert judgement or worst case scenarios. More mechanistic approaches are required to these problems, so that the species and communities at risk can be better defined and also the critical rates and magnitudes of climate change can be defined.

A unified understanding is needed to reconcile the differences in responses to elevated CO_2 that have been observed among species and situations, so that the responses of vegetation can be assessed at a habitat and regional scale, and put into context with the impacts of other drivers of change.

4.5 PRINCIPAL IMPLICATIONS FOR OTHER SECTORS AND THE UK ECONOMY

Change in the flora, fauna and landscapes per se will have little impact on the UK economy, although it may affect local tourism. However, changes in conservation policies may have some impact on agriculture and forestry, possibly imposing greater restrictions, but the overall effect will probably be small.

However, widespread and noticeable change of any kind in the natural environment will heighten public concern about the quality of life and the sustainability of our current way of life. Concern will be raised by, for instance, the death of trees in response to droughts, the loss of high-profile species (eg birds, orchids, butterflies) and the dramatic loss of (or change in) conservation areas by fire or drought. This heightened concern may be given political expression and lead to changes in Government commitments under the Climate Change Convention, increased energy costs and other strategies to restrict greenhouse gas emissions.

4.6 RESEARCH AND POLICY ISSUES

4.6.1 Future research effort

There is a continuing need to develop a predictive capability to be able to forecast the effects of given scenarios of climate change on species survival, abundance and migration, and on the fates of communities and landscapes.

The Core Model framework developed by the Department of the Environment (Parr and Eatherall, 1994) has been able to link biological, hydrological and chemical responses to projected changes in climate, and to assess the effects of potential climate change on the distributions of plant and animal species, taking into account habitat distribution and land use. The Core Model provides a framework in which new information can be assimilated so that questions posed by policy makers and planners can be answered with ever greater precision and certainty. The current need is to improve the formulation of the model (incorporating more of the processes that respond to climate, including CO_2) to improve the databases on which they depend, and continually to test predictions against observations in experiments and at monitoring sites in the field.

Many controlled environment experiments have been conducted in recent years to explore the responses of species, communities and associated invertebrate fauna to changes in temperature, CO_2 and other variables. As mentioned, a variety of sometimes conflicting results have been obtained, and it has not always been easy to extrapolate from controlled environments to field conditions. There may be continuing uncertainty until it is possible to conduct ecosystem-scale, long-term studies in the field, including treatments with elevated CO_2.

An Environmental Change Network has been set up in the UK to monitor change in environmental variables at a range of sites. This network and associated statistical research needs to be maintained in order to detect early warnings of the impacts of climate change against the background of large annual and interannual fluctuations.

4.6.2 Policy issues

Policy issues need to be addressed concerning the conservation of the existing flora and landscape and the risks of invasion or spread of alien or harmful species.

Conservation

The existing conservation strategy, embodied in the Wildlife and Countryside Act (1981) and Amendment (1985) and the Environment Protection Act (1990), is to protect and conserve species and habitats in the UK. The Government's statutory nature conservation advisers, have powers to conserve National Nature Reserves (NNR) and Sites of Special Scientific Interest (SSSI). This strategy assumes that the main driver for change is human interference and 'damaging operations' such as changes in land use. The Acts give powers to curtail such human interventions. By contrast, there is no provision for positive management of SSSIs, nor for the possibility that habitats may change in response to climate change, and that the natural ranges of species and communities will change. The main challenge for the future is to develop a strategy that recognises that species distributions may not be static and that the conservation values of NNRs and SSSIs may change. The 'best guess' is that an increase if 1°C would significantly alter the species composition in over half of the statutory protected areas.

Also, conservation in the UK is based mainly on the protection of isolated sites, each representing some particular habitat. Most of the species that live on these sites are marooned there, with no opportunity to move elsewhere if conditions become unfavourable. The major conservation need for the future may be to modify landscapes so that organisms can spread along habitat corridors from one region to another as conditions dictate. Many existing nature reserves may become less valuable as species die out, and the only way that these local extinctions can be compensated may be to enhance the possibility of species colonising other areas.

Pests and spread of alien species

Regulations which lessen the chances of introducing damaging pests and animals to the UK may need to be strengthened and more rigorously enforced. A policy may also be needed to prevent large amounts of pesticides being used rather than adapting to change. However, there is probably no requirement to take further measures to control the spread of alien plant species, most of which already occur in the UK, other than to stop the planting of clonally spreading exotics in bottomlands and private woodlands, and to establish some coastal stretches that are free of Hottentot fig. Established alien species such as sycamore and *Rhododendron ponticum* are unlikely to become worse problems following climatic warming, and it may be impossible to keep alien plant species away from invadable coasts and rocks (Hill *et al.*, 1994)

5. Agriculture, Horticulture and Aquaculture

SUMMARY

- Agriculture in the UK is generally sensitive to climatic factors which contribute significantly to year-on-year variability in output. Changes in the intensity, annuality and distribution of precipitation, together with prevailing temperatures and the incidence of extreme weather events would have the greatest effects upon production.

- Alterations of rainfall, temperature, and the incidence of extreme weather events will impact upon all components of the sector. The north-south gradients of climate variables (in particular summer temperatures and rainfall) will also be of major significance in terms of impact. Elevated atmospheric CO_2 will stimulate plant productivity but by a variable amount depending upon species, location and management.

- Spatial variation in inputs will affect specific production systems in different ways. Limitations caused by reduced water availability in the south and east UK, coupled with higher temperatures and increased evapotranspiration may shift potential production of arable and other field crops northwards and westwards as well as placing extra pressures on water for irrigation.

- Grassland productivity in the wetter north and west would be sustained by warmer winter temperatures, and the boundaries of forage maize cultivation may continue to move northwards. Stock damage to wetter pastures could reduce the advantages of increased grassland productivity.

- Trout farming is likely to show the greatest sensitivity within aquaculture systems. This would be most marked in the south due to warmer temperatures and to the increased likelihood of drought restricting river flows.

- Adverse effects on soils and increased incidence of pests, weeds and diseases could reduce or negate any yield increases attributable to climate change.

- Warmer and drier summers will increase cultivation opportunities for novel crops including industrial and perennial biomass crops (e.g. *Miscanthus*).

- The sensitivity to climate change of the whole sector, and the industries which it supports, is likely to be less than that of specific production systems. Altered patterns of land use will help to sustain overall performance. The effects of climate change on global production will, however, impact at the sectoral level in the UK in terms of changes in market price and in the UK's altered competitive position with respect to other food-exporting regions of the world.

5.1 INTRODUCTION AND BACKGROUND

This chapter considers the potential effects of climate change on agriculture, horticulture and aquaculture. This sector of the UK economy has a gross output of some £14 billion and is the major provider of raw materials to the food manufacturing sector which accounts for 2.9% of GDP, with a gross output of £60 billion.

Agriculture is the largest user of land within the UK, covering about 77% of the total land area. Grassland makes up more than 70% of this total (c. 13 million hectares). The remainder is used for a range of arable and horticultural enterprises. River and coastal aquaculture occupy only a small area. Although less than 3% of the labour market are employed directly within this sector, it is the major generator of revenue in rural areas, and forms an essential part of the infrastructure supporting tourism, recreation, forestry and the water industry.

The 1991 CCIRG report concentrated upon the direct effects of major climate variables on the processes involved in primary production. It concluded that there would be significant direct effects on agricultural systems, but that these could also generate novel opportunities. Particular areas of uncertainty included the significance of indirect effects upon soils and upon pests and diseases. The scope of this part of the second CCIRG report has been widened to consider, in addition to primary production processes, the implications of climate change on the UK industries based upon them, which, in turn, function as components of global economic activity.

The main sources of new information have been: the relevant IPCC Working Group II documents (Reilly *et al.*, 1996); increased fundamental knowledge on biological responses to climate change at the system, organ and cell level (Lawlor and Mitchell, 1991; Long, 1991); the refinement of predictive models for crop growth and yield (Porter, 1993); and the progressive extension of such modelling into global economic activity (Rosenzweig and Parry, 1994). In addition, the enhanced resolution of general circulation models of climate has resulted in more refined projections (Chapter 2) which have been used as the basis for this discussion. A number of coordinated research programmes have been initiated since the 1991 report (Table 5.1) and the outputs from these will become available over the next few years and will lead to further refinement of impact forecasts.

It is important to bear in mind that significant changes at one level in the hierarchy of activities from farming process to global economy may have very different consequences at another. For

Table 5.1: Current or recent UK research programmes on the biological effects of climate change.

Funding Agency	Remit
NERC	Ecosystem responses to global climate change
BBSRC	Mechanisms of adaptation to climate change
MAFF	Responses and contributions of agricultural systems to climate change
SOAFD	Spatial and system responses to climate change
DANI	Monitoring environmental change

example, increased use of forage maize is a visible indicator of significant changes in agricultural land-use patterns within the UK. However, these changes have had little impact upon world-wide livestock agriculture which is much more sensitive to factors affecting global demand. At all levels, political considerations affecting subsidy have resulted in marked changes in agricultural land-use, the make-up of the industry, and its impact on similar industries in other countries. The integration of such factors into projections of climate change is not straightforward. However, the IPCC Working Group II concluded that, on a global scale: i) the effects are likely to be highly discontinuous; but ii) that global agricultural productivity is expected to be broadly sustainable under equilibrium climate scenarios based upon double current effective CO_2 concentrations. The complex interactions at the crop level between elevated CO_2, temperature, rainfall and nutrient status which determine primary productivity are not, however, fully understood, and may modify these broad projections.

5.2 ESTIMATED EFFECTS OF CLIMATE CHANGE AND SEA LEVEL RISE

5.2.1 Sensitivity of agriculture, horticulture and aquaculture to weather and climate

Primary crop productivity in the UK is determined by prevailing temperature, incoming solar radiation, water and nutrient availability. The growth and development of the canopy determines the light interception and ultimate productivity, and this is strongly affected by temperature and nutrient status. Canopy photosynthesis is driven by incoming solar radiation, but may be affected by constraints imposed by water and nutrient availability, or by temperature stress. All systems will respond to changes in these variables, but sensitivities vary markedly between systems and in time and space within systems. For example, crop establishment in the spring is strongly influenced by temperature, whereas production in the summer is affected more by light interception and water availability. In some cases, these responses have been quantified and modelled, and specific predictions are possible (Lawlor and Mitchell, 1991). Increasingly, the potential effects of climate change can also be included in such models and sensitivities identified. In general terms, as suggested by the 1991 CCIRG report, most crop production systems are likely to show significant responses to changes in the mean values of temperature, CO_2 concentration and rainfall, together with potential evapotranspiration and windspeed. It would appear, however, that major reductions in overall European crop productivity under 'steady-state' alterations in mean climatic values are unlikely (Reilly et al., 1996). What is less clear are the effects of altered frequency of extreme weather events. Modelling analysis suggests that changes in variability without changes in the mean do have significant effects upon predicted yields (Semenov and Porter, 1995). Also unclear is the interaction between primary responses and both secondary effects on soils, pests and diseases and the interactions with other anthropogenic stresses (NO_x, O_3, UVB, etc.) (Allen, 1990).

Aquaculture in the UK is largely tied to the farming of salmonids (rainbow trout, Atlantic salmon), mussels and oysters (Williamson and Beveridge, 1994; Muir et al., in press). Salmonids in particular are highly susceptible to climatic conditions. Temperature influences growth and, among salmon, the timing of sexual maturity. In addition, the farming of molluscs can only be carried out in areas whose microbiological quality is acceptable under the conditions specified by the European Community Shellfish Directive.

The sensitivity of agricultural industries to localised changes in primary production will be much lower than the sensitivity of primary production because of their proven ability to adapt (Section 5.3). However, they, in turn, will be sensitive to global patterns of food trade and population needs. These will be affected by climate change, but the overall sensitivity is difficult to estimate. Global estimates of cereal production under 2 x CO_2 climate change scenarios have been integrated with models of world food trade to generate predictions of price and risks of hunger for developed and developing countries. This study suggests that the global production response is fairly low, but that the production in developing countries would decline with a concomitant increase in the potential risk of hunger in these regions (Rosenzweig and Parry, 1994). Such projections suggest that UK agriculture would be sensitive to global perturbations, but that the effects would be positive in terms of both increased output and increased demand. These predictions are not, however, consistent with current trends towards reduced real costs of food (Alexandratos, 1995) or with forecasts of continuing decline in such costs. It must be pointed out, however, that such forecasts are not universally accepted (Bongaarts, 1994).

5.2.2 Effects of climate change

Elevated CO_2 will generate a positive and variable increase in productivity for annual C_3 crops, which is likely to be larger than that for perennial crops including grassland. The magnitude of potential changes has been established in a range of studies, and typical annual increases in gross productivity of about 30% at 700 ppm CO_2 are realistic where other factors are not limiting. The range of responses would, however, be large (Reilly et al., 1996; Table 5.2). The effects of elevated CO_2 on crop performance interact strongly with the changes in temperature and precipitation, which are themselves directly attributable to increases in greenhouse gas. For annual crops, increased temperature speeds development, hastens maturation and may reduce yield if the period of useful growth is shortened (Porter, 1993). A 4°C increase in ambient temperature over the whole growing season decreases yield of winter wheat by about 20% (Mitchell et al., 1995). However, cultivar variation in temperature requirements in crops is high, and it is also likely that, for some crops, breeding adjustments will be feasible. In others, critical developmental phases may show extreme sensitivity to temperature, and these sensitivities may not be so susceptible to amelioration by breeding. There are also critical thresholds in certain crops which affect distribution (e.g. threshold temperatures and accumulated growth temperature in forage maize). Here, climate changes will also affect potential cultivation acreage. Thus elevated temperatures are predicted both to alter the yields of existing crops and to improve the suitability for cultivation of alternatives (Parry and Duncan, 1995).

The changes in precipitation indicated in the 1996 CCIRG scenario are given in much more detail than in the 1991 CCIRG report. The predicted seasonal and latitudinal variation will impact upon agriculture, particularly in southern England. The potential sensitivity of many crops to lower precipitation is high (Figure 5.1), and it is probable that greater levels of irrigation will be needed to sustain yields in the south of England. The 1996 CCIRG scenario indicates alterations in other variables, of which the sea-

Table 5.2: Relative yield increases with upper and lower limits for a range of agricultural and horticultural crops under double ambient CO_2 conditions (from Kimball 1993).

Crop	Mean Yield Increase	Lower Limit	Upper Limit
Lettuce	1.35	1.19	1.53
Tomato	1.20	1.12	1.28
Wheat	1.37	1.11	1.69
Soybean	1.27	1.01	1.60
Potato	1.64	0.95	2.81

These observations were derived by Kimball (1993) from the primary literature and have been normalised to permit comparison. The estimates do not take into account interaction with temperature, water stress, nutrient availability, etc. They indicate, however, the general positive nature of the response and the variability between crops and experiments.

sonal and geographical variation in winter wind speed and summer solar radiation are likely to have marked effects on primary products. Increased wind speed could affect fruit crops, if it occurs prior to harvest or leaf loss, and these effects are predicted to be maximal in the southern half of England. Reduction in irradiance caused by increased cloud cover would have direct effects on photosynthesis. The magnitude of this effect will depend upon when in the growing season it is manifest. Finally, increased frequency of extreme high-temperature events and the decreased frequency of frosts will also im-

Figure 5.1a: Predictions of yield of forage maize using mathematical models driven by meteorological data from 93 English and Welsh Sites (From Davies et al., 1995). Baseline, climatic conditions.

Topographical and soil quality overlays were used to exclude land unsuitable for cultivation. The models were re-run using data amended to give a 2°C rise in temperature and either a 10% increase or a decrease in precipitation. The altitude overlay for projections of increased mean temperature was raised from 180m to 500m.

Figure 5.1b: Predictions of yield of forage maize using mathematical models driven by meteorological data from 93 English and Welsh Sites (From Davies et al., 1995). Temperature +2°C, Precipitation - 10% of baseline conditions.

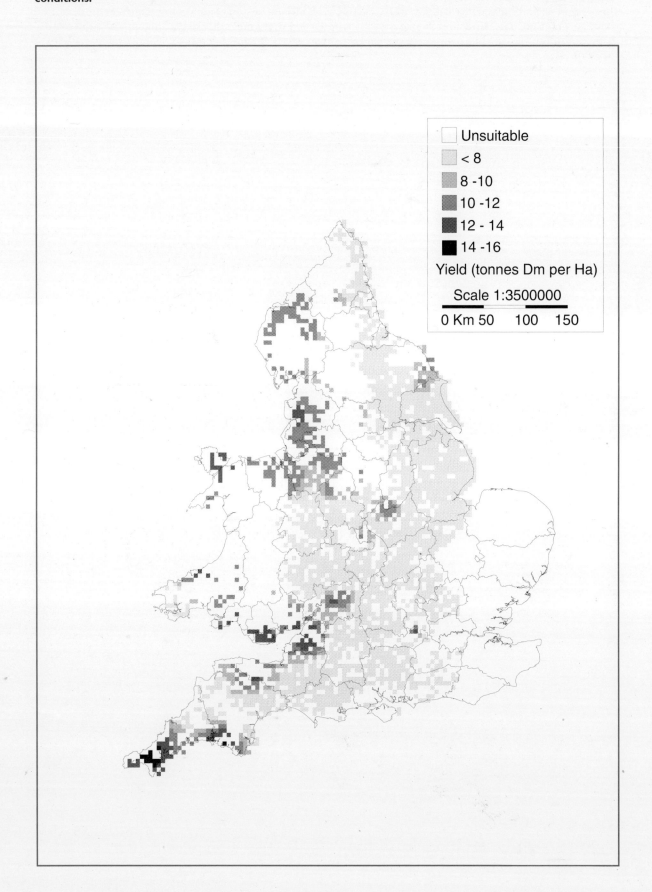

Figure 5.1c: Predictions of yield of forage maize using mathematical models driven by meteorological data from 93 English and Welsh Sites (From Davies *et al.*, 1995). Temperature +2°C, Precipitation + 10% of baseline conditions.

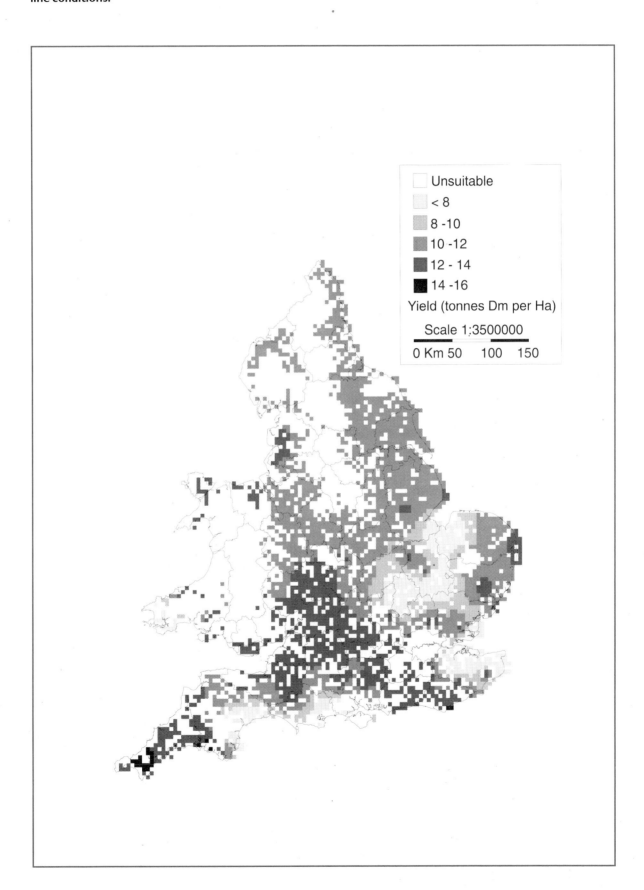

pact significantly upon crop performance and survival. It is known, for example, that short duration exposure to high temperature can adversely affect both reproductive development (Wardlaw et al., 1989) and starch deposition in cereal grains (Jenner, 1991) and warming in early spring can sharply increase vulnerability of perennial forages and winter-sown cereals to subsequent frosts (Pollock and Eagles, 1988).

Despite these observations, the IPCC Report (IPCC, 1996b) rates the general direct vulnerability of European agricultural systems to climate change as being low. This is attributed to: i) the small proportion of the GDP represented by primary agriculture; ii) the small number of people employed on the land; and iii) the capacity of the system to adapt (see below). Individual sectors, however, will be affected differentially, and marginal or niche enterprises (e.g. Pembrokeshire early potatoes) could be very sensitive to changes in productivity elsewhere. By the same token, limited opportunities may arise for alternative cultivation systems better able to exploit warmer or drier climates (e.g. grain maize and sunflower). Perennial biomass or industrial crops may also be better able to exploit warmer summers where water is limiting, since canopy development does not depend upon current photosynthesis.

Arable systems

These include both determinate and indeterminate crops, and there have been a number of relevant studies on the responses of such systems to elevated CO_2 and temperature. What is now becoming clear is the degree of interaction between these responses and changes in nutrient and water availability. The extent of acclimation of the primary photosynthetic system to elevated CO_2 depends strongly upon both phenology and the other prevailing limitations to growth, and this is a major cause of the variations in yield increase observed in different studies (Long, 1991; Lawlor and Mitchell, 1991; Nie et al., 1993).

Decreases in summer rainfall would be particularly significant in affecting arable production. This could have the effect of shifting the area of potential maximum production northward and westward away from the traditional areas of cultivation (Brignall and Rounsevell, 1995; Figure 5.2). Warmer, wetter winters may cause increased incidence of pests and diseases (Harrington and Woiwod, 1995) and lead to problems of waterlogging which would affect cultivation of winter cereals (Brignall et al., 1995).

A range of growth models exist for the major temperate arable crops, some of which have been used to predict responses to climate change scenarios (Mitchell et al., 1995). In general, they have not been adequately tested with experimental data relevant to climate change conditions, although some comparative studies have been carried out (Grashoff et al., in press). There are also significant gaps in the range of models for minor crops, and most models estimate production potential and do not take into account losses due to pests and diseases. There are also significant discrepancies between the outputs of different models for the same crop, so the reliability for future projections may be questionable (Davies et al., 1994).

Horticultural systems

The majority of horticultural cultivation is unprotected, or partially protected, and is thus subject to similar constraints to arable systems. Increased temperatures could bring benefits in reducing time to harvest for high-value overwintering crops, but once again the interactions with rainfall, nutrient availability and incidence of pests and diseases has not been well studied. Elevated temperatures can cause reductions in both quality and appearance in some vegetable crops (Wurr, 1994). Pressures on water supplies for irrigation associated with the projections of geographic shifts in winter rainfall would be partially offset by increased water use efficiency, but the magnitude of this amelioration is likely to be variable between species. For perennial fruit crops and some vegetables,

Figure 5.2: The effects of changed temperature and precipitation on the distribution of winter wheat suitability (From Brignall and Rounsevell, 1995). (a) Temperature +1°C; (b) Temperature +2°C; (c) Temperature +2°C, Precipitation -10%; (d) Temperature +2°C, Precipitation +10%.

Suitability was predicted based upon the frequency of occurrence of drought stress and the length of time in the autumn when soil moisture contents were suitable for tillage.

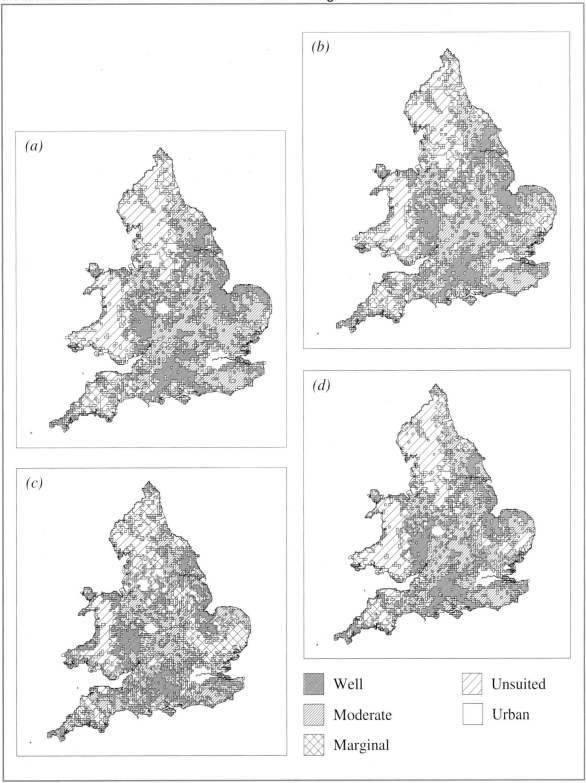

exposure to chilling temperatures is necessary to trigger further development, so warmer winters could adversely affect phenology. Fruit crops are also sensitive to late frosts and autumn wind damage, so increased incidence of extreme events could be damaging. Projected reductions in summer irradiance in northern Britain, associated with increased cloud cover, could also lower yield in some cases. Increased demand for irrigation in the south and east will increase cultivation costs and may result in altered patterns of horticultural land use, favouring the wetter north and west. Increased temperatures will alter the energy inputs required for protected horticulture and may affect the range of crops grown under cover.

Predictive models exist for some horticultural crops, and there have been a restricted number of studies which have incorporated climate change scenarios. Similar limitations apply to these studies as to those discussed above.

Livestock systems

Grassland-based livestock systems will be affected by warmer, wetter winters. Increased temperatures, particularly at night, will increase respiration at a period when photosynthesis is restricted by low irradiance and short photoperiods (Thomas and Norris, 1977). Increased growth during this period will also render the sward more susceptible to damage by pests and diseases and by wind; it will also reduce the persistence of white clover in mixed swards. By contrast, warmer summers will enhance grass production, particularly if sward height is optimised by good management. Risk of drought stress may increase. Annual forages (particularly maize) will be favoured by increased temperatures, but water availability may limit cultivation in the east (Figure 5.1; Davies et al., 1994). The spread of other crops northward and westward under the influence of climate change will impinge upon lowland grass systems which favour alternative cultivation practices.

Animal production will be less affected by current UK climate change projections than will forage production, although outdoor grazing in winter may bring increased risks of health problems, such as lameness, and may increase damage to wetter pastures. A greater move towards extensive animal production systems will reduce the extent of such damage. The higher mean temperatures projected are not high enough to cause health or welfare problems or to affect the efficiency of conversion (Reilly et al., 1996). Animal housing may have to be of a higher construction standard if the magnitude of extreme weather events increases significantly (Chapter 11) and tree planting to provide shelter belts may become more widespread.

Predictive models of grassland production systems do exist, and have been applied to climate change scenarios. They do not take into account secondary factors such as soils, pests and diseases. There have been few long-term mechanistic studies at the system level, although there is some work using open-topped chambers and FACE (Free Air Carbon Dioxide Enrichment) systems (Pollock et al., in press). Integrated models of animal production systems have been used in SOAFD-funded studies on impact assessment.

Aquaculture systems

Trout farming is likely to be the most affected activity within this sector. Changes in precipitation patterns, especially in the southern UK, will affect water availability and cost. More episodic rainfall will affect carrying capacity, suspended solids loads and acidic surface run-off. Higher summer temperatures may affect water quality, at a time when effluent controls are likely to become stricter (Beveridge et al., 1994). This may result in a northward shift in trout production. Milder winters may improve production in the north, although higher UVB levels may require greater use of shading (Bullock, 1988), or a move from pond to tank-based systems. Effects on parasite life cycles are not known.

Mariculture is less likely to be affected. Higher sea temperatures may exacerbate problems of premature sexual maturity, although controls for this problem are being explored. Changes in sea temperature and carbon dioxide levels will affect plankton production and hence that of shellfish. Microbial quality in shellfish-growing areas may deteriorate, requiring additional procedures to be adopted. Any effects of climate change on fishmeal prices would be very important in determining production costs for salmonids.

There have been few attempts to model effects of climate change on aquaculture, although the use of GIS techniques is being explored (Ross *et al.*, 1993).

5.2.3 Effects of sea level rise

Projections in this report are for a slower rise in mean sea level than that proposed in the 1991 CCIRG Report. Nevertheless, any significant probability of flooding will put at risk areas similar to those identified in the previous report (Chapter 17). Fifty-seven per cent of all Grade 1 land lies below the 5 metre contour line, particularly in the Humber Estuary and in East Anglia. Although there is some protection by sea defences in these areas, this land must be regarded as 'at risk' (Whittle, 1990). These rises would not significantly affect UK aquaculture.

5.3 ASSESSMENT OF POTENTIAL ADAPTATION

Both historically and currently, UK agriculture and horticulture have shown themselves to be strongly adaptive. Adaptation has been almost entirely directed, and has occurred at the level of altered land use patterns, primary system performance, industry targeting and general efficiency increases. Both internal changes (new varieties, crops, agro-chemicals, managements, etc.) and external changes (new markets, industrial and biomass crops, altered competition, altered customer demands) have driven this adaptation which has been the principal engine for the relatively low current cost of food. There is, however, little opportunity for adaptation in the UK aquaculture sector, which is largely tied to the production of a few species and which is dependent upon a few key resources. There may be some opportunities for altered distribution of fish farms to cope with changes in water supply and quality, and improved production methods may help combat parasite problems. It may be possible to produce modified grain crops which will reduce dependency upon fishmeal.

Adaptation is rarely a response to a single defined environmental variable, although the appearance of, for example, specific exotic animal diseases would generate a significant direct adaptive response by the livestock industry. It seems likely that the sector will adapt via a combination of altered management practices, patterns of land use, exploitation of new opportunities and awareness of new markets. Improvements in variety production (assisted by the increasing use of biotechnological methods), cultivation, harvesting, storage and transport will all help to mitigate any deleterious effects of climate change, and the underpinning research base may be applied as necessary to specific problems associated with climate change. Strategies for crop improvement may improve tolerance to environmental factors over those exhibited by current cultivars, but it is by no means certain that overall yields will be sustained. On a global scale, there is debate about whether the adaptability of agriculture is adequate to cope with increasing pressures (including a changing climate and population increase), and about whether the social, scientific and governmental structures always provide an adequate basis for large-scale adaptation at national level. The IPCC Second Assessment Report identifies a range of responses in terms of research, extension work, general education, transport, distribution, marketing and support which, if implemented, would increase adaptability (IPCC, 1996b). Most of these responses do occur in some form or another in the UK, and it seems probable that the general adaptability of this sector is adequate in the short term. Attempts have been made to integrate global and regional estimates of adaptation in terms of altered land use (Parry *et al.*, in press) and these confirm both the sensitivity to other factors driving adaptation and the effects of climate change upon geographical patterns of land use.

What is less clear is the relationship between short-term adaptability of specific systems (e.g. via additional inputs) and their long-term sustainability. Sustainability is now a major goal for agriculture in many developing and developed countries and it seems likely that this will restrict the options for longer-term adaptation to altered climate. The research implications of sustaining and extending adaptability are discussed below.

5.4 UNCERTAINTIES AND UNKNOWNS

Significance of secondary effects

There is still very little hard evidence on the effect upon primary production of secondary factors which are themselves sensitive to climate change. This is particularly true of soils (Chapter 3), pests and diseases. In particular, warmer, wetter winters could markedly affect the overall patterns of pest and disease overwintering which, in turn, would affect the subsequent development of epidemics and modify cultivation methods (Treharne, 1989). The more rapid life cycle of such organisms will also amplify any effects of UVB radiation on genetic change within the population. Although specific studies on a restricted range of pests and diseases have been performed which generally support the above view, the pattern of response of individual pests and diseases is likely to vary markedly, and it will be difficult to predict where the major constraints will occur.

Effects of climate change on the quality of agricultural products

There is a close interaction between the metabolism of carbon and that of nitrogen in crop plants. This interaction is a primary determinant of quality in a variety of forage, horticultural and grain crops, and there is increasing evidence of the effects of climate change upon various indicators of quality (Tester *et al.*, 1995; Williams *et al.*, 1995). Changes in forage and grain quality would also affect animal production (Nie *et al.*, 1993).

Timing and distribution of extreme events

The increased detail presented in Chapter 2 compared with the 1991 CCIRG Report has given strong support to the view that the climate is likely to become more variable, at least in certain key respects. From the standpoint of agriculture, horticulture and aquaculture, the timing and distribution of these events is of paramount importance (Morison and Butterfield, 1990, Semenov and Porter, 1995, Seligman and Sinclair, 1995). Biological responses to stress can be both adaptive and acclimatory (i.e. a degree of tolerance can be inbuilt, and it can also increase in response to a range of environmental cues). If the relationship between environmental stress and the cues that regulate induced tolerance is disturbed, the effects can be dramatic (e.g. damage caused by late frosts after hardiness has been lost). There is genetic variation in the nature of the environmental signals associated with acclimation which could be used in the improvement programmes, but clear indications of major targets will be needed, and this will require further refinement of the methods used to predict extreme events and their impacts upon agricultural production.

5.5 PRINCIPAL IMPLICATIONS FOR OTHER SECTORS AND THE UK ECONOMY

The major beneficiary from agriculture, horticulture and aquaculture is the food industry, Any alterations in land use which might occur as a result of altered practice will also impinge upon forestry, tourism, the natural landscape and the water industry. In turn, the projections of increased recreational land use (Chapter 16), reduced availability of irrigation water (Chapter 7), increases in salination (Chapter 17) and alterations to soil structure (Chapter 3) have significant implications for agriculture. The inherent adaptability of the sector suggests that overall changes are likely to be gradual, and thus manageable. However, the collapse of niche enterprises, or their replacement by others, will have significant local effects. The broad conclusions would, however, be that the effects directly attributable to climate change are only part of the more widespread changes within the sector which will impact on the UK economy over the foreseeable future.

5.6 RESEARCH AND POLICY ISSUES

5.6.1 Future research effort

A number of specific research needs can be identified:

- Continued spatial and temporal refinement of predictive climate modelling.
- Further studies on the responses of fast-growing organisms (particularly pests, weeds and diseases) to climatic change. This should involve increasing the range of organisms which have been studied, and increasing the detailed knowledge of primary responses and interactions within systems.
- Integrated systems studies (involving soil science, where appropriate) to assess long-term responses (particularly with perennial crops). Such studies should involve predictive modelling and include socioeconomic considerations.
- Fundamental studies on the interaction between elevated CO_2, temperature rise and other climatic factors and pollutants (NO_x, O_3, UVB, etc), for major animal and plant systems.
- Investigation of the interactions between sustainable agriculture and increased tolerance of climate change.
- Experimental analysis of responses of agricultural systems to variable and /or extreme conditions.
- Studies on a wider range of crops of the interactions between climate change and quality.

5.6.2 Policy issues

It is inappropriate to define a suite of policy requirements which can be wholly assigned to the problems caused by climate change. What is appropriate is a coherent sectoral response to its development as part of the UK economy through the 21st Century. Such a response needs three broad strands of understanding, backed by a policy environment which facilitates the application of such knowledge. The three main strands are:

i) Prediction e.g.

- What is the place of this sector within the UK and the global economy?
- How will the balance alter between food supply and food demand?
- How will that affect the industry in different regions?
- How will the major environmental constraints affect primary production?
- How significant will the extensive (sustainable) production systems become regionally and globally, and what effect will this have on both total production and on efficiency?
- What effects will increasing wealth have on consumer behaviour?

ii) Amelioration e.g.

- What improvement strategies for various components of this sector can be implemented in response to the predictions above?

- What are their costs and what are their benefits?

- Do any of these (e.g. those involving genetically manipulated organisms) lead to consumer resistance?

iii) Alternatives e.g.

- To what extent can the sector direct its efforts to the economic production of alternative crops and non-food products, and how will these impinge upon the management of landscape?

In all cases, climatic change is an important, but not unique, component of the problem and it should be addressed in proportion in formulating research needs and in moving towards robust solutions.

6. Forestry

SUMMARY

- Expansion of UK forestry earlier this century was intended to reduce reliance on imports. Future changes in area will also be policy-driven, reflecting shifts in the balance of agricultural support, social policies and conservation values, rather than affected by responses to changes of climate.

- The timber trade in the UK, supplying only 15% of domestic wood product demand, is dependent on conditions in exporting countries (Scandinavia, N. America) and thus relatively insensitive to changes in British supply. Changes in overseas supply would alter production economics here, thus affecting expansion and species choice.

- UK forests, uniquely in Europe, are based on introduced species, which are more productive and in greater demand by wood-using industry than native species. Most forests are young, have short rotations and thus could allow introduction of species (or origins) better adapted to the altered climate. More rapid early growth of commercial species, already occurring, may be attributable to increased ambient CO_2, nitrogen deposition and higher temperatures. Further increases due to future changes of climate would generally improve the financial return on afforestation, encourage forest expansion and reduce reliance on imports.

- Yields of the main commercial species in central and northern UK may increase by about 25% (equivalent roughly to 1.0 M m^3 yr^{-1}) by 2050 in response to increased mean and accumulated temperatures (> 5.6 °C) wherever water availability is not limiting.

- In southern UK decreased precipitation and increased evapotranspiration would reduce general productivity and drive sensitive species (e.g. beech) from marginal sites affecting wood supply, amenity, recreation and conservation values. Urban trees may be particularly stressed and substitute species should be sought now.

- Areas subject to more frequent drought might expect increased insect pest damage, fire hazard and poorer wildlife habitat.

- Adaptation to climate change will require positive management intervention. Existing species populations cannot migrate and would therefore require deliberate replacement with better adapted material while management strategies would need to alter to accommodate changes in growth patterns and frequency of damaging impacts.

6.1 INTRODUCTION AND BACKGROUND

The forestry sector involves a wide range of activities from the establishment, manipulation and protection of forest stands to the harvesting and marketing of wood and other products. Forests are managed mainly for their timber products but also provide a variety of other goods and services. These latter include the maintenance of wildlife habitat and biodiversity, landscape, amenity and recreational facilities. Forests are also a means of sequestering CO_2 in themselves or in the long-term storage of their wood products. The balance between different objectives of management reflects the perceived needs of society and, recently, greater emphasis has been placed on the non-timber aspects of forest management.

Forests currently occupy 10% of the land surface in the UK having increased from a low of 3% at the turn of the century. This expansion of forest cover has been achieved by a vigorous afforestation policy conducted by both State and private agencies, the latter assisted by fiscal incentives. However, recent new planting at about 15k ha is less than half the Government's annual target rate (33k ha) The current distribution of the forest estate (Table 6.1) illustrates the preponderance of afforestation with coniferous species, particularly in Scotland, and the much greater proportion of broadleaved species in England. Coppice working is almost entirely confined to the south of England. The distribution largely reflects the availability of marginal agricultural lands for afforestation, particularly between 1945 and 1990, and historical aspects relating to land use and ownership.

The species composition of UK forests (Table 6.2) is the result of a combination of ecological and economic factors. Firstly, the native tree flora, which invaded post-glacially, includes only one productive conifer, Scots pine, which is limited ecologically to freely draining soils in relatively low rainfall areas. Secondly, the land available for afforestation has been predominantly in the uplands with higher precipitation, poorly drained and often impoverished soils. These conditions favour the use of adapted exotic species, such as Sitka spruce or lodgepole pine, and are unusable for most broadleaved species. The productivity of the introduced species (Table 6.2) considerably exceeds that of the native tree flora making them attractive economically. The shorter rotation of productive conifers, as well as their greater yield, greatly enhances profitability compared to the low yielding broadleaved species.

The long time scale of forest production, with rotations of 40-160 years according to species, means that climatic changes proposed in the 1996 CCIRG scenario for the 2050s will occur well within the life span of many existing tree stands. In addition the rate of change may exceed possible migration rates of most tree species (IPCC, 1996b). However, the dissected nature of land use in the UK and its dependence

Table 6.1: Total forest area of Britain (at 31/3/94) (thousands of hectares). (Source Forestry Commission 1994)

	High Forest		Coppice	Total Productive Woodland	Other Woodland
	Conifers	**Broadleaves**			
England	382	450	38	870	103
Wales	168	65	2	235	13
Scotland	966	100	0	1066	90
Great Britain	1516	615	40	2171	206

Other woodland consists of areas where timber production is not a main objective. It includes areas managed chiefly for amenity and public recreation.

Table 6.2: Productivity and rotation lengths of major species groups in British forestry. (Source: FICGB Year Book 1994)

Species group	Average MAI[1] m^3 ha^{-1} yr^{-1}	Rotation length years	Forest area %
Pines	9	45-75	22
Larches	9	45-55	8
Spruces	13	40-70	34
Douglas fir	14	45-60	2
Oak	5	120-160	9
Beech	6	100-130	4
Ash	5	60-80	4
Birch	5	40-60	4

[1] MAI is mean annual increment

on socio-economic factors makes natural migration difficult for tree species, if not for their associated organisms.

The demand of the wood-using industries is overwhelmingly for coniferous timber for almost all end uses, although there is a good market for high-grade hardwoods in relatively small quantities. Of the 15.7 million m^3 of wood (round and sawn) imported in 1992 only 1.7 million m^3 was of hardwood (Forestry Industry Committee of Great Britain, 1994). The UK reliance on imports for some 85% of its wood product requirement, at an annual cost of about £6 Bn, and the recent large investments (£1.3 Bn) in wood manufacturing industry are major factors behind the increased area of productive forest. On the other hand, the need to maintain open land areas for landscape and conservation, constrains the possibilities for afforestation.

For the above reasons UK forestry is unique in Europe, being very largely based on introduced species from many of the temperate regions of the world. British forests are also unusual because most have been established recently. Apart from the small fraction of residual native woodlands, the majority of planted areas are less than 50 years old and in their first rotation. Though apparently acclimated to current climates these stands are not in equilibrium with their associated fauna and flora or their soils but are in a transitional stage in the development of potentially novel ecosystems which may be only partially analogous to other temperate or boreal forests.

Further increase in the land area devoted to forestry depends on European policies for elimination of agricultural surpluses. In the lowlands, imposition of 'set-aside' policies and production quotas is leading to expansion of mainly broadleaved woodland, primarily for amenity and recreational benefits. Expansion of forest in the uplands is strongly influenced by the relative balance between Government support for hill-grazing systems, incentives for afforestation and social policies to maintain viable communities in more remote areas. These socio-economic forces are likely to influence land-use changes as much as any of the currently projected changes in climate.

This section considers the possible effects of potential climatic change on the distribution and development of UK forests, their production and the utilization of forest products, taking into account likely changes in the availability of imported wood products. The First CCIRG Report (1991) dealt only briefly with forestry, suggesting that increasing temperature and CO_2 levels would enhance tree growth rates, reduce soil water availability and increase fire and possible pest hazards while it noted the importance of wind and storm frequency on the stability of upland forests. These and other effects are further discussed here.

6.2 ESTIMATED EFFECTS OF CLIMATE CHANGE AND SEA LEVEL RISE

The 1996 CCIRG climate scenario for the period up to the 2050s in Britain (Chapter 2) suggests a SE-NW gradient in both mean summer and winter temperatures, the largest effects being in SE England (+2.0 °C) and the least (+1 °C) in NW Scotland. Mean seasonal temperatures are important in several ways for the distribution and function of tree species, but it is the frequency of extreme seasonal conditions (e.g. summer of 1976) which test the stress tolerance or survival of particular species. The frequency of anomalous warm summers or mild winters is projected to increase with concomitant reductions in the frequency of air frost days, again particularly in southern Britain. The occurrence of frost in the growing season is another critical factor for reproduction and survival of many tree species.

The projected patterns of precipitation to the 2050s are less secure but suggest an increase in winter precipitation in southern Britain diminishing northwards to present values in northern Scotland. The pattern is reversed for summer precipitation (decreasing in south; increasing in north) with parallel changes in rainfall intensity and relative humidity while mean incoming solar radiation in summer shows differences from the present of +12 to -12 (Wm^{-2}) along the SE-NW gradient across the UK. These changes when expressed as potential evaporation indicate enhanced drying effects in the south-east but reduced evaporative stresses in northern areas which would have strong impacts on the availability of soil water in both regions.

Changes in mean wind speeds described by the 1996 CCIRG scenario show little difference in summer but modestly increased windspeeds in southern areas in winter which will tend to reduce existing differences across the country. However it is the frequency of damaging wind storms (> 36 m s^{-1}) that most influences forestry, because potential rotation lengths are curtailed by windthrow. Although the occurrence of such wind storms cannot yet be predicted, the projected increase in the frequency of gales in upland Britain would affect the orderly flow of timber products to markets, increase the proportion of small roundwood and thereby reduce the return on investment.

6.2.1 Sensitivity of forestry to weather and climate

Initially it might appear necessary to separate the climatic responses of native from introduced tree species. Some 30 species (Mitchell, 1981) of tree or large shrub are considered native. With the exception of Scots pine all native trees are broadleaves, some of which are close to their northern limits in the UK. None is endemic and all have Continental distributions spanning wider latitudinal ranges than found in the UK. It is often claimed that native species are best adapted to existing climatic conditions, their populations having persisted through climatic changes equivalent to those now projected, albeit at a slower rate of change. However, past management of many semi-natural woodlands resulted in the introduction of species of continental origins (e.g. oak in Scotland). These populations therefore may not be best suited to existing conditions. With the resurgence of broadleaved planting in recent years, home-collected seed has been insufficient to meet the demand so that imports from Europe continue to be used.

Introduced species, which considerably outnumber natives and are, in forestry terms, far more important, were initially imported on an *ad hoc* basis but subsequently their natural distributions have been screened for those origins that are well adapted to British conditions. As a result of extensive provenance testing it has been possible to identify ecotypes that are phenologically suited (Lines, 1987). In many cases the genetic variability within the major exotic species is better understood than in the native species.

Tree species, in common with other plants, show considerable genetic variability *within* populations which in natural conditions allow them to adapt to quite large variations of climatic factors. This variability is often greater than that found *between* populations within a region, although marked ecotypic differences can occur across well-defined environmental boundaries. In managed forest there is a tendency to reduce variation as a result of selection for productivity and qualitative characteristics (e.g. stem form). Tree improvement programmes attempt to take this into account by either assembling selected populations covering the range of climates that exist now or by testing genotype x environment interaction across a number of sites. However, the current selection of species or genotypes for forest production could encounter problems with changes in the frequency of extreme events, such as droughts or wind storms.

Temperature relations

The distribution of species is often described in relation to isotherms, for example, the altitudinal limit for Scots pine in Scotland is thought to occur where summer temperature exceeds 10 °C (Pears, 1967). However, distributions may be better described if species occurrences are plotted against the mean temperatures of the warmest and coldest months which provides a 'climate-envelope' for the species. Clearly there needs to be some rationale for the variables chosen to define the 'climate-envelope' in terms of the performance of the species (Parr and Eatherall, 1994). Recently Jeffree and Jeffree (1994) have suggested that species distributions, when plotted against these temperature axes, fall within an ellipsoidal shape, with its principal axis inclined to the ordinate of winter temperatures, within which 75% of the distribution can be delineated, whether the species distribution is geographically continuous or disjunct. The projected changes in mean temperatures for individual localities, where European beech occurs (Figure 6.1), indicate improved conditions for Swedish and Norwegian populations while those in France and Italy become more marginal. UK stands are little affected in relation to temperature change. These techniques, while defining species temperature tolerances for naturally-occurring populations do not, of course, take account of the artificially extended distribution of many species. For example, beech stands planted in northern Scotland have survived and regenerated over two centuries.

The genetic variation within the distributions of most tree species, commonly occurring in Britain, is probably sufficient to accommodate the mean temperature changes projected in the 1996 CCIRG scenario.

Many tree species may be damaged by out of season frost, either after dehardening in spring or before hardening their tissues in autumn. Although the projections in Chapter 2 suggest a considerable reduction in frost days ($T_{min} < 0$ °C), it is possible that tissue susceptibility will be adversely affected by generally increased mean temperatures in winter and increased growing season lengths.

Atmospheric carbon dioxide

The effect of increased CO_2 concentrations in the atmosphere is variable, young trees apparently being more responsive than mature. Experimental acclimation of young trees to elevated atmospheric concentrations (x 2 ambient CO_2) indicates modal increases of 30-40% in growth rate (Jarvis, in press). Strong interactions with nutrition show that with adequate nitrogen supply increased growth rates in response to enhanced CO_2 concentrations do not lead to changes in structure or physiology. On the other hand, in poor nutritional conditions root production is stimulated rather than above-ground growth, while water use efficiency may be increased and phenology of bud-burst and bud-set altered (Murray *et al.*, 1994).

Figure 6.1: Projected trends of temperature changes for coldest and warmest months for locations in western and northern Europe. Arrows indicate direction of temperature change relevant to the 78% probability region of beech (*Fagus sylvatica*) occurrence. The start and end points of the arrows represent the period of projection (1990-2050).

Arrow clusters labelled S (Sweden); N (Norway); F (France): I (Italy) and UK. Data from United Kingdom Meteorological Office transient model (UKTR). Source: C E Jeffree unpublished data.

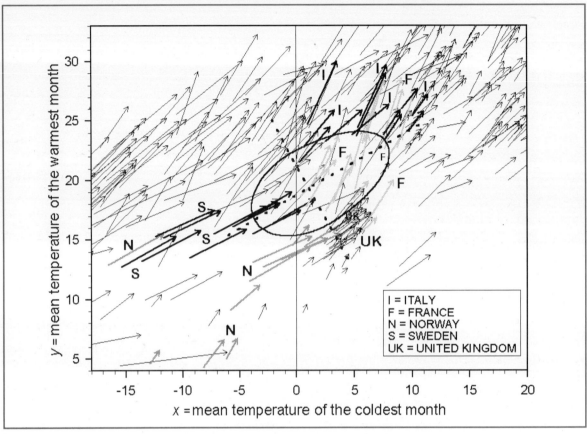

Wind relations

The relatively high windspeeds recorded in the UK particularly in the north and west, affects tree growth both physiologically and by mechanical damage. Exposure to consistently high winds reduces growth rates through cooling of tissues and interference with normal gas exchange. Mechanical damage leads to deformation of immature tissues, leaf damage, shoot breakage and ultimately windthrow of the whole tree. Windthrow is a major problem in the management of stands established at higher elevations on soils with limited rooting depths, restricting rotation lengths and severely affecting economic returns. In the present regional wind climates, the anticipated height of the tree stand, at which windthrow occurs, can be estimated from the combination of elevation, topography and soil type of the site (Quine *et al.*, 1995). Regional changes in the frequency of damaging gales could increase the areas at risk, making some upland areas less suitable for timber production, because of the reduced tree sizes obtainable and increased costs.

Soil relations

Most of the younger forest stands, in the uplands, have been established with the aid of soil disturbance, either through disruption of existing organic horizons or their aeration by drainage. This cultivation is designed to improve soil rootability and stimulate decomposition and mineralisation of organic matter thus making mineral nutrients available. This process is often assisted by the addition of phosphorus fertiliser where this element is lacking. The initially enhanced tree growth rate is maintained by the drying effects of the developing stand and the deposition of litter. The decomposition rate of this litter is, however, closely tied to soil temperature and moisture availability. In cold (high elevation) or excessively wet (anaerobic) conditions the availability of nutrients may become insufficient to support the continued growth of old tree stands. Where soil temperatures increase and there is adequate, but not excessive, moisture availability, nutrient cycling should be enhanced.

Productivity of forest trees

The productivity of the main coniferous species used in UK forestry is strongly related to climatic variables and it can be estimated quite closely by indices of temperature and windiness that integrate differences between sites in altitude and distance from coast. For example, Worrell and Malcolm (1990) accounted for about 80% of the variation in Sitka spruce productivity in upland Britain, using only accumulated temperature (day-degrees $> 5.6°C$) and wind run measured by flag tatter rates. Soil moisture factors are considerably less important in the uplands where precipitation in the growing season is usually adequate. At lower elevations (< 250 m) soil water deficits may limit productivity (Jarvis and Mullins, 1987). These known relationships allow the estimation of future productivity levels in altered climates and can be used to calculate elevational limits for productive afforestation as well as returns on investment.

Associated organisms

Insect populations tend to increase more rapidly at higher temperatures and are more damaging to host tree species when the latter are stressed physiologically because of drought, high temperature or poor nutrition. Climate change may also result in the invasion of species from continental Europe hitherto unable to adapt to existing climatic regimes. At higher temperatures and humidities some fungal leaf pathogens may be more damaging. The projected changes in climate would lead to new balances between parasites and their hosts and, for some species combinations, could become seriously damaging.

Forestry operations

In the UK forestry operations are carried on throughout the year and so are little affected by temperature but mainly by frequency of wet conditions for working. Worker productivity decreases with increasing rainfall and incidence of rain days, so that all operational costs increase along with the difficulty and cost of machine operation. Harvesting operations are particularly disadvantaged on wet soils.

6.2.2 Effects of climate change

Forests are complex ecosystems which express the interaction of many biotic and abiotic influences. The interaction is dynamic, adjusting to variations in climatic factors and disturbance by management. The marked changes in climate projected for the 2050s (Chapter 2) would therefore lead to multiple related effects which may be beneficial or detrimental to overall forest productivity.

Increased carbon dioxide concentrations

The increase in ambient concentrations of CO_2 since 1850 may already be influencing tree growth rates. Reports from Scandinavia and Central Europe indicate increases in current volume increment of up to 50% (Sterba, 1995) in Norway spruce and Scots pine compared to similar stands a century ago. In the UK several recent species-site studies suggest that younger stands are growing faster than older stands did at the same stage (e.g. Tyler *et al.*, in press). Other factors, such as higher temperature or nitrogen deposition, may be involved but the effect is likely to increase with ambient CO_2 concentrations. Enhanced growth rates in younger stands will be important for forest management, resulting in shorter rotations and increased return on investment.

Increased temperature and productivity

Given adequate water supply, increased summer temperatures will result in higher growth rates for most tree species. The increase projected (Chapter 2) for the growing season, expressed in degree-days ($T_{mean} > 5.5°C$), across the country is in the order of 25-30%. Such a change would in itself result in appreciable gains in productivity of major tree species. For example, an average stand of Sitka spruce at 300 m elevation in mid-Argyll, currently with about 1350 degree-days, would show an increase in volume growth rate of about 3 m^3 ha^{-1} yr^{-1} for a 25% increase in accumulated temperature, assuming no change in mean windspeeds in northern Britain (Worrell and Malcolm 1990). This increase in productivity, could, for spruce alone, be up to 1.0 M m^3 yr^{-1}. This would greatly improve the financial return of much of the productive forest of central and northern UK. The upper elevational limit for commercial planting in the UK (usually taken to equal a yield of 8 m^3 ha^{-1} yr^{-1}) could also be extended from about 500 m to 650 m away from coastal areas.

The possible effects of temperature increases on a range of species have been outlined by Cannell *et al.* (1989). Higher summer temperatures might encourage the planting of some productive exotics such as *Pinus radiata* and possibly *Eucalyptus* spp. although sharp fluctuations in winter temperatures might still control their use (Evans, 1986). Many species can be damaged, including otherwise hardy natives (birch, hawthorn, etc.), when dormant shoots deharden in response to mild spells in early spring followed by a return to sub-zero temperatures.

Changes in precipitation

The increased evaporative stresses projected (Chapter 2) for the south of England, coupled with decreased rainfall would markedly increase soil water deficits. By the 2050s these changes would adversely affect growth rates in many species and reduce the available sites for those that need moist conditions (e.g. beech and ash). Mortality of species like beech, which suffered severely in the 1976 summer, would be increased and a series of local extinctions could be expected for species on marginal sites. Similar problems can be anticipated for trees in the urban environment where they already operate in conditions of enhanced temperature and drought stress. On the other hand, the scenario for northern Britain suggests that moisture limitations for some species (e.g. spruce) would be relieved allowing their use on sites presently considered too dry.

On wetter sites, in the north, increased precipitation and reduced evapotranspiration would encourage the enhancement of peat forming processes in upland areas, thus greater emphasis on surface drainage would be required to maintain forest productivity and to avoid early windthrow. Wet snow falling at near zero temperatures is a common cause of serious damage to upland forests and this phenomenon might become more frequent with marginally higher winter temperatures and an increase in precipitation.

The projected drier conditions in the southern UK in summer will inevitably increase fire hazard, particularly on freely drained soils which are already occupied by flammable stands (e.g. pine). The areas at risk can be expected to increase if moisture demanding species are replaced by those which are more drought resistant.

Forest 'decline' in Europe is now thought to bear some relationship to consecutive years of below normal precipitation, exacerbated by the effects of atmospheric pollution. Such events could become more widespread under the scenarios of future altered climate.

Phenology

It is generally accepted that buds flush in response to increasing temperatures following a necessary period of chilling. On the other hand cessation of growth in autumn is mainly a response to reducing photoperiod. Both phenomena are considered adaptations to avoid out-of-season frosts when tissues are not hardened adequately. Cannell *et al.* (1989) showed that early flushing could lead to an increased probability of spring frost damage. Although it is unlikely that adequate chilling will not be attained even with increased mean winter temperatures, the pre-flushing thermoperiod may be shorter leading to early flushing, commonly seen in northern continental species when moved south. A potentially longer growth period in autumn would be truncated by the existing adaptation to photoperiod although some gain could result from modest transfers northwards of southern provenances (Beuker, 1994). However Kramer (1995) suggests on the basis of clones of seven European species transferred to various phenological gardens, that individual genotypes display considerable plasticity to variable climate conditions. Nevertheless, differences in their growth-period response may make individual species more or less competitive than at present and thus affect establishment and native woodlands.

A further phenological effect is the frequency of flowering; which in most species is enhanced by higher summer temperatures and water stress in the previous year. Seed production might therefore be greater in many species in the south but might become less frequent in the north. This would influence the current trend to rely more on natural regeneration of forest stands to establish the next rotation, but reproduction of maladapted stands might be undesirable.

Pests and pathogens
An overall increase in mean temperatures would be expected to lead to increased damage from insect pests by better overwintering conditions (e.g. *Elatobium* on *Sitka* spruce) and by increased moisture stress of species susceptible to bark beetles (e.g. *Dendroctonus* and *Ips* on spruce and larch) while conditions might favour further invasions from Europe. Current periodic control programmes of pests (e.g. *Bupalus* on pines) might have to be more frequent.

Higher humidities in the north and increased winter temperatures would enhance conditions for some pathogenic fungi such as *Brunchorstia* on pines or *Botrytis* in forest nurseries. The reverse conditions would be experienced in the south.

In general, the effects of the projected changes in climate will not alter the species composition in upland forests, where there should be substantial increases in productivity. Deleterious effects will be most marked in southern, lowland, Britain. Drought and high temperatures will exclude sensitive species from some sites and risks from pests, pollution and fire will increase together with an overall reduction in productivity.

6.2.3 Effects of sea level rise

Only a few coastal forests exist on areas of sand dune. Increased coastal erosion might destroy the dune system and disrupt these areas but generally sea level rise would have little impact on forestry.

6.3 ASSESSMENT OF POTENTIAL ADAPTATION

The ability of forests and woodlands to adapt to the proposed changes in climatic variables is severely limited. Neither the time for adaptation nor the potential for migration is generally available. Changes in the species composition of existing forests will, therefore, be mediated deliberately by management. Declining performance and increased mortality of existing tree stands would lead to early exploitation and replacement by better-adapted species or provenances. This implies an awareness of the causes of declining performance and the availability of suitable replacement genetic material. The most critical problems can be expected in the southern UK with warmer drier conditions. Low water availability can be partially offset by maintaining lower stand densities but at a cost in production and quality of product.

Adaptation is probably best sought through an expansion in tree improvement programmes to encompass a wider range of species and the inclusion of more southerly provenances or selections in breeding populations. The use of more drought tolerant species, such as Douglas fir and Corsican pine could be expected to increase in southern UK. Additional provenance studies in these and other candidate species need to be commenced now. Existing breeding populations should be expanded to include a greater proportion of southerly provenances or selections (e.g. Oregon/Washington in Sitka spruce). Effective results and the availability of suitable material requires a minimum of about 25 years with longer periods for adequate testing.

Adaptation to climate change in forestry is inevitably slow as areas planted now will only be maturing in 50 years time. Thus assumptions about projected climate scenarios must inform current decisions if adaptation of species distribution and cultural techniques were to be appropriate in the 2050s. Opportunities to restructure existing plantation forests will exist in the next decades, and are being taken now for other reasons. Mixed species stands, enhanced conservation and biodiversity measures and stand structures better able to withstand climatic hazards, such as windthrow, are being adopted generally.

At a national scale reductions in production because of less favourable conditions in the south might be compensated for by increased productivity in central and northern UK. In addition changes in land use induced by decreased demand for agricultural land might allow further expansion of the forested area and thus an increase in total wood supply. Nevertheless, the availability of some products, for example high quality hardwoods like beech, might diminish while substitution of broadleaves by more drought resistant conifers would have large contingent effects on landscape, recreation and nature conservation values. Such deliberate changes would probably be unacceptable socially so that substitute origins of broadleaves should also be sought.

Major wood-using industries are already concentrated in the northern UK closer to the main supply areas so there is unlikely to be a need for relocation. Specialist users of threatened species currently located in the south might be disadvantaged to the extent that they rely currently on home-grown supplies.

6.4 UNCERTAINTIES AND UNKNOWNS

Much is known of the ecophysiology of many of the tree species used in UK forestry and considerable research effort is being devoted to modelling the effects of climate change and increased CO_2 concentrations. Educated estimates of species tolerances, adaptability and probable future productivity can be derived from given scenarios of future climate. However, there remains a need for more information on the growth cycle, reproductive biology, phenology and partitioning of assimilated carbon to provide the parameters of models describing species responses to climatic variables.

Insect populations are expected to increase but the extent to which they may be held in check by predators or require management control is not known. Hitherto non-damaging species might reach pest levels and invading species could be dangerous. Conditions favouring the spread and epidemic status of pathogens are also poorly understood. Interactions with wildlife generally and effects on biodiversity remain obscure.

6.5 PRINCIPAL IMPLICATIONS FOR OTHER SECTORS AND THE UK ECONOMY

The potential decline and mortality of some tree species, particularly in the heavily populated areas in the southern UK would have major impacts on the flora, fauna and landscape

(Chapter 4). The recreational value of woodlands and urban trees would decline with knock-on effects on the quality of life for human populations. Increased precipitation, cloudiness and rain days would adversely affect forest work in northern areas and lead to lower profitability of forest production and decreased recreation possibilities (Chapter 16).

Increased wind damage to forests in the northern UK through changes in the areas of poorly drained soils and frequency of severe wind storms could disrupt supplies to wood-using industries, reduce wood prices, imbalance orderly harvesting procedures and generally increase costs.

Decreased production of quality hardwoods might lead to greater reliance on imports and increasing costs for furniture and similar trades.

The timber trade does not operate in a free market and UK prices are subject to conditions in the countries of supply. These markets are not particularly stable with resulting fluctuation in the market for home-grown timber products. Reliance on other temperate or boreal sources of supply for manufactured wood products (paper and panels) could be reduced slightly if UK forests become more productive, or could become more difficult if supplying nations suffered reductions in productivity (e.g. hardwood pulp from Mediterranean or sub-tropical sources).

The direct effects of climate change per se may have less influence on the expansion and production of UK forests than the economic factors and political attitudes it induces.

6.6 RESEARCH AND POLICY ISSUES

Continued refinement of climatic models is a prerequisite for the prediction of future forest health and production. Better estimates of extreme occurrences and climatic variability in general are required at a regional level.

6.6.1 Future research effort

Areas of research that should be pursued include:

- Physiology and genetics to resolve effects of climate and atmospheric changes on the growth, development and adaptation of key species;
- Process-based models of forest stand development with particular reference to the effects of increased ambient CO_2. Scaling-up of tree and stand models to regional levels to match with climatic models;
- Elucidation of the biology of potential pest and disease organisms in relation to host physiology;
- Analysis of existing records (mensurational plots, provenance and species trials) in relation to climatic factors with a view to developing better yield-site relations;
- Development of monitoring techniques to detect the effects of environmental change with emphasis on changes in productivity in existing forests.

Programmes of research in many of these areas already exist but there is a need to integrate them with other more general ecological studies. The timescales involved militate against large scale field experimentation and results are urgently required if forest management is to respond adequately.

6.6.2 Policy issues

Up to now projections of climate change have been insufficiently precise to estimate detailed effects and therefore to convince policy makers to divert scarce resources to a revision of incentives for land use change. The importance of woody vegetation for carbon sequestration is now better appreciated and support for short-rotation coppice systems for biofuel has the dual effect of utilising surplus agricultural land and reducing dependence on fossil fuels.

An increase in productivity of commercial forests in the northern UK with financial benefits for their owners, should encourage expansion of the area under forest vegetation. This would also increase C sequestration and reduce reliance on imports, while sustaining wood-using industries.

The long-term nature of forestry investment makes allocation of priorities difficult but policy decisions need to be taken now if UK forestry is to respond fully to the projected future climate.

7. Water Resources

SUMMARY

- The high degree of sensitivity of the water industry and water users in the UK to climate variability has been illustrated by the droughts of 1988-92 and 1995, and by the floods of 1993, 1994 and 1995.

- Under the 1996 CCIRG scenario there would be an increase in river flow in winter and a decrease, especially in the south, during summer. This would adversely affect abstractions of water from rivers in summer (for water supply, irrigation and cooling water) as well as instream uses such as navigation, recreation and ecosystem maintenance. The effects of changes in the volume and timing of flow on managed water systems will depend significantly on the size and number of reservoirs and the degree to which different sewers are linked.

- Increased winter rainfall and wetter catchment conditions are likely to increase the frequency of riverine flooding.

- It is not yet established whether groundwater recharge would increase in future wetter winters, because the recharge season would be shortened by increased autumn and spring evaporation. It is therefore quite possible that recharge would be reduced.

- Higher water temperatures will speed up self-purification processes and improve water quality, but could also increase the risk of algal blooms in rivers and lakes. Furthermore, lower flows in summer would reduce the dilution of effluents and pollutants. The effects on water quality of changes in temperature and flow may be less important than the effects of changes in land use and agricultural practices.

- Sea level rise would have an easily avoidable impact on both coastal aquifers and freshwater intakes located close to the tidal limit.

- Climate change might add an additional 5% onto the predicted 12% increase in demand for public water supplies in southern England between 1990 and 2021, largely due to increased usage in gardens. The increase in peak demands could be much greater. This has important implications for the reliability of supply to domestic consumers and may require the redesign of parts of the distribution networks. Demands for spray irrigation are predicted to rise by 69% by 2021 without climate change, and could rise by 115% if temperatures were to rise.

- Changes in the frequency of drought and restrictions on water users depend not only on changes in water availability and demand, but also on catchment geological conditions and the volume of storage available.

- Aquatic habitats will be affected by the increase in water temperature and changes in river flows. Most fish species in Britain are unlikely to be significantly affected, but exceptions include cold-water lake fish and, most importantly, the native brown trout, which are both likely to be adversely affected by a rise in temperature. Changes in flow regimes might also affect salmon migration patterns.
- Water management has always been an inherently adaptive process, responding to and anticipating many changes. Possible climate change is one extra pressure, with a longer time horizon than most other changes. Current management techniques can be used to adapt to change, but this may involve economic, social or environmental cost, and some systems are better able to adapt than others.

7.1 INTRODUCTION AND BACKGROUND

The 1990s have seen extremes of both flood and drought in Britain. River flows, groundwater levels and reservoir contents in many areas reached record lows during the drought of 1988-1992 (Marsh *et al.*, 1993). At the peak of the drought, 18 million consumers were prevented from watering their gardens, and farmers in parts of East Anglia were not permitted to irrigate their crops. Severe restrictions were imposed in summer 1995, when high temperatures led to increases in demand and the lack of rainfall meant flows in some rivers and stocks in some reservoirs became very low. During the winters of 1994 and 1995, however, many parts of the UK experienced their worst flooding for decades; property and land were flooded, and transport routes disrupted. Taken together, these events illustrate the effect of climate on the UK economy, and indicate the potential implications of climate change in which summers will tend to be warmer and winters wetter. This chapter explores these implications in some detail.

Taken as a whole, the UK appears well supplied with water. However, at the regional scale there are considerable pressures on resources in drought years. The wetter west is generally well-supplied, although there are occasional problems in summer in the south west due to large numbers of visitors. Some reservoirs in the north and west of England emptied much more rapidly than usual in the dry, hot summer of 1995. Eastern UK is drier than the north and west, and has both the greatest demand and the fastest rate of increase in demand. Water resources are under particular stress in parts of the Thames and Trent basins, East Anglia and in southern England, although the degree of stress locally depends on the volume of storage available.

The water industry in the UK affects three groups: suppliers, users and regulators. The *suppliers* are the publicly-quoted and privately-owned water utilities in England and Wales, and the Regional Councils in Scotland (three water authorities after April 1996). These provide water to users, and some of them also discharge treated effluent back into the environment. *Users* include not only households, industry and agriculture, but also hydropower generators, river navigators and aquatic and riverine ecosystems. Some users who abstract their own water are capable of degrading the resource for downstream users. The *regulators* include the National Rivers Authority (NRA) in England and Wales (the Environment Agency from April 1996) which licenses abstractions from and discharges to rivers and aquifers. The NRA, together with the Scottish River Purification Boards (the Scottish Environmental Protection Agency from April 1996), also have powers to protect against flooding and pollution incidents, and have a duty to protect and enhance the environment. The Office of Water Services (OFWAT) controls the prices and operational standards of water companies in England and Wales. All three groups are likely to be affected by climate change, to varying degrees and in different ways.

There are additional pressures on the water industry. Water is generally being used more efficiently than in the past, but demand has been increasing in the south and east, due largely to an increasing concentration of population and industry and growth in spray irrigation. Between 1990 and 2021, demand for public water under a medium growth scenario would increase by 27% in the Anglian Region, by 23% in South Western Region and by 15% in Southern Region (NRA, 1994); in the north and west increases would be smaller. There is an increasing recognition of the demands placed on water resources by aquatic and riverine ecosystems, and there are changing legislative pressures, including European Union directives. Public expectations are altering too, especially since privatisation, and users increasingly expect reliable and good quality supplies. Lastly, changes in land use affect hydrological characteristics, and hence water resources. Particularly important are agricultural changes - such as set-aside - which result in changes in the application of agricultural chemicals with consequent effects on stream and groundwater quality.

Over the last decade there have been a number of studies into the possible effects of climate change on river flows, both in the UK and elsewhere. There has been considerably less work on water quality, very little on groundwater and nothing on river flooding. There has also been little research on specific water supply systems and operational consequences. However, implications for changes in demand have been investigated (Herrington, 1995) and a review has been conducted for the National Rivers Authority (Arnell et al., 1994). A similar distribution of studies is described for the world scene in the IPCC Second Assessment Report (IPCC, 1996b).

7.2 ESTIMATED EFFECTS OF CLIMATE CHANGE AND SEA LEVEL RISE

7.2.1 Sensitivity of water resources to weather and climate

Suppliers, users and regulators will all be affected by climate change, to varying degrees. Table 7.1 summarises the key impact areas in the water sector, together with the specific aspects of each impact area sensitive to change and the activities affected by such changes.

Impacts on the three groups which comprise the water sector are felt through the consequences of climate change on the amount of water available at different times in rivers, reservoirs and aquifers. Such changes would affect not only the supply of water to domestic, industrial and agricultural users, but also hydropower generation, river navigation, recreation possibilities and aquatic ecosystems. A change in water qual-

Table 7.1: Sensitivities to climate change in the water sector

Impact area	Aspects sensitive to climate change	Activities
Volume and timing of water	River flows Groundwater recharge and levels Reservoir contents	• Domestic, industrial and agricultural water supply • Power generation • River navigation • Aquatic ecosystems
Quality of water	River water quality Groundwater quality Saline intrusion into aquifers Saline intrusion into estuaries	• Domestic, industrial and agricultural water supply • Pollution control • Power generation • Aquatic ecosystems
Society's demand for water	Domestic demand Industrial demand Agricultural demand	• Domestic, industrial and water supply • Distribution of water to users
Flood frequency	Riverine flood risk Coastal flood risk	Flood management
Riverine, lake and wetland environments	Temperature River flows Groundwater levels Lake levels	Fisheries Conservation Recreation

ity, due to changes in water temperature and river flows, would alter water supplies and affect pollution management. Rising sea level could lead to saline intrusion into coastal aquifers, and also increased penetration of salt water along estuaries, threatening low-lying freshwater intakes. Thermal power generation and some industrial processes use river water for cooling; higher water temperatures would reduce the efficiency of water cooling systems. Changes in the demand for water would affect water supply and distribution systems; the distribution network is particularly sensitive to peak demands. Increased incidence of high river flows and sea levels has implications for flood protection standards. Finally, changes in water temperature, river flows and groundwater levels may affect instream, river corridor and wetland ecosystems, with implications for fisheries, conservation and recreation.

7.2.2 Effects of climate change

Virtually all impact assessments have followed a sequential approach. Climate change scenarios are first used to perturb the climatic inputs to a model of the hydrological system, and hydrological effects of change are simulated. The consequential changes for water resources are then explored using other models or, in many cases, expert judgement. There are several very significant uncertainties in this process, as outlined in Section 7.4; here it is necessary only to note that the major uncertainties lie in the definition of the climate change scenarios and the assumed response of water managers to hydrological change. The hydrological models used to convert climatic inputs into river flows and groundwater recharge are relatively reliable, but water quality models and aquatic ecosystem models are less well developed.

Chapter 2 has presented scenarios (1996 CCIRG Scenario) for changes in rainfall, temperature and potential evaporation across the UK. No published studies of water resources have yet used this scenario, but some simulations were conducted specifically for this review, and results of earlier studies can be approximately adjusted to account for differences between scenarios.

Volume and timing of supply

Approximately one third of public water supply in the UK is taken directly from unregulated rivers, one third is drawn from reservoirs or rivers regulated by upstream reservoirs, and the remaining third comes from groundwater. There are very large regional variations; in the south and east of England a larger proportion comes from groundwater, and metropolitan areas are often supplied from a variety of conjunctively operated sources.

River flows and groundwater recharge will be affected by changes in rainfall, temperature and evaporation; changes in catchment vegetation due to the altered climate and enhanced CO_2 concentrations (Chapter 4) may also affect catchment-scale evaporation and the water balance. Arnell and Reynard (1993) investigated the effects of climate change on river flow regimes in Britain, using a daily rainfall-runoff model and a range of climate change scenarios. The same exercise was repeated with the 1996 CCIRG scenario. Figure 7.1 shows the percentage change in average annual runoff by the 2050s (from the 1961 to 1990 baseline) across the whole of Britain. Annual runoff totals would increase in the north and decrease in the south. The effect, however, would vary month by month.

Figure 7.2 shows average monthly runoff under current conditions and the 1996 CCIRG scenario, for six representative catchments. In southern Britain (the Harper's Brook and Medway catchments) river flows would be substantially reduced during summer (by up to 50% in some months) and might even be lower in winter than at present. Further north and west (for example in the Greta, Conwy and Nith catchments), runoff is expected to increase in winter under the 1996 CCIRG scenario, and show little difference in most summer months.

Figure 7.1. Percentage change in average annual runoff by the 2050s under the 1996 CCIRG scenario.

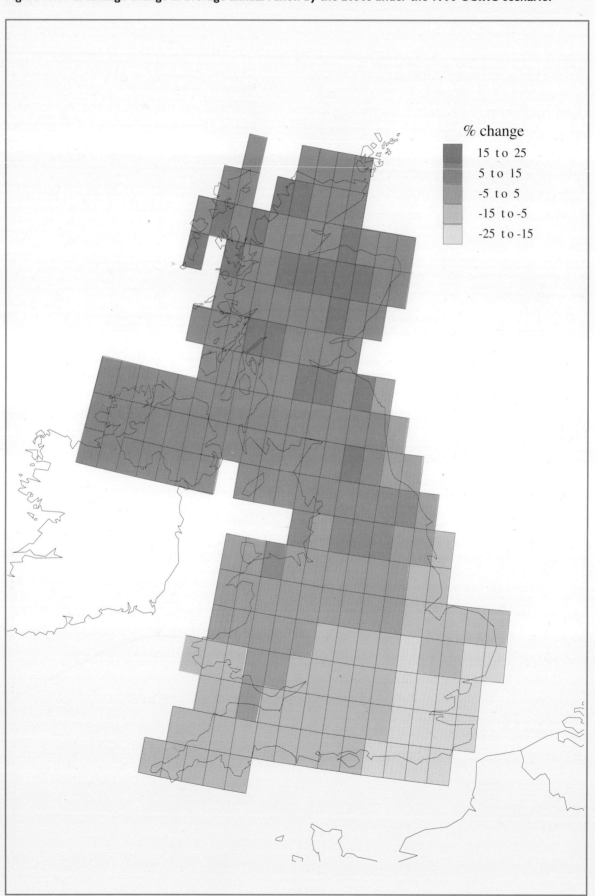

Figure 7.2. Monthly average runoff at present and in the 2050s in six representative catchments, under the 1996 CCIRG scenario.

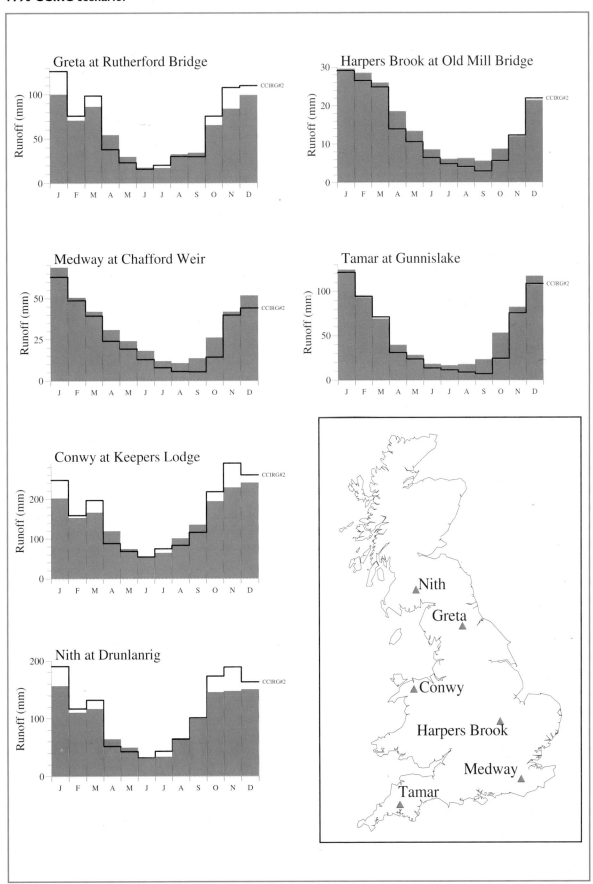

Snowmelt generally makes only a minor contribution to river flows in the UK, and with a temperature rise by the 2050s of between 1°C and 1.5°C would be virtually eliminated from most of the country. This effect on discharge is shown for the Greta, Conwy and Nith catchments in Figure 7.2, a case where snowmelt has some significance under current conditions. Spring flows in this catchment (especially in April) would decline considerably because winter precipitation would no longer be stored as snow and melted in April; winter runoff is correspondingly increased.

The general decline in summer river flows under the 1996 CCIRG scenario, at least in the southern UK would cause problems for direct river abstractions, used for a third of public supplies, by many large industrial users and by many irrigators. Other supplies, however, are taken either from on-line reservoirs, from off-line reservoirs filled by pumping from rivers, or from rivers regulated by upstream reservoirs. The effect of climate change on these systems depends on the size and characteristics of the reservoir (Hewett et al., 1993) and how it is filled. Cole et al. (1991) showed that a reservoir with a small storage volume (relative to the total annual runoff) would be much more sensitive to change in inflows than a larger reservoir, and that a reservoir with high yield (relative to average flow) would be more sensitive than one with a low yield. If climate in the UK were to change in line with the 1996 CCIRG scenario, then the change in reliable yield would depend on whether the reservoir could store enough of the extra winter runoff to compensate for reduced summer inflows; the larger the reservoir, the greater the likelihood that this will be the case. Increased winter runoff over a shorter period could lead to reservoirs filling more quickly, but overflowing more frequently. From the supplier's perspective, such water would be wasted. In general, an altered seasonal distribution of reservoir inflows could have implications for the operation of the reservoir (in terms of volumes released at different times of the year), and for the availability of water in the reservoir at different times in the year.

There have been very few studies into possible changes in groundwater recharge anywhere in the world. Groundwater recharge in the UK occurs mostly during winter, after soil moisture deficits have been removed during autumn and before they return as evaporation increases in spring. The effect of climate change on recharge depends on the extent to which extra rainfall during the recharge season compensates for a shorter recharge season, and whether all of the extra rainfall contributes to aquifer recharge; some may go rapidly to increased river flows. Under scenarios used by Cooper et al. (1995), similar to the 1996 CCIRG scenario, the effect of the shorter season was simulated to outweigh the increased rainfall, and recharge would thus reduce. Hewett et al. (1993), however, reported that simulations using independently-calibrated recharge and groundwater models of the North Kent Chalk block showed net gains in recharge and groundwater levels, using scenarios similar to the 1996 CCIRG scenario.

In the most populated areas of the UK, much of the water used is withdrawn from several sources, which include inter-basin transfer, direct river abstraction, reservoirs and groundwater withdrawals. In principle, the greater the interconnections within a supply system, the better the system can withstand periods of shortage. Adjustments to change should also be easier, although changes may be necessary to the transmission and distribution network. Changes in the reliability of a complicated integrated system, however, are difficult to assess without a realistic system simulation model. Bunch and Smithers (1992) examined the Lancashire Conjunctive Use System, and although their results are case specific, the general conclusion is that changes in climate would lead to changes in operating costs: capital investment may also be needed in some cases.

The effects of climate change on users would not be apparent in most years, but would be seen through changes in the frequency of occurrence of droughts and restrictions on use, especially if low supplies are associated with high demands (see below).

Quality of supply

The quality of water influences the use to which that water can be put. In the most general terms, streamwater and groundwater quality will be affected by changes in water temperature, runoff and recharge processes and river flow patterns, but changes in inputs of chemicals, due largely to changes in agricultural practices, will also be important and perhaps dominant in much of the UK.

Except in very small headwater streams, river water temperature closely follows air temperature, with a given increase in air temperature resulting in a slightly smaller increase in water temperature (Webb, 1992). By the 2050s, river water temperature could therefore be between 1°C and 2°C higher than at present, with the greatest increases in large, wide, open channels, and the smallest increases in groundwater-dominated catchments and where the watercourse is shaded. Such a change in water temperature would affect aquatic ecosystems (as discussed below), and also affect the rate of operation of biogeochemical processes within the river.

In general, biogeochemical processes (which act to purify water) operate faster in higher temperatures, but sensitivity to temperature varies between processes. For example, denitrification processes are more temperature sensitive than nitrification processes (Jenkins et al., 1993) thus, with no change in river flows or chemical inputs, higher temperatures would result in the removal of nitrate. However, the temperature effects are frequently outdone by flow discharge effects, and a reduction in flow would reduce dilution levels and tend to increase nitrate concentrations. Other factors are also important; higher temperatures would increase the mineralisation of organic nitrogen in the soil, and increased rainfall intensity in autumn could lead to increased flushing of nitrogen into the river network.

Higher water temperature would increase both the rate of biological activity within the stream (and hence de-oxidation) and the rate of re-aeration, but the effect on de-oxidation is greater, thus oxygen contents would fall (Jacoby, 1990). This decline could be offset by an increase in flows, but would be exaggerated further if flows were to decline, as the biological oxygen demand would be more concentrated. Sewage treatment works are more efficient at higher temperatures as the biogeochemical treatment processes work faster, thus the biochemical oxygen demands from treated effluent could fall, leading to improved water quality.

Higher temperatures could lead to an increase in the growth of algal blooms on lakes and in rivers (Arnell et al., 1994); a reduction in flow (and hence increase in residence time and nutrient concentrations) would stimulate even further the growth of algal blooms in rivers. Algal blooms cause oxygen depletion (occasionally leading to fish-kills) and some are toxic. Bloom growth is also stimulated by the input of phosphates, so a trend towards lower agricultural inputs, independent of climate change, may ameliorate the situation.

The quality of groundwater resources depends largely on the quality of water recharging the aquifer, and a change in the quality of recharging water may, after a delay of perhaps several years, affect groundwater quality. Increased mineralisation of organic nitrogen, for example, might lead to increasing nitrate concentrations in winter recharge, but there have been no studies into potential changes in groundwater quality as a result of climate change; indeed, the effects of current land use practices on groundwater quality are far from understood. Changed flow pathways, due for example to increased cracking in the soil, might mean that pollutants could reach groundwater more readily (Soils, Chapter 3). Higher groundwater temperatures can in principle lead to increased solution in carbonate aquifers (such as chalk) and hence a rise in calcium concentrations.

Demand for water

Demand by the public for water is expected to increase by 12% in England and Wales between 1990 and 2021, under a medium growth scenario and *without* climate change (NRA, 1994). Most of this growth is expected in the five NRA regions in the south and east. Climate change will lead to additional increases.

Herrington (1995) predicted an increase in per capita domestic demand of 21% between 1990 and 2021 - in the absence of climate change - due largely to a projected increase in use of dishwashers and other appliances. A predicted increase in the use of water for garden watering and personal showering due to climate change would add another 5% onto this increase, and would increase peak usage by a greater proportion: the large increases in peak demand during the hot, dry summer of 1995 emphasised the sensitivity of demand to climate. At present, approximately 5% of the water delivered over the year to domestic consumers in the south and east is used for garden watering: by 2021 this could be increased to 11%. Estimates of future demand are, of course, highly dependent on policies which may be introduced to curb increases in demand or encourage efficiency of use. The predicted increase in peak demand has particularly significant consequences for the water distribution network, which distributes water from source to user; capacities may need to be increased, which will be very expensive. Provision for investment in upgrading distribution networks is generally not made in current water company Asset Management Plans.

Spray irrigation in England and Wales is concentrated in East Anglia, and is undertaken to guarantee the high quality and yields, particularly of vegetables, demanded by supermarkets (Chapter 5). Demand for spray irrigation is predicted to increase by 69% between 1990 and 2021 without climate change (Weatherhead *et al.*, 1994), and by 115% with climate change (Herrington, 1995), although this increase is concentrated in a few areas. Most irrigation water is taken directly from rivers or groundwater, although irrigation reservoirs filled in winter are increasingly required by the NRA before irrigation licences are awarded. Substantial increases in the demand for irrigation would place additional pressure on resources, particularly during the summer low flow season. However, predicted demands for irrigation are very much influenced by future prices of water and agricultural produce, and even at present, irrigation demands were not fully satisfied during the particularly dry summers of 1990 and 1991.

The total industrial demand for water in Britain has fallen over the last few decades as heavy industry has declined and water is used more efficiently. Significant demands for cooling water still exist, particularly from power stations; they will be affected by changes in both the volume of water available and the temperature of the water, although there is a trend in the electricity generating industry towards coastal locations and more efficient generating plants. There have been no studies of sensitivities in the UK.

Almost a quarter of the water put into the public supply system is lost through leakage (Department of the Environment, 1992). Any efforts to reduce this loss (which will be independent of climate change) would offset to an extent the effects of demand growth and climate change.

Riverine flood risk

There have been no quantitative studies of potential changes in riverine flood risk in the UK. Most floods occur during the winter season which become wetter under the 1996 CCIRG scenario. Hence it can be inferred that floods are likely to become more frequent. An increased intensity in summer convective storms may also increase summer flooding. However, the effect of any change in rainfall characteristics will depend on the physical properties of the catchment (such as size, topography and geology), and are difficult to predict. The system contains a strong non-linearity; thus, a small change in conditions that generate floods could lead to major changes in the frequency with which certain critical thresholds, or design standards, are passed.

The aquatic environment

The aquatic environment comprises instream habitats, the river corridor, wetlands, ponds, lakes and reservoirs. It has not only an ecological significance but is also important for fisheries and recreation and has great cultural value.

Instream habitats will be affected by changes in water temperature and changes in the seasonal pattern of river flows. Higher water temperatures will affect growth and survival rates of some fish species, although in the UK most species are unlikely to be significantly affected as few are near their thermal limits (Arnell *et al.*, 1994). Exceptions include some cold-water lake fish found in northern England and Scotland and, more significantly, the native brown trout, whose growth rate slows as temperature rises. Higher water temperatures may also increase the outbreak of fish diseases and fungal infections. A change in river flows could alter habitat suitability and the availability of breeding sites, and such changes in habitat may be more significant for fish and aquatic populations than an increase in water temperature. Changes in seasonal flows may also influence migration patterns; lower summer flows may mean that summer salmon runs in some rivers might disappear. Very little is known about the sensitivity of British aquatic ecosystems to changes in climatic variables, thus any assessments of change are very uncertain (see also Chapter 4).

River corridor and wetland ecosystems are generally maintained by water levels either in a river or in shallow groundwater. Experience with such human interventions as field drainage, river channelization and water abstraction in the UK and elsewhere suggests that such ecosystems can be very sensitive to changed hydrological conditions, although there have been no studies in the UK of the potential effects of global warming. Minor water bodies, such as village ponds and urban watercourses, can have high cultural and amenity value, thus change could have very significant local consequences.

7.2.3 Effects of sea level rise

The most obvious effect of a rise in sea level would be an increase in the risk of coastal flooding and the overtopping of coastal defences. A small change in mean sea level can have a major effect on the risk of water levels exceeding a particular threshold (CCIRG, 1991), and this risk would be further increased if storm frequencies increase (Chapter 2). The most significant implication is that coastal defences would need to be enhanced: this, as outlined in Section 7.3, is already under way. The Thames Barrier protects London from coastal flooding, and its design allowed for a rise in sea level of around 0.8m per century over the notional 50-year design life. This trend is an extrapolation of the historical rise in sea level, and may accomodate a certain amount of sea level rise due to global warming, particularly if the isostatic effects are less pronounced than in earlier periods. However, it is possible that the probability of a storm overtopping the Barrier will increase, and that the Barrier would need to be closed more frequently (Kelly, 1991).

A rise in sea level results in increased saline intrusion into coastal aquifers, and hence a potential reduction in groundwater quality. Most of the coastal aquifers potentially at risk are along the south and east coasts. Clark *et al.* (1992) have shown that a rise in sea level of 0.6m would lead to reductions in yield of only 1 or 2% from a few coastal boreholes, which could easily be compensated for by altered management practices.

A higher sea level would also mean increased penetration of saline water up estuaries, causing a potential threat to the 35 public supply abstraction points located close to the current tidal limit. Studies in a number of estuaries (Clark *et al.*, 1992; Dearnaley and Waller, 1993), however, have indicated that any change would be well within the current variability between tidal cycles, and that a reduction in freshwater inflows to the estuary would have a greater effect on the position of the saline front. Few intakes would be directly threatened by a modest change in saline intrusion along estuaries. However, if the amount of freshwater flow needed to control estuary salinity is affected by rising sea level there could be substantial economic consequences.

Some lowlying parts of the coastline are drained by pumping out accumulated surface water at low tide. A rise in sea level would probably raise the costs of pumping, but the magnitude of the impact is uncertain: it may be very easy to adjust pumping schedules.

7.3 ASSESSMENT OF POTENTIAL ADAPTATIONS

Water management has always been an inherently adaptive process, adjusting to new demands, new technologies, new data, new legislative arrangements and new expectations. The effects of the drought of 1988-1992 were less severe than they might have been, largely because of the lessons learnt and infrastructure installed by the water industry during the 1976, 1984 and earlier droughts, and as a result of emergency responses implemented through the course of the drought. The water industry in the UK is currently adapting to new institutional arrangements (following privatisation), new legislative requirements (principally EU water quality directives), and increasing demand. The consequences of many of these changes are uncertain in their magnitude and timing. Adaptation will entail economic, financial, social and environmental costs and benefits, and the water industry and its regulators need to develop an acceptable balance between these factors.

Table 7.2 summarises the impacts of climate change on producers, users and regulators in the water sector, and lists some of the feasible adaptation strategies - some of which are currently already being adopted for reasons other than climate change.

In the managed part of the water sector (water supply, flood defence and so on) the choices are between 'business-as-usual', maintaining current standards or reaching a new, economically-optimum, standard (Arnell and Dubourg, 1995). 'Business-as-usual' assumes no explicit response to climate change, and may result in standards of service declining or improving as environmental conditions change. Maintaining current standards will involve work to continue to meet current levels of service; this may involve changes in the operation of a water supply and distribution network, or the construction of new sources or flood protection works. A new economically-optimum standard of service would be based on an assessment of the costs and benefits of providing services to different standards; it would, however, require a sophisticated analysis of economic and environmental costs and benefits, and assumes a water sector run on economic, rather than service, principles. In past practice, much water management in Britain is based around fixed standards of service, set by the suppliers themselves and regulators, so response to climate change is most likely to take the form of maintaining current standards at the minimum cost. However, new financial imperatives are now being introduced into the equation following privatisation.

Some parts of the water sector, however, cannot adapt to climate change. Although it is possible

Table 7.2. Impacts on suppliers, users and regulators, and feasible adaptations

	Impact	Adaptation
Suppliers	Change in the reliability of resources and demands for resources	- new sources - integration of sources - increased efficiency of exploitation of sources - new technologies - demand management - improvements in distribution network (eg leakage control and improved peak capacities)
Users	Change in resource availability	- increased efficiency - substitution
	Change in river basin characteristics	- manage flows and water temperature
Regulators	Change in amount and quality of available water	- redefine and reallocate licences and discharge consents
	Change in risk	- upgrade protection schemes - accept change in risk

to artificially control water levels and water temperature to maintain desirable aquatic ecosystems, this could be very expensive and, perhaps, undesirable in an ecological sense: the managed communities would be unsustainable without outside intervention. Water management activities may be able to minimise the effects of climate change, or at least not to exaggerate them, but it will not be feasible to prevent them completely. Aquatic ecosystems will adapt to global warming by changing their composition. This may involve the loss of some species from some areas, and their replacement by other species.

Suppliers can continue to adapt to changes in the reliability of resources, water quality and the demand for water in several ways. New sources may be introduced (which may be expensive or politically difficult) or existing sources could be more fully integrated and run more efficiently. Operating rules may be altered. New control technology could be introduced, and improved leakage control may release further resources. Changes in the distribution network may be needed to cope with increased peak demands, which would be very expensive. Demand management involves the manipulation of demand through fiscal incentives and other encouragements. There is currently a major debate within the UK water industry on the merits, techniques and effects of demand management, which include the setting of tariffs for water use. The major factor discouraging proactive adaptation is relationship between the uncertainty in future changes, the magnitude of any trend over the planning horizon, and its size relative to the very considerable year-to-year variability which already exists.

Water users can adapt to a change in resources available either by using water more efficiently (as is continually being done), or by substituting some other resource (such as using 'grey' water for some purposes). Those users which discharge effluent into the river network might be obliged to upgrade treatment facilities to maintain downstream water quality.

The regulators could adapt to changes in the amount and quality of water available by redefining licences, and possibly reallocating licences amongst different users. This, however, could be difficult for legal or other reasons, and would be expensive. Catchment plans are being increasingly used by the NRA, and provide a vehicle for reconciling competing water uses. Changes in the risk of extremes can be either ignored (at the risk of increasing losses) or responded to by alterations to protection schemes. The NRA is already designing coastal flood defences which can easily be increased in the future if necessary due to increased sea levels (Arnell *et al.*, 1994).

7.4 UNCERTAINTIES AND UNKNOWNS

There is a cascade of uncertainties in the translation of global warming into impacts on the water sector. First, there are uncertainties in the simulations of global climate models (Chapter 2), particularly in their simulation of regional precipitation, the most important single driver of the hydrological system. Second, there are considerable uncertainties in the downscaling of GCM simulations to the finer spatial scales necessary for hydrological simulation. Many techniques have been proposed (Arnell, 1995), including simple interpolation, relationships between large-scale climatic features and point climate, and the use of nested regional climate simulation models. All, however, essentially add uncertain detail to uncertain global model simulations.

The third stage is the translation of climate change into hydrological, water quality, ecological or water resource impact. Current hydrological models are reasonably accurate, and represent the most reliable element in the cascade. There remain, however, difficulties in simulating high flow extremes. Ecological climate-response models are currently in very early stages of development, and the ecological effects of a change in climate inputs are therefore difficult to assess. Models of water resource systems are relatively reliable, and indeed are used in operational water management.

The fourth and final stage is the estimation of the response of components of the water sector to changed conditions, and it is these responses which will determine the magnitude and cost of any impact of climate change. Responses will be affected by social, managerial and legislative factors, as well as non-climatic pressures and the effects of climate change on other sectors.

7.5 PRINCIPAL IMPLICATIONS FOR OTHER SECTORS AND THE UK ECONOMY

The water sector affects many other sectors within the UK economy, and a change in water resources would therefore have consequent effects on large parts of the UK economy.

Irrigation, for example, is currently used in cereal and vegetation production, particularly in south east England: changes in demand for water and the availability of water will feed through to the agricultural sector (Chapter 5). Industry uses water as a raw material, and again changes in the availability and price of water might, in some industries, have a significant effect on costs (Chapter 10). Power generation requires cooling water, although as noted in Chapter 8, technological changes are likely to have major impacts on demand. River navigation (Chapter 12) relies on levels in rivers and canals, and water-related recreation (Chapter 16) is primarily influenced by water quality. Changes in water quality could also have health implications (Chapter 15), again largely through contact during recreation. Changes in flood risk will affect the insurance industry (Chapter 13). Finally, changes in river flow regimes, water quality and the uses of the river environment will impact upon instream and riverine aquatic ecosystems (Chapter 4).

7.6 RESEARCH AND POLICY ISSUES

7.6.1 Future research effort

There are several areas where research is essential:

- Defining credible scenarios for changes in rainfall characteristics, at hydrologically-appropriate spatial and temporal scales (daily, at a spatial resolution of the order of 1000km^2);

- Developing scenarios for change in potential evaporation, considering explicitly the effects of changes in plant characteristics and catchment land use, at similar resolutions to the rainfall scenarios. It is particularly important to understand the effects of changes in plant physiology and growth due to increased CO_2 concentrations on water use *at the catchment scale* (Chapter 4);

- Refining estimates of the effects of climate change on river flows, groundwater recharge, water quality and supply system yields;

- Developing accurate models of the flood generation process;

- Developing accurate models of aquatic ecological processes;

- Understanding potential responses of actors within the water sector to uncertain climate change, including the feedbacks between sectors and actors.

In addition, it is necessary to undertake investigations of the sensitivity of *real* water supply and management systems to climate change, in order to determine the potential magnitude of impact and feasibility of different adaptation options. Such studies must place climate change in the context of the other, non-climatic, changes that are affecting water management.

7.6.2 Policy issues

Climate change is just one of the pressures currently facing the water industry, and is also one of the most uncertain. Different components of the water sector must develop policies, to meet different objectives over different time scales, and it is not appropriate to make prescriptive proposals. However, it is possible to make two general recommendations.

i) Initiate and continue monitoring programmes to check for evidence of climate change. This monitoring is needed to maintain the credibility of any response to climate change.

ii) Adopt a precautionary approach: it is not necessary to make direct adaptations now, but it would be prudent when designing a new water management system to (a) investigate the sensitivity of the proposed system to feasible climate change, and (b) ensure that the system is sufficiently flexible to be adapted to altered circumstances (perhaps by openly including climate change with all other factors for which a combined planning margin is needed). The second action is already current good practice in the water industry; the first can be encouraged by the regulators in the sector requiring producers to consider the effects of 'approved' climate change scenarios. The NRA has already adopted the precautionary principle in water resource management (NRA, 1994).

8. Energy

SUMMARY

- The impacts of climate change on the energy sector are diverse and few in themselves will be of major significance. The more important climate impacts relate to energy demand and the availability of natural resources. The technological potential for adaptation is high because the lifetime of most assets in the energy sector are shorter than the timescales associated with projected climate change. Rapid climate change and unusual weather conditions such as those experienced in recent years could however lead to a degree of mismatch between investments and climate conditions for which they were not designed.

- Globally, the energy sector is the largest source of anthropogenic emissions of CO_2, the most important greenhouse gas. The direct impacts of climate change on the energy sector will be much lower than those resulting from market developments, technological change or mitigation policies designed to constrain greenhouse gas emissions. There have, for example, been major changes in the electricity sector, in terms of both organisation and technological choice, even since the first CCIRG report (1991).

- Climate change could reduce energy demand for space heating in the UK by about 5% below the level which it would otherwise have been by the 2050s, and by about 3% by the 2020s. This would benefit householders who, on average, currently spend 4.8% of their budget on home heating. Recent warmer winters have reduced space heating demand below expected levels. Climate change would increase the use of existing air conditioning systems and could stimulate the wider adoption of new systems. Energy use for air conditioning would increase most rapidly in southern parts of the UK which is already warmer and where a greater temperature rise is projected.

- Climate change could affect the availability of renewable energy resources. In the UK hydro-electricity is currently the most important renewable energy source but the use of wind energy and biomass could increase in the future. Photovoltaic energy may become important at the global level but is less sensitive to climate. Hydro-electric potential would be affected by both the timing and volume of run-off. Wind energy resources could increase. Higher CO_2 concentrations could lead to higher biomass availability as long as water availability is adequate.

- All of the UK's petroleum refineries, all of its operating nuclear power stations and 70% of its conventional fossil-fuel fired power generation capacity are located on coasts or estuaries. Requirements for further protection against sea level rise are highly site-specific but are unlikely to cause any major difficulties.

- Energy sector operations may be affected in a number of minor ways by climate change. Increased storminess would affect offshore oil/gas exploration and production operations. The availability of cooling water could affect power generation. Overhead electricity cables and various renewable energy systems are vulnerable to storms and high winds.

8.1 INTRODUCTION AND BACKGROUND

The energy sector covers a variety of related activities including: extraction of oil and gas; coal mining; conversion of fossil fuels to electricity in power stations; petroleum refining; primary electricity generation (nuclear and renewable energy sources); energy transport and transmission; and distribution and marketing. The UK is the only EU country to be self-sufficient in energy. The energy industries account for about 5% of UK gross domestic product (Central Statistical Office, 1995). Energy is utilised in a wide range of final demand sectors (Department of Trade and Industry, 1995a) including: transport (33%); households (29%); industry (25%); commercial services and public administration (12%); and agriculture (1%). Climate impacts on the demand side of the energy market are likely to be at least as important as those on the supply side.

The energy sector is currently undergoing major changes in terms of both patterns of supply and institutional arrangements. The production and utilisation of coal are declining rapidly while the use of natural gas is growing. These developments are associated mainly with changes in preference for fuels and technology in electricity generation. The largest institutional changes have occurred in the 'network' industries, electricity and gas, where privatisation and the introduction of competition has had a profound effect.

The energy sector is the largest source of anthropogemic emissions of carbon dioxide (CO_2), the most important greenhouse gas (GHG). Policies to mitigate climate change would have a significant impact on the energy sector. The Department of Trade and Industry has recently published scenarios of energy demand and CO_2 emissions to the year 2020 which take account of uncertainties about economic growth and world energy and which incorporate assumptions about the impacts of the national climate programme (Department of Trade and Industry, 1995b). These foresee continuing change in the energy sector (Table 8.1). Without additional CO_2 mitigation action, primary energy demand could increase by 15-35% between 1990 and 2020. The transport sector and service sectors would be the most important sources of growth while demand in industry and households would grow much more slowly.

Few published energy scenarios look as far ahead as 2050, though the IPCC 1992 global scenarios look forward to the year 2100 (IPCC, 1994). Globally, the range of uncertainty about future energy demand is very wide. Under the IPCC scenarios, CO_2 emissions, mainly associated with energy use, range from 35% below to 400% above 1990 levels by 2100. Whatever the global outcome, CO_2 emissions in an industrialised country such as the UK are unlikely to grow significantly.

Table 8.1: Scenarios for UK energy demand excluding climate impacts (million tonnes of oil equivalent)

	1990	2020 Low	2020 Central	2020 High
By Sector:				
Transport	38.8	72.5	87.2	94.5
Industry	41.6	42.8	47.9	50.9
Household	40.8	45.6	47.4	47.4
Other	18.1	23.9	25.9	27.2
Conversion†	81.5	68.4	74.8	78.2
By Fuel:				
Coal	69.5	51.0	20.5	19.5
Oil	77.8	88.3	128.1	138.4
Gas	55.8	107.0	128.2	133.8
Nuclear/imports	17.3	4.1	4.1	4.1
Renewables	0.5	2.7	2.3	2.3
TOTAL	220.8	253.2	283.2	298.2

Note: † = computed as a residual

'Low' corresponds to low growth, high prices; 'central' to central growth, low prices; 'high' to high growth, low prices.

Source: Department of Trade and Industry (1995b)

The UK's position of energy self-sufficiency is unlikely to endure during the first half of next century as North Sea reserves become depleted. Proven plus probable oil reserves in 1995 would last 14 years at current rates of production while gas reserves would last 23 years (Department of Trade and Industry, 1995d). Further exploration and development in new oil provinces, west of the Shetlands for example, may however extend the period of self-sufficiency.

The direct impacts of climate change will have a much lower impact on the energy sector than will market developments, technological change or mitigation policies designed to constrain GHG emissions. By the 2050s, the energy system may have changed radically and could exhibit a very different set of sensitivities to climate. Mitigation measures are not within the scope of this report. Nevertheless, it is important to bear in mind that mitigation measures could have major economic impacts on energy supply companies on much shorter time-scales than those associated with climate change itself. CO_2 reduction policies rather than climate impacts have attracted the attention of corporate strategists. Aggressive mitigation strategies, for example the promotion of renewable energy, could significantly change the nature of the energy system with which climate variables would interact.

The literature assessing the direct impacts of climate change on the energy sector in the UK is sparse. The first CCIRG report (CCIRG, 1991) for example relied heavily on sources which addressed sensitivity to weather. Few of the cited sources addressed the issue of climate change. There has been little development of the literature other than that relying extensively on the CCIRG report itself (Skea, 1992; Skea, 1995). Internationally, the literature has expanded since 1991 but much of the original work is focused on one country, the United States. Several national studies have considered the energy sector (Smith and Tirpak, 1989; Nishioka et al., 1993; Mundy, 1990) but in little detail. A large proportion of the work has covered the impact of climate change on energy required for space heating and air conditioning. Other issues covered include: offshore oil/gas production at high latitudes; hydro-electric potential; impacts on the availability of other renewable energy sources; and water availability for thermal power generation. This chapter relies to some extent on interpreting overseas literature in a UK context.

8.2 ESTIMATED EFFECTS OF CLIMATE CHANGE AND SEA LEVEL RISE

8.2.1 Sensitivity of the energy sector to weather and climate

The impacts of climate change on the energy sector are extremely diverse and few are, in themselves, of major significance. Climate impacts relate to: i) *markets for energy*, especially those associated with space heating and air conditioning; ii) *the availability of energy resources*, especially renewable energy sources which are intimately related to climate; and iii) *operations* within the energy sector such as power generation, transmission or petroleum refining. Climate variability also has implications for energy demand and resource availability. At the same time, the energy sector will be affected by sea level rise and by a wide range of climate variables other than temperature and precipitation, including humidity, windiness, insolation and the frequency of extreme events. Table 8.2 provides a synopsis of the sensitivity of different components of the energy sector to specific climate variables.

8.2.2 Effects of climate change

Energy markets
Space heating and air conditioning are the most climate-sensitive uses for energy. However, irrigation pumping in agriculture, refrigeration and lighting also display a degree of climate

Table 8.2: Synopsis of climate sensitivities in the energy sector

	Temperature	Precipitation	Windiness	Frequency of extreme events	Water availability	Sea level rise	Other
Renewables	more evaporation from reservoirs, shifting seasonal run-off	hydro-electric potential	wave potential, reservoir evaporation wind potential	many renewable systems vulnerable- especially wind turbines, solar systems	hydro-electric potential	design of tidal, wave systems	insolation affects solar potential
Biomass	biomass availability	biomass availability		damage to biomass crops	biomass availability		
Energy demand	less space, heating, more air conditioning		space heating				more humidity, more air conditioning
Energy extraction	open-cast coal mining	open-cast coal mining		offshore oil and gas		offshore oil and gas	
Energy conversion	slightly less efficient thermal generation				cooling water availability	coastal power stations, refineries	
Energy transport/ transmission	lower capacity of power lines	icing of power lines		effects on power lines			

Key: ▓ = significant impact requiring adaptive response at a strategic level
      ▒ = modest impact requiring adaptive response
      □ = minor impact

Note: This table identifies impacts and their degree of significance. The direction of impacts and uncertainties are discussed in the text.

sensitivity. Space heating currently accounts for approximately 35% of UK final energy demand while air conditioning accounts for less than 1% (Hardcastle, 1984; Herring *et al.*, 1988). Energy demand for space heating is virtually static while air conditioning demand has been growing rapidly, notably in commercial buildings. The dominant fuel used for space heating is natural gas, though significant quantities of electricity, coal and oil are also used. Air conditioning demands are met entirely by electricity.

Space heating and air conditioning are intimately tied to building design and operation and are discussed also in Chapter 11 (Construction). This chapter focuses specifically on the impacts of climate change on the markets in which energy supply companies must operate.

Space heating

The annual level of demand for natural gas is significantly affected by average temperature conditions over the heating season. Recent warm winters, for example in 1988-89 and 1994-5, have reduced gas demand with a negative impact on both revenues and profits in the gas industry (British Gas, 1995).

Energy suppliers have analysed in detail the relationship between weather conditions and energy demand. In the case of the electricity industry, the variables taken into account include cloudiness, precipitation and wind as well as temperature (Parry and Read, 1988).

The use of energy for heating buildings is closely related to 'degree-days' which measure the extent to which mean daily temperatures fall below a base temperature, conventionally set at

15.5°C. Degree-day statistics can be manipulated to take account of temperature rise and hence derive an approximate indication of the impact which climate change would have on energy demand for space heating. Under the 1996 CCIRG climate scenario, space heating demand could fall by 11% by the 2020s and 17% by the 2050s. Columns (2) and (3) of Table 8.3 show the impacts on energy demand in different sectors. The biggest impacts would be felt in the household and service sectors where space heating accounts for a high proportion of demand.

In practice, building occupiers would take the opportunity provided by climate change to improve comfort levels. Many householders cannot afford to heat their homes to the temperatures which they might desire (Boardman, 1991). Statistical relationships based on historical associations between temperature conditions

Table 8.3: Possible impacts of climate change on UK final energy demand (reductions attributable to space heating)

	Degree-Day Method		Statistical Method	
	2020s	2050s	2020s	2050s
Household	7%	11%	5%	7%
Services	7%	11%	7%	9%
Iron & Steel	<1%	<1%	-	-
Other Industry	2%	3%	2%	2%
Agriculture	-	-	-	-
Transport	-	-	-	-
TOTAL	**3%**	**5%**	**3%**	**4%**

and energy demand can take account of this factor. Columns (4) and (5) of Table 8.3 show the results of feeding the 1996 CCIRG scenario into the temperature dependent models developed by the Department of Trade and Industry (Department of Energy, 1989). The projected changes in energy demand are slightly lower than those projected using the degree-day method, reflecting the fact that behavioural changes have been taken into account.

The shorter heating season means that annual energy demand for space heating will decline more than peak energy demand. In principle, the resulting lower utilisation of space heating equipment could have implications for fuel choice (e.g. electricity versus gas). In practice, the effect is likely to be too small to make an appreciable difference. Although climate-induced changes in space heating energy demand may appear large, other factors, including market changes, technological change and policies to reduce greenhouse gas emissions and hence energy demand, could have a much large impact over a period of 30-50 years.

Air conditioning

The current climate of southern England is marginal with respect to the need for air conditioning, although a quarter of office space is now air-conditioned (Herring *et al.*, 1988). Air conditioning is an automatic choice for new office buildings, especially those located in London and the south east or in noisy and polluted areas. Single-unit room air conditioners of the type which are found in parts of the US, Japan and parts of continental Europe are not commonly used.

Higher temperatures would increase electricity use associated with existing air conditioning systems and could accelerate the rate at which new markets for air conditioning might develop. Cooling degree-days, conventionally calculated from a base of 18°C, would increase sharply with higher temperatures. Energy use for air conditioning is affected by humidity, internal heat gains and solar heat gains as well as by temperature. Cooling degree-days are therefore rather unreliable as an indicator of air conditioning demand. The Building Research Establishment has estimated that a temperature rise of 4.5°C, far higher than that now projected for next century, would double the average 'full-load' usage of a typical system from 1,250 hours/year to 2,500 hours/year (Milbank, 1989). This estimate does not take account of the fact that relative humidity in summer could decline by up to 6 per cent. A summer temperature rise of 1.6°C by the 2050s would induce a perceptible increase in air conditioning demand.

No published UK studies have assessed the impact of climate change on the rate of installation of new air conditioning systems. A Japanese study found that the sale of air conditioning units increased by 40,000 for every degree-day in excess of 30°C (Sakai, 1988). It is not possible to extrapolate this result to UK conditions but a sensitivity to hotter climate conditions is clearly demonstrated.

Impacts on energy supply

Climate-induced changes in energy demand have implications for supply investments. It is certain that the market for fossil fuels will be smaller than it otherwise would have been as a result of climate change. Fuels used for space heating in the household and service sectors will be affected to the greatest extent. If current patterns of energy use were to continue, the largest impact would be on natural gas markets.

The impact of climate change on electricity demand will be the net result of two opposing effects. Demand for space heating will fall, while demand associated with air conditioning and refrigeration will increase. The use of electricity for space heating still greatly exceeds demand for air conditioning, although the latter use is growing and will be proportionally affected to a greater degree. Under the 1996 CCIRG climate change scenario, it is likely that there will be a net decline in electricity demand as a result of climate change. However, this greatly depends on changes in energy demand patterns, including the degree to which electricity is used for space heating and air conditioning purposes in the next century.

Currently, peak electricity demand in England and Wales in winter, approximately 46,000 MW exceeds peak demand in summer by about 10,000 MW (NGC Settlements, 1995). It is not likely that changes in demand for air conditioning will be sufficient to induce a switch from a winter peaking regime to a summer peaking regime such as is experienced in parts of Japan and the US. However, insufficient work has been carried out in this area to reach a definitive conclusion. A levelling out of the summer/winter demand pattern would have operational implications for the electricity supply industry.

Energy resources

Most renewable energy resources are directly dependent on climatic resources and are hence sensitive to climate change. Renewable energy, mostly hydro-electricity, currently accounts for only 2% of UK primary energy supply (Department of Trade and Industry, 1995a). The government has estimated that production could increase up to ten-fold by 2025 under 'severe pressures of need and economics' (Department of Trade and Industry, 1994). Energy crops and wind energy account for the majority of possible future renewable contributions. Promotion of renewable energy forms part of the mitigation element of the UK's climate strategy.

Hydroelectricity

Hydroelectric potential will be affected by changes in patterns of precipitation, the rate and timing of snowmelt, the ratio of run-off to reservoir storage and various factors (temperature, insolation, wind, humidity and local vegetation cover) which influence rates of evaporation (Cohen, 1987). Hydrological issues are discussed in more depth in Arnell and Reynard (1993) and in Chapter 7. Most UK hydro-electric generation takes place in Scotland and there have been no published studies showing the possible impacts of climate change. Work carried out in other parts of the world shows that it is virtually impossible to reach general conclusions concerning the impact of climate change on hydro-electric potential. Several studies have shown that hydro-electric production could increase rather than decrease. For example, Singh (1987) showed that the potential in the James Bay region of Quebec could rise by 7-20% depending on the climate scenario used. Fitzharris and Garr (1995) and Lettenmaier et al. (1991) found that climate change would result in reduced snow accumulation and a shifting of peak production from spring to winter, resulting in a better match between demand and supply.

Any future development of hydro-electric resources in the UK is likely to take the form of small run-of-the-river projects. As discussed in Chapter 7, the availability of water resources for hydroelectric production is linked to the management of storage and flood defences.

Biomass fuels

The UK Renewables Energy Advisory Group believes that there is a significant potential for electricity generation from biomass or 'energy crops' (UK Renewables Energy Advisory Group, 1992). Several projects were financed under the third tranche of the electricity industry's non-fossil fuel obligation (NFFO). In the UK, the most cost-effective biomass route is the coppicing of high-yield poplar and willow trees. Harvested wood would be gasified and burned using advanced generation technology (Department of Trade and Industry, 1994). Cereal crops could in principle be used to derive ethanol for use as a transport fuel.

The availability of biomass from coppices would be subject to the same sensitivities as other forestry products. Higher CO_2 concentrations could lead to faster tree growth, though this depends on adequate water availability and it is not yet certain whether the benefits would be permanent (Fajer and Bazzaz, 1992). Under the 1996 CCIRG climate change scenario, the UK winter precipitation is projected to increase but there is no consensus about even the direction of summer precipitation. Southern parts of the UK may become drier while northern regions could be wetter. Species selection and further breeding could help to mitigate any negative implications of climate change.

Biomass is potentially vulnerable to storms, pests and frost. Coppiced stands of the type suggested as appropriate for UK conditions are relatively resistant to storms. The frequency and severity of frosts (see Chapter 6) would decrease in a warmer world.

Wind energy

There is a considerable potential for the use of wind energy in the UK. Under the 1996 CCIRG climate change scenario mean winter wind speeds could increase by up to 6% in Wales and southern England by the 2050s. Since available energy increases as the cube of wind speed (Cavallo et al., 1993), climate change could significantly raise the UK's wind energy potential. Higher wind speeds could also make viable sites which would not otherwise have been developed. There are 7200 km^2 of land in the UK with wind speeds above 8 m/s and a further 4500 km^2 with wind speeds above 7.7 m/s (Chester, 1988).

Other renewable energy

Tidal and wave energy production is technically feasible but does not currently appear to be economically viable (Department of Trade and Industry, 1994). A tidal barrage such as that proposed at the Severn would have a very long life (about 120 years). It would be essential to take into account sea level rise and altered storm activity if tidal barrages or wave energy were to be developed.

Solar energy, in the form of either solar thermal or photovoltaics (PV), is a promising renewable energy source globally but may be relatively less attractive in the UK given its latitude. PV cladding of buildings represents a possible new means of electricity generation in the longer term. The 1996 CCIRG scenarios foresee little change in incoming solar radiation in winter, but an increase of up to 10%, associated with reduced cloud cover, in summer in southern England by the 2050s. In Scotland and Northern Ireland, solar radiation could decrease. In principle, these changes could make PV or solar water heating more attractive in southern parts of the UK.

Operation of energy facilities

The operation of energy facilities could be affected in a variety of relatively minor ways by climate change. Most of the sensitivities described here could be overcome by simple adaptive responses.

Offshore oil and gas

Offshore oil and gas operations could be affected negatively if storminess were to increase. The number of days on which it would be possible to service rigs would decrease. There is little confidence attached to statements concerning changes in storminess as a result of climate change, but heavy rain events may become more frequent (IPCC, 1996a).

Coal mining

Issues relating to coal mining are covered in Chapter 9 which deals with minerals extraction.

Conventional power generation

Traditional thermal power generation based on steam turbines, whether fossil or nuclear-derived, depends on the availability of cooling water. Water availability may become a constraint for power stations situated on rivers. Two types of cooling are available. Once-through systems abstract considerable quantities of water but return almost all of it to the river, albeit at a higher temperature. With 'evaporative' cooling (cooling towers), less water is abstracted, but all of the make-up water is 'consumed' in the process (Miller *et al.*, 1992).

Low levels of river flow can limit the output of power stations if: thermal pollution of the river becomes excessive; water consumption in evaporative systems becomes excessive in relation to river flow; or, in extreme cases, river levels fall below the level of the water input pipe. In France and the US, these factors have led to curtailments of plant output under drought conditions (Miller *et al.*, 1992). In the UK, fossil plants reliant on river cooling tend to be older and smaller than coastal power stations. Many are being closed. The UK's only non-coastal nuclear plant, Trawsfynydd, has recently closed.

Most new power stations are of the combined cycle gas turbine (CCGT) design and have much lower cooling water requirements per unit of output than do traditional steam turbine designs relying on coal or oil firing. The electricity industry's reliance on cooling water is therefore being reduced.

The efficiency of thermal power generation could be reduced slightly due to higher ambient temperatures (Ball and Breed, 1992). Existing technologies are however deployed in much warmer climates and adaptation therefore presents no problem.

Renewable energy

Climate change would have an impact on the performance of various renewable energy technologies. Wind turbines, solar thermal and PV systems are also subject to severe weather damage and would need to be strengthened if the frequency of extreme events were to rise (Jensen and Van Hulle, 1991; Kelly, 1993). Drier weather can result in increased soiling and reduced efficiency of wind machines, solar thermal systems and PV systems (Lynette and Associates, 1992; Radesovich and Skinrood, 1989; Goossens *et al.*, 1993). Changes in rainfall patterns and changes in local vegetation can also influence siltation, and hence the capacity, of reservoirs used for hydro-electric generation (Ball and Breed, 1992).

Energy transport and transmission

Overhead electricity cables are vulnerable to icing and storms. A more distributed pattern of electricity generation involving a greater use of combined heat and power (CHP) and renewables could reduce future vulnerability to supply interruptions. While icing is likely to decrease with climate change, the frequency of storms could increase. As noted in the First CCIRG Report, the major storms of 1987 and 1990 caused great damage to the English electricity transmission and distribution system. The October 1987 storm led to an average of 250 minutes loss of supply for customers in England and Wales with those living in the south east affected even more significantly (Electricity Council, 1988).

The capacity of electricity transmission lines is reduced at higher temperatures because they sag to a greater extent. A 400kV line currently has capacity of 2190 MVA in summer and 2720 MVA in winter (Eunson, 1988). Climate change would therefore have a perceptible effect on capacity but this could be overcome, at a cost, by re-dispatching plant or, in the longer term, reinforcing the transmission network. This issue has not been discussed in the published literature.

Compressors on gas pipelines would have to work slightly harder at higher temperatures. Higher temperatures could ease winter weather conditions, reducing problems associated with moving coal and oil by road or rail.

8.2.3 Effects of sea level rise

The energy supply industry, like other economic sectors, includes facilities which are located in coastal zones and are hence potentially vulnerable to sea level rise and/or increased storm activity. Petroleum refineries and some power stations for example have been preferentially located on the coast in order to take advantage of access to sea-borne supplies of fuel and, in the case of power stations, to gain access to cooling water. Currently, all of the UK's 14 petroleum refineries, all of its nuclear power stations and 70% of the major fossil fuel-fired electricity generation capacity (excluding CCGTs) are located on coasts or estuaries (Table 8.4).

Petroleum refiners are effectively 'locked into' existing locations. The economic advantages of ready access to supplies of crude oil are considerable and existing refineries are connected to an existing UK products pipeline network. The economic viability of specific refineries will be influenced much more by market conditions, or environmental controls relating to product quality, than by sea level rise. There may be a need for additional protection against sea level rise but requirements will be very site-specific. Refiners face considerable capital investment programmes over the next decade or so and any additional expenditure on site protection is unlikely to be significant by comparison.

On the other hand, technological change in power generation is having a significant effect on location. It is not necessary to locate CCGT stations on the coast. If coal were to become again the fuel of choice for power generation in the early decades of next century, new stations could however be advantageously located at coastal sites to gain access to sea-borne coal supplies.

Sea level rise could have some impacts on offshore oil and gas production and exploration. Shell have added one-two metres to the height of a production platform to take account of sea level rise over the lifetime of the facility (National Academy of Sciences, 1992). The incremental cost was however only 1% of the total investment. Most new exploration and produc-

Table 8.4: Location of major energy facilities in the UK

	Refineries (m tonnes distillation capacity)	Nuclear Power Stations (MW)	Conventional Fossil Stations (MW)
Coastal/lower estuary	92.9	12,462	13,860
Upper estuary	-	-	13,180
River	-	-	11,420
Other	-	-	480
TOTAL	92.9	12,462	38,940

Source: derived from Department of Trade and Industry, 1995c

tion facilities are floating rather than fixed and sea level rise is less relevant. Support facilities for offshore oil and gas production could also be affected by sea level rise.

8.3 ASSESSMENT OF POTENTIAL ADAPTATION

Specific research work on climate adaptation in the energy sector is virtually non-existent and the literature that does exist has often been judgmental and general in its nature. The majority of a panel convened by the US National Academy of Sciences concluded that the sensitivity of the energy sector to climate change is low in relation to that of natural systems and agriculture while adaptability is high. One panel member however dissented from this view, believing that insufficient note had been taken of interdependencies between the energy sector and other parts of the economy (National Academy of Sciences, 1992).

At the technological level, the capacity to adapt to climate change depends partly on the relationship between the rate of climate change and the rate of replacement of equipment and infrastructure. Many assets employed in the energy sector (see Table 8.5) have lifetimes of only a few years and will be replaced several times over the timescales assumed under the 1996 CCIRG climate change scenario. Little existing plant will remain by the middle of next century and even medium-life assets such as conventional power stations will have been retired. Only very long-lived assets, such as some residential and public buildings and dams, tidal barrages and infrastructure which locks certain activities into specific locations (petroleum refineries, ports, roads, rail links) will remain.

Some forms of adaptation will be undertaken by individuals and organisation without the need for any specific policy actions ('autonomous adaptation'). To take a simple example, central heating thermostats will allow householders to adapt to climate change while hardly being aware that they have done so. On the whole, it is likely that successful autonomous adaptation will take place in the energy sector. Problems might arise however in relation to longer-lived assets which may need to function in a climate for which they were not designed. In practice, much infrastructure is modified during its lifetime while retaining the same location and basic function. Re-building or upgrading offers a useful opportunity to adapt to changed climatic conditions. Autonomous adaptation will however take place most successfully if individuals and organisations are adequately informed and have the institutional and financial capacity to act.

8.4 UNCERTAINTIES AND UNKNOWNS

Table 8.5: Asset lifetimes in the energy sector

Conventional light bulb	weeks up to 3 years
Electric white goods	5-10 years
Central heating boilers/systems	10-15 years
Motor vehicles	10-15 years
North Sea oil field	10-30 years
Gas supply contract for combined cycle gas turbine	15 years
Life of renewable energy project	20 years
Conventional power plant	40-45 years
Housing stock	50 years, but some very long
Infrastructure (roads/rail/ports)	50-100 years
Tidal barrage, dam	100 years +

Source: Skea (1995)

Most sensitivities to climate within the energy sector, such as space heating demand, are in principle relatively easy to quantify. Given perfect information about changes in climate variables it would, in principle, be possible to quantify most climate impacts. The unavailability of reliable scenarios for climate change at the regional level is therefore a barrier to further impacts analysis. There are two areas in which further work is merited:

- The relationship between climate change and the adoption of *new* air conditioning is not well understood although the performance of *existing* air conditioning systems under a changed climate is relatively simple to project.
- Sensitivities within the energy sector are quite well understood but the interaction between interdependent economic sectors displaying various degrees of climate sensitivity - water, energy, agriculture, industry - is not.

8.5 PRINCIPAL IMPLICATIONS FOR OTHER SECTORS AND THE UK ECONOMY

Energy is an essential input to economic life and many other sectors will be indirectly influenced by climate impacts on the energy sector, although these impacts are unlikely to be large:

- Users will be affected by changes in energy needs and energy prices. Energy accounts for 4.8% of the average household budget and 10.7% of the budget of low income households (Central Statistical Office, 1994). Since most household energy is used for space heating, the impact of climate change would be beneficial. Energy accounts for 2.7% of total industry production costs, though this can rise to about 10% in industries such as water and paper (Department of Trade and Industry, 1995c). Industrial energy demand is relatively insensitive to climate.
- The building industry will need to take account of climate change and the impacts on energy demand in designing new buildings. This issue is discussed in Chapter 11.
- Higher temperatures will lead to increased demand for air conditioning equipment.

Those studies which have attempted to attach a monetary value to the impacts of climate change (Cline, 1992; Titus, 1992) have generally taken into account only space heating/air conditioning impacts in relation to the energy sector. Superimposing the projected 2050s climate on the current pattern of energy demand would result in a cost reduction equivalent to approximately 0.3% of GDP. The actual impact on GDP in the 2050s would depend on the pattern of energy demand/supply and the relative importance of energy within the economy. There would be other economic consequences arising from the impacts of climate change on the energy sector, but these are less easy to quantify. Some of these impacts, such as the vulnerability of power lines to more frequent storms, could be negative rather than beneficial

8.6 RESEARCH AND POLICY ISSUES

8.6.1 Future research effort

Little research into the physical impacts of climate change on the UK energy sector has been conducted. There is probably little merit in conducting a substantial amount of research unless more reliable projections of climate change at the regional level become available. The impact of climate change is likely to be small in relation to other pressures. Results are unlikely to be useful unless they can be presented with a much higher degree of confidence than they currently enjoy. Work in the following areas could be justified:

- Improving the reliability of regional climate scenarios so that a better indication of the wide range of climate variables which may influence the energy sector can be obtained.
- Understanding better the relationship between climate variables such as temperature and humidity and the

adoption of new air conditioning systems. This work would need to be carefully tied to work on new building designs.

- Understanding better the impact of climate change on the availability of renewable energy resources in the UK, especially biomass, hydro-electricity and wind.

- Developing techniques (such as input-output modelling) which would enable an exploration of the way in which interdependencies between energy and other economic sectors can transmit climate sensitivities throughout the economy.

8.6.2 Policy issues

It is likely that the energy sector can, for the most part, adapt to climate change without the need for specific policy actions. Government may have a role to play in two areas:

- Providing, or stimulating the provision of, information which will enable organisations to anticipate better the nature of the climate in which infrastructure and facilities will have to operate. Both a better characterisation of current climate (means and variabilities) and more reliable projections of future climate, at the regional level, would be helpful.

- Acting more directly in defining planning codes for especially long-lived assets such as dams, tidal barrages or coastal infrastructure which are vulnerable to climate or sea level rise.

9. Minerals Extraction

SUMMARY

- The UK minerals extraction industry is accustomed to continuously and quickly adapting to changes in production locations, strong environmental protection pressures for onshore and offshore operations, the implications of sustainable development, privatisations and changes in market conditions. Over the last 30 years significant changes have included the rapid development of the offshore oil and gas industry and the reduction in the coal industry.

- The UK is in a strong position for the production of energy, construction materials and certain industrial minerals but is heavily dependent on imports for base metals, iron ore, coking coal for steel production and precious metals.

- The historical record of successful adaptation of the UK minerals extraction industry suggests that the projected changes in climate in relation to temperature and rainfall do not create any significant technical problems for the oil, gas, construction materials, industrial minerals and coal industries.

- More importance will be attached to information about increases in the frequency and intensity of storms and the rise in sea level These factors could cause inland and coastal flooding, coastal erosion, tidal surges, hazards for shipping and increased salinity. The impact of these changes would be to raise the costs of extraction and, possibly, would result in the relocation of production, transport and processing facilities associated with minerals extraction. Although technical solutions are likely to be developed, there is no guarantee that the producing activity would be economically viable.

- Current estimates about the rate of increase in the sea level suggest that coastal investment associated with minerals extraction should allow adequate time for current capital investments to be recovered. An increased frequency and intensity of storms could reduce the time periods before action is required in relation to decisions about improving sea defences or relocating operational facilities.

- Decisions to improve sea and river defences would require increased use of rock armour, large sizes of hard rock and sand from marine dredging. It is likely that this would require imports of rock armour to supplement UK production and to limit overland transport of this material.

9.1 INTRODUCTION AND BACKGROUND

This section describes the location, techniques of extraction and contribution to the UK economy of the minerals extraction industry. It demonstrates the ability of the industry to adapt to changes in production, strong environmental pressures on onshore and offshore operations, the implications of sustainable development, privatisations and changes in market conditions. In particular it shows how the offshore oil and gas industry has met and continues to meet the challenge of successful exploitation in increasingly hostile climatic conditions. The historical record of the UK minerals extraction industry suggests confidence in its technical ability to adapt to climate change.

The total value of mineral production in the UK in 1994 is estimated at £17,000 million which is 2.5% of Gross Domestic Product. The major contributions are from oil and natural gas liquids (57%) and natural gas (22%). Industrial and construction minerals provide about 11% and slightly less is supplied by coal at 10%. Oil made a positive contribution to the balance of payments and this is estimated at over £4,000 million in 1994. Government revenue from oil and gas was £1,600 million in 1994/95 (Department of Trade and Industry, 1995b [DTI]). The total of £17,000 million in 1994 is relatively low compared with the 1985 figure for GDP contribution of £27,500 million for the value of minerals production. In 1985, before the major fall in the price, oil contributed about £20,000 million.

During the last five years the minerals extraction industry in the UK has accommodated a significant contraction in the size of the coal industry and a continuing expansion in the production of natural gas. These changes have occurred in response to the Government's policy as stated in The Prospects for Coal. Conclusions of the Government's Coal Review;

'The aim of the Government's energy policy is to ensure secure, diverse and sustainable supplies of energy in the forms that people and businesses want, and at competitive prices. The aim needs to be pursued in the context of the Government's economic policy as a whole, of other Government policies, especially on health, safety and the environment, and of the United Kingdom's European community and other commitments.' (DTI, 1993).

With the privatisation of British Coal Corporation (BCC) in December 1994 as part of the above policy and the prior closure of unprofitable mines, a new structure for fossil fuel production emerged in the UK.

Like farming and forestry the minerals industry is an important user of land, but of a far smaller amount. The total area of land used for all types of surface mineral workings in England in 1988 was just over 96,000 hectares. This equalled just 0.75% of the total land surface and, after its temporary use for mineral extraction, most of it will be restored to other beneficial use (Confederation of British Industry, 1994 [CBI]).

Oil

In 1994 offshore production of oil was 114.3 million tonnes, land production was 4.6 million tonnes and natural gas liquids production was 7.8 million tonnes. This total of 126.7 million tonnes was a near record, falling just short of the 127.6 million tonnes produced in 1985 (DTI, 1995b). Prior to the 1970s the UK was almost wholly dependent on imports for its oil supplies; the only domestic supplies came from a few land-based oilfields.

There were 73 offshore fields and 20 onshore fields producing crude oil and condensate on 15th March 1995 and 11 new development projects were approved in 1994 (DTI, 1995b). Production from most fields is controlled from production platforms of either steel or concrete which have been built to withstand severe weather, including gusts of wind up to 260 km/hour (160mph) and waves of 30m. Design criteria vary significantly in the UK offshore sector. Values of around 50% of those indicated above are used in the shallow waters of the southern North Sea (Grant, *et al*., 1995). With cumulative production of oil to date of 1,786 million tonnes the total UKCS remaining reserves are estimated to be in the range 1,884 - 7,034 million tonnes (DTI, 1995b).

The industry has now reached the stage where several of the earlier platforms are reaching the end of their production life and methods of decommissioning these installations to avoid danger to shipping and fishing in the North Sea are the subject of a major domestic and international debate. To the necessity of avoiding environmental damage has been added the extra requirement to organise the industry to incorporate the concept of sustainable development. Both the costs associated with meeting these requirements and the ultimate responsibility to pay them remain unresolved at the present time.

New oil field developments are moving to the Atlantic west of the Shetlands. Decisions are required regarding the production techniques for use in these deeper waters. There is a choice between Floating Production Storage and Offloading (FPSO) in conjunction with shuttle tankers and the alternative of submarine pipelines. The economic viability of an oil pipeline depends upon the discovery of a sufficient amount of oil, not yet achieved to the west of the Shetlands. The cheaper system preferred by the United Kingdom Offshore Operators Association Limited (UKOOA) is the use of FPSO and shuttle tankers. The discussion about the two systems centres on profitability and the environmental damage which could result from accidents to the floating platforms and shuttle tankers as a result of severe weather conditions in the area. The UKOOA has a data base of wind, wave and current information taken from daily measurements over a four year period. Operators continue to adapt design criteria to meet new circumstances of the type envisaged for the first field, Foinaven, west of Shetland (Grant, *et al*., 1995).

The production of onshore crude oil and natural gas liquids (NGLs) was 4.6 million tonnes in 1994 the majority of which came from Wytch Farm in Dorset (Central Office of Information, 1995 COI). The principal land based installations of the oil industry are the refineries and the onshore pipelines.

Oil pipelines brought ashore about 80 per cent of offshore oil in 1993. Some 1,930 km (1,199 miles) of major submarine pipeline brings oil ashore from the North Sea oil fields. Crude oil onshore pipelines link harbours, land terminals or offshore to refineries. Also, onshore pipelines carry refined products to the major marketing areas. In 1993 the refinery sector processed 96.3 million tonnes of crude and process oils and about 80 per cent of the output is in the form of lighter, higher value products such as gasoline (COI, 1995).

The location of the oil fields, offshore pipelines, oil terminals and refineries are shown in Figure 9.1.

Gas

The UK is the world's fifth largest gas producer and British natural gas accounts for a quarter of total primary fuel consumption in the country.

In 1994 the production of natural gas from offshore gas fields was 69,728 million cubic metres and 241 million cubic metres from onshore fields, (DTI, 1995a). Converting this into oil equivalent tonnes gives an offshore and onshore production of 59 million tonnes for 1994. This is a significant increase from 32 million tonnes in 1984 (British Petroleum, 1995). Indigenous supplies in 1994 provided 97% of UK natural gas consumption (Central Statistical Office, 1995 [CSO]).

With cumulative production to date of 982,000 million cubic metres of gas from the UKCS, the remaining reserves are estimated to be in the range of 1,218,000 - 3,815,000 million cubic metres (DTI, 1995b).

Figure 9.1: Location of oil installations in the UK (Source: An Official Handbook 1995, by Central Office of Information, HMSO)

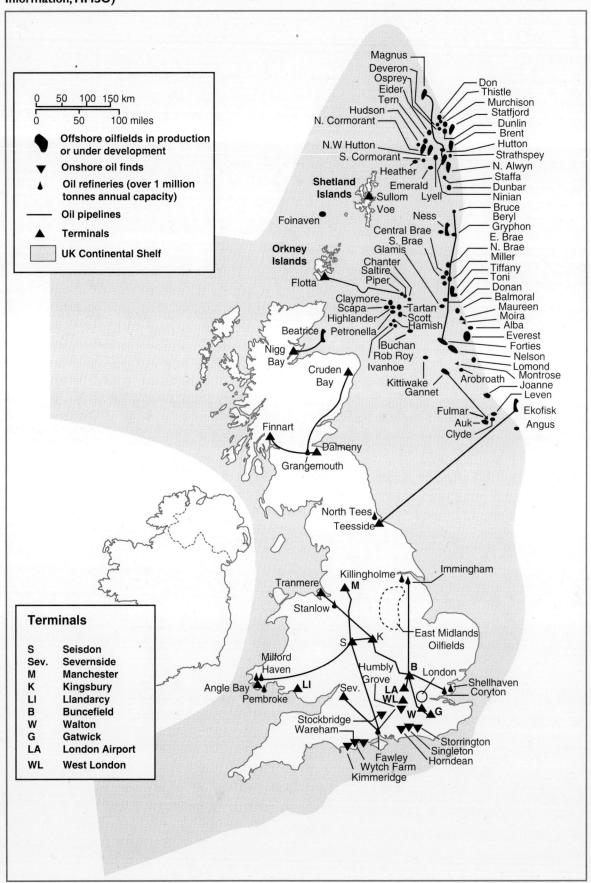

Figure 9.2: Location of gas installations in the UK (Source: An Official Handbook 1995, by Central Office of Information, HMSO)

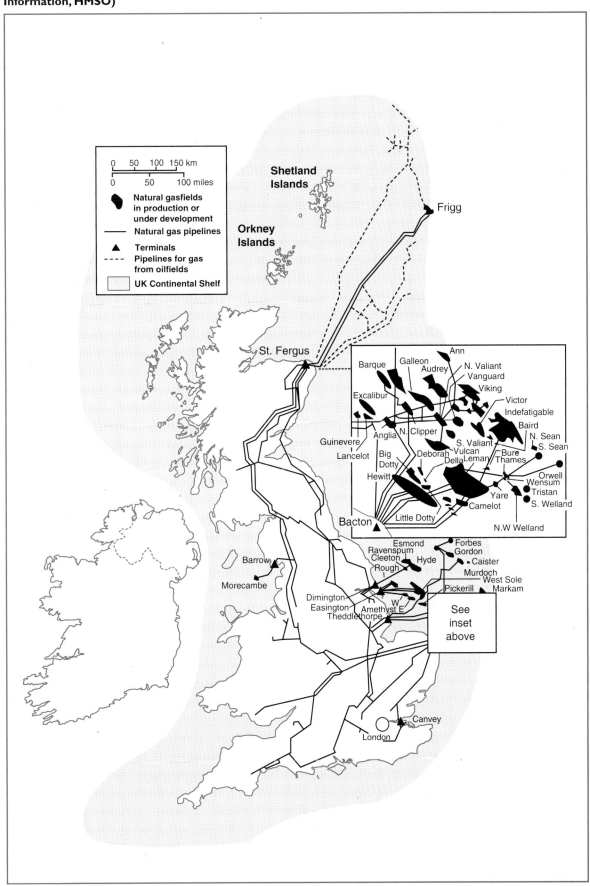

The total number of gas fields was 56 on 15 March 1995, 53 offshore and 3 onshore (DTI, 1995b). Significant contributions of associated gas from North Sea oil fields are produced in addition to supplies from gas fields. The British Gas national and regional high pressure pipeline system of some 265,000km (164,660 miles) transports natural gas around Great Britain. It is largely supplied from four North Sea shore terminals and one in Barrow in Furness (COI, 1995).

The locations of natural gas producing fields, terminals and pipelines are shown in Figure 9.2.

Coal

The estimated UK coal output of 48 million tonnes in 1994 was produced by a combination of the former British Coal Corporation (BCC) mines and the original private sector coal mines. Deep mined output was 31.1 million tonnes, opencast production was 16.6 million tonnes and small amounts were produced from slurry recovered from dumps (DTI, 1995a). Prior to the privatisation of BCC, the private sector produced about four million tonnes per annum. Total inland consumption of coal in the UK in 1994 was 81.7 million tonnes of which domestic production supplied 59%, net imports were 17% and reduction in stocks supplied 24% (DTI, 1995a).

An indication of the major changes in the industry since 1955 and the ability of the industry to adapt to changed circumstances are shown by the fall in output from the former BCC underground mines of 211.3 million tonnes with a labour force of 740,000 at 850 underground mines. In March 1994, the BCC employees totalled about 19,000.

For the purposes of privatisation the coal mining assets of BCC were divided into five regions, with 14 operating underground mines, 2 underground mines being developed and 32 open-pit mines (Table 9.1).

Three separate private sector companies purchased the coal mines in the three regions shown in Table 9.1. The company which bought the mines operating in England and North Wales is comparable in size to the largest private sector coal mining operations in the world.

In addition to the mines listed above an additional seven underground mines, which had been closed by BCC, were offered for sale and bids were received for five of them. The total number of underground coal mines in the UK is increased by a further six because they were taken over from BCC by the private sector in

Table 9.1

	Underground Mines	Open Pit Mines	Number of Regions
Scotland	1	9	1
England + North Wales	15*	14	3
South Wales	0	9	1
Total	16	32	5

* including two mines being developed.
Source: N.M. Rothschild & Sons Limited. Preliminary memorandum issued on behalf of HM Government, April, 1994, Coal Mining Activities of British Coal Corporation.

1993 on a lease/licence basis. Some of these additional mines are operated by the three companies mentioned above and some by other private sector companies.

The underground mines which were included in the privatisation programme are in sharp contrast in size to the very small underground mines operated by the previous private sector under licence from BCC. All of the open-pit mines are small by international standards with average lives of about five years. The known technically recoverable reserves of coal in the UK are very large but a new assessment is required after the extensive mine closures. Also economically recoverable reserves will have been reduced significantly in response to the need to reduce costs to meet internationally competitive prices. A new public body, the Coal Authority, has been formed to take over the ownership and licensing of coal production from BCC. The Coal

Authority became effective on 31st October 1994.

Future output levels of UK coal will depend upon the private sector's ability to compete on price and quality with alternative fuels while at the same time meeting increasing environmental protection requirements. There is possible scope for replacing some of the UK coal imports, which totalled 15.0 million tonnes in 1994, with domestic production and perhaps increasing exports above the 1.2 million tonnes in 1994. Import substitution will be principally connected with steam coal because the UK does not have significant economic reserves of high quality coking coal required by modern steel works.

The location of coal and lignite deposits is shown in Figure 9.3.

Industrial and construction minerals

Sand, gravel and crushed rock are the most extensively used of all mineral products in the UK. Traditionally they come from land-based sources, but over the last 20 years marine-dredged sand and gravel has become increasingly important (CBI, 1994). The estimated output of non-fuel minerals in the UK in 1994 was 337.2 million tonnes (British Geological Survey, 1994 [BGS]). The bulk of this total is provided by construction aggregates, i.e. crushed rock (limestone, igneous rock and sandstone) and sand and gravel. This sector of the industry has also demonstrated its ability to adapt to market changes. There has been a threefold increase in the production of aggregates over the past 30 years and UK consumption of aggregates per capita was 4.2 tonnes in 1992 (CBI, 1994).

Annual demand for construction aggregates is forecast to rise to between 370 to 440 million tonnes in 2011. The balance between producing adequate supplies of construction materials and environmental protection will be a continuing challenge. This has resulted in an increased interest in developing very large coastal quarries in Scotland, each capable of producing about five million tonnes of aggregates per year from reserves of 150 million tonnes, and transporting it by sea to areas of high demand in England such as the southeast region. One operation of this type is in production at Glensanda at Loch Linnhe in western Scotland; a planning application has been made for another at Lingarabay on the Isle of Harris in the Outer Hebrides. Other possible sources of aggregates are coastal sites in Norway and Northern Spain. Research indicates that port facilities for the importation of this material into the southeast region could be located along the Thames Estuary. Significant wharf and port development would be required to accommodate large volumes of material. In addition, secondary and recycled materials are expected to increase with Government targets of 40 million tonnes per year by 2001 and 55 million tonnes per year by 2006 (Department of the Environment, 1994).

Marine dredged sand and gravel production contributed about 20% of total UK production of 99,000 tonnes in 1994. The Government is aware of the special need for marine aggregates in soft coastal defence schemes where it is often impossible to make use of material from non marine sources (Department of the Environment, 1994). Marine dredging from 20 km offshore is being used to supply sand for the Lincshore coastal defence system in Lincolnshire. This project involves the largest beach recharge scheme ever undertaken in the Anglian region of the National Rivers Authority (NRA). Covering nearly 20 km between Mablethorpe and Skegness along some of the most vulnerable coastline in Britain, the defences protect more than 35,000 people over 15,000 homes and 18,000 residential caravans as well as extensive agricultural, commercial, industrial and service related industries on 20,000 hectares of low lying land (NRA, 1995).

Other sectors of the non fuel minerals industry which make a contribution to the economy at regional and national level are china clay, potash, salt, chalk, ball clay and fluorspar. The UK is the world's second largest producer of china

Figure 9.3: Location of coalfields and lignite resources in the UK (Source: United Kingdom Minerals YearBook, Harris 1995, British Geological Survey.)

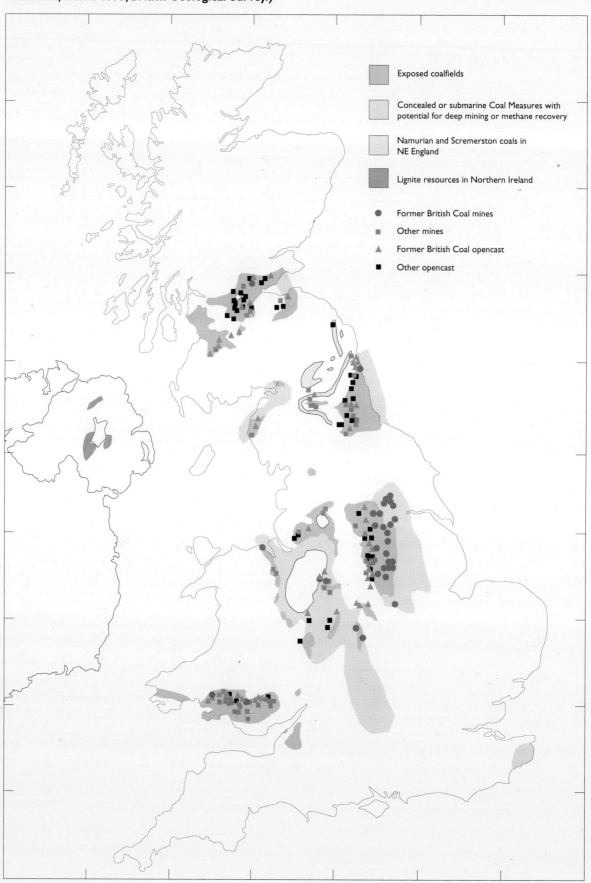

Figure 9.4: Location of some metals and industrial minerals in the UK (Source: An Official Handbook 1995, by Central Office of Information, HMSO)

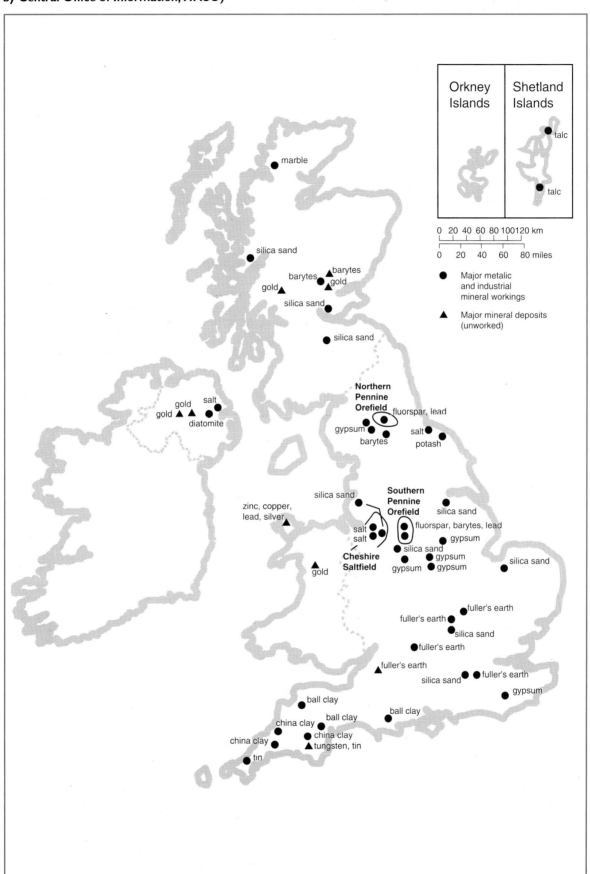

clay and estimated sales in 1994 were 2.64 million tonnes of which 87% were exported. The paper industry accounts for approximately 73% of total sales, ceramics manufacture 19% and the remainder for other industrial applications including a fill in paint, rubber and plastics (BGS, 1994).

The location of these industrial minerals is shown in Figure 9.4.

Metals

Metal production in the UK in 1994 consists of 1,900 tonnes of tin in concentrate and small amounts of lead and iron ore. Low prices have reduced tin output in Cornwall. The Parys Mountain polymetallic copper/lead/zinc/silver deposit in Anglesey remains undeveloped pending an upturn in metal prices. Minor amounts of lead and zinc are produced as by-products in the Pennine fluospar operations (BGS, 1994). Two small gold mining operations in Scotland and Northern Ireland recently obtained planning permission.

The location of metallic deposits in the UK are shown in Figure 9.4.

The UK is dependent upon large scale imports of copper, iron ore, aluminium, nickel, zinc, lead and the other metals and minerals required for an advanced industrial economy.

Employment

There is no definitive survey of the employment levels in the oil and gas related activities in the UK, but UKOOA has estimated that direct and indirect employment could be about 300,000. A survey by the Scottish Office and Scottish Enterprise estimated that in 1993 some 77,000 jobs were accounted for by companies involved mainly in oil-related activities. In September 1994, the Inland Revenue estimated that the total offshore workforce was 27,000 of whom 92% were UK nationals (DTI, 1995b).

In March 1994, prior to privatisation of the coal mining assets, there were 19,000 employees of BCC. There were an additional 4,500 employees in other sectors of the coal extraction industry (e.g. open-pit mines) operated by companies under contract to BCC at that time and also employees in small mines which were already in the private sector. The employment in the extraction of non-fuel minerals in the UK is 32,000 (BGS, 1994).

9.2 ESTIMATED EFFECTS OF CLIMATE CHANGE AND SEA LEVEL RISE

This section deals with the estimated effects of global warming on the techniques for the extraction of metals and minerals on land and offshore UK.

9.2.1 Sensitivity of minerals extraction to weather and climate

The historical performance of UK offshore oil and gas extraction gives a sufficient indication that the projected temperature and rainfall changes under the 1996 CCIRG scenario would cause no major difficulties. More importance needs to be attached to any problems which might result from increased frequency and intensity of storms and sea level change associated with climatic change.

9.2.2 Effects of climate change

The problems which could occur in relation to temperature and rainfall changes can be classified more accurately as inconveniences and irritations which, although causing higher costs, would not be sufficient to significantly alter the financial and competitive position of the industry. It is likely that the climate change projected by the 1996 CCIRG scenario could lead to improved conditions in some mining operations in the UK.

The mining industry in the UK, particularly the surface mining sector, has much experience in dealing with the problems associated with noise, dust, traffic congestion and restoration. Many of the surface coal mining operations are located near built-up areas, thus environmentally acceptable techniques must be demonstrated before planning permission is granted.

Hot, dry summers are likely to increase environmental problems due to increased volumes of dust. Additional dust suppression will be required, particularly for surface mining. Stockpiles of coal and waste at deep mining operations would have to be kept to minimum levels. Dry periods resulting in water shortage would detrimentally affect coal washing facilities, with consequent production losses and increases in costs. Overcast skies can affect the strength of blast noise due to reflection from the cloud base.

Many UK underground mines work seams located in strata that contain quantities of methane under pressure. Under normal operating conditions, gases are gradually released from these strata into the ventilation airstream, which is designed to dilute them satisfactorily. Normally the pressure difference is relatively stable but when barometric pressure falls rapidly, this can result in a dramatic increase in the gas emission rate which would temporarily interrupt production. Current trends in underground coal mines are towards working fewer faces at very high rates of production with high availability time for machine operation. Interruptions in continued production on these faces can result in increasing costs which affect the financial returns from the capital intensive equipment. This problem can be alleviated by methane drainage from the strata direct into pipe systems. This technique is widely practised in UK mines and the methane is utilised on colliery premises or elsewhere for power generation and space heating.

Surface mines for all minerals would be adversely affected by increased wet periods; flooding of the pits and pollution control difficulties would be the main consequences. Large underground mines are currently working below the water table and the possibility of flooding is minimal. The effects of increased pressure on shaft linings would need to be studied. Some small underground coal mines of the type currently owned by the private sector would be more detrimentally affected.

For the quarrying of construction materials, the most important impact would be movement of the water table. A raised water table would increase pumping costs at a hard rock quarry. In the case of sand and gravel, excavation under dry conditions is favoured but water is essential for processing. A higher water table could complicate restoration on quarry sites as soils used for restoration can only be moved in dry periods. The stripping of soils prior to extraction could be severely impeded by unseasonable rain.

Oil and gas from the Northern Basin of the North Sea is produced in some of the most challenging climatic conditions found anywhere in the world. Oil and gas exploration and production technologies have improved rapidly with a shift into more difficult operating regions such as the move from the Southern Basin to the Northern Basin of the North Sea and into the Atlantic west of the Shetlands. It is unlikely that technology would fail to keep pace with the projected climate change scenario but the main question, which cannot be answered on the basis of current knowledge, is whether the production activity would be economically viable (Taylor, Personal Communication).

The minerals extraction industry will also have to respond to any changes in construction standards which are developed to deal with the effects of climate change (CCIRG, 1991).

9.2.3 Effects of sea level rise

The effects of sea level change incorporate flooding, coastal erosion, tidal surges exacerbated by storms and increases in salinity. Considerable experience already exists in the United Kingdom for dealing with coastal erosion and storm surges.

For the offshore oil and gas industry a rise in sea level would effectively reduce the air gap between the crest of the design wave and the underside of the lowest level of fixed offshore structures. Regular reviews and updates of the metocean design conditions for existing structures would highlight potential problems and appropriate action could be taken (Grant et al., 1995).

It is likely that the oil and gas terminals and refineries could be affected by sea level change due to their coastal location (Figures 9.1 and 9.2). Also, the impact on the development of large coastline quarries in Scotland needs to be examined in relation to associated port facilities. Direct effects on marine dredging would be manageable but could involve increased processing costs, if chloride levels rise and extra sedimentation lead to extra contamination by clays.

Any rise in sea level or higher storm tides may affect the extraction of minerals close to sea level, and special measures would be required to reduce the effects. For example, the surface extraction of minerals, such as gravels, in river flood plains would be affected as the high tide would reach further up river than previously and higher up the banks of rivers already tidal. This would involve a change in mining technique from, possibly, land based draglines to floating dredges. The land, post dredging, would also be affected by the higher tide. Underground mining, principally of coal in low lying areas, already has restrictions on the amount of subsidence allowed; this has effectively meant that only one seam may be worked. In Germany and other countries subsidence has been reduced by stowing waste material in the void created by mining, however, this technique is very expensive. (Richards, Personal Communication).

The indirect effect of a rise in sea level on the minerals extraction industry could be very significant. If extra sea defences are required there would be an increased demand for rock armour to build the necessary barriers. The problems associated with higher quarry output in England have already been mentioned. Strong sea defences require large stones which generally are provided by coastal quarries. Imports from outside the UK may be necessary to supplement UK production and to avoid overland transport.

9.3 ASSESSMENT OF POTENTIAL ADAPTATION

The technical ability of the minerals extraction industry in the UK to adapt to changing marketing and physical operating conditions is demonstrated by recent events.

The offshore oil and gas industry developed rapidly over the last 30 years and the coal industry has selected the appropriate techniques to meet a variety of geological conditions. Both industries incorporate into their operations increasingly stringent environmental protection measures. As well as ideas to deal with new circumstances being generated within the UK, there is a vast amount of information available from the rest of the world for extracting minerals under a wide range of climatic conditions.

Possible changes in the location and installation of coastal defences due to sea level changes alone are likely to be achieved in the time periods without affecting the recovery of investment in current installations. The ability to adapt without too much inconvenience will depend upon information being available which is consistent with the investment planning time horizons. An increased frequency and intensity of storms combined with sea level rise could reduce the time periods before action is required in relation to decisions about improving sea defences or relocating operational facilities.

Adaptation to the possible effects of climate change and associated sea level change does not require a dramatic change of culture for the minerals extraction industry. The offshore oil and gas industry has continuously been adapting to different climatic conditions as shown by the development of a new metocean design for the deep water harsh environment off the North West Shelf of Europe. This design incorporates information relating to wind, waves, currents, water levels, air temperatures, sea temperatures, ice and snow and marine growth (Grant, et al., 1995).

Oil and gas extraction, mining and quarrying operations are carried out in a wide range of climatic conditions throughout the world. Appropriate techniques have been developed to deal adequately with particular problems and this information is transferred easily between countries. In some situations, such as the extraction of oil and gas from offshore fields in deep water, the UK can be considered a leader in this technology.

As a result of increased environmental protection measures in relation to dust from mineral stockpiles (e.g. coal and waste material from mines), research on the impact of weather conditions is being carried out. This research will assist in developing appropriate techniques for adapting to climate change. For several years, the on land minerals extraction industry has been subject to rising standards of environmental care and has been required to mitigate the effects of noise, vibration and dust from both the workings and associated heavy lorry traffic.

Atmospheric pressure variations cause pollutant levels in mines to fluctuate, for example, the release of methane from mine workings. Rapid fall in pressure is the main concern as this can lead to rapid increases of gaseous pollutants in mine ventilation airstreams. Weather warning systems, in some countries, warn mines and other subsurface facilities of rapid falls in barometric pressure; this allows remedial action to be undertaken prior to the event occurring. The majority of mines are exhaust ventilated, i.e. the mine is a depressurised zone. If the frequency or magnitude of rapid falls in barometric pressure increases, a simple, relatively low cost, method of reducing the effects of these events is to force ventilate the mine, i.e. to make the mine a pressurised zone. This is already undertaken to a large degree in the case of single entry headings with large gas makes where forced auxiliary ventilation is used and in the civil tunnelling industry where compressed air pressurised working is undertaken to reduce pollutant inflows. Other subsurface facilities, for example, nuclear waste repositories, require a much greater level of ventilation control than in mines to avoid contamination of airflows. In such situations it is usual to both force ventilate and exhaust ventilate the facility to achieve the required level of airflow and pressure control. Such systems are less affected than mines by changes in barometric pressure (Tuck, Personal Communication).

A survey of the literature dealing with climate change and minerals extraction shows a long term awareness of the factors involved. Two examples are indicative of the information and continuous assessment of the situation.

"The exploitation of minerals from the sea bed especially in deep water requires an analysis of weather conditions, seasonal variation and their impact on the ship, floating structure or submersible from which the operators are conducted" (Hughes, 1978).

"Gas explosions in coal mines and the idea that changes in barometric pressure have played a role in many major explosions world wide is far from new. If major disturbances are expected, education of personnel to test regularly for gas and take rigorous precautions during gas alerts when barometric pressure falls and investment in methane monitoring systems, as essential courses of action" (Fauconnier, 1992).

9.4 UNCERTAINTIES AND UNKNOWNS

In order to assess the impact of climate change and associated sea level rise on the technical capacity of the UK minerals extraction industry, information is needed on the following: sea defence policies, frequency and intensity of storms and tidal surges, modifications to construction standards, frequency of barometric pressure changes and lengths of periods of dry and wet weather on a regional basis.

9.5 PRINCIPAL IMPLICATIONS FOR OTHER SECTORS AND THE UK ECONOMY

The products of minerals extraction make a direct and essential contribution to all aspects of good material living standards. Without domestic production it would be necessary to import and any reduction in output would adversely affect the balance of payments. The production of oil, gas and coal gives the United Kingdom independence in energy supplies which, in turn, provides a strategic advantage over other countries.

The analysis of the effects of climate change and sea level rise on the minerals extraction industry does not suggest any insurmountable problems regarding the technical ability of the industry to function and maintain its contribution to the UK economy. The costs of the offshore minerals extraction industry could rise in response to increased frequency and intensity of gales and this would affect the national economy through higher energy costs. If international competition precludes price rises, then the Government's revenue from the industry would fall if profits are squeezed and employment levels in the industry would be reduced.

Significant increases in the extent and size of sea level defences could involve increased domestic production, imports of rock armour and additional requirements for marine dredging of sand. Careful planning will be required to reduce the environmental impact of these developments.

9.6 RESEARCH AND POLICY ISSUES

9.6.1 Future research effort

Future research effort should be directed towards the impact of sea level rise on coastal and estuary sites where there are facilities for mining processing, refining and storing metals and minerals. Movements of the water table should also be investigated. A study of the availability of materials and the costs for improving sea defences is also important.

Research needs to continue into the performance of floating platforms/shuttle tankers for extraction in deep level offshore oil fields under severe weather conditions.

Research is needed into the economically viable methods for stowing waste materials in underground mine openings to reduce subsidence so that minerals may be extracted without affecting land usage.

9.6.2 Policy issues

Policies on the extent to which sea and river defences should be strengthened and the environmental risk associated with offshore oil and gas extraction are required for planning the future development of the minerals extraction industry.

10. Manufacturing, Retailing and Service Industries

SUMMARY

- Fewer severe winters would be beneficial, reducing disruption at all stages from the supply of raw materials through processing to markets for the finished goods. A parallel increase in hot, dry summers, which has been shown to disrupt sectors with a large water requirement, is unlikely to have a substantial impact over the long-term. The water supply industry is expected to adjust appropriately, although costs may increase. Industry has demonstrated the ability to economize in both water and electricity use.

- Overall sales volumes will be dictated by the economic environment, but buying patterns would be influenced in part by climate change, affecting the type of goods required from the manufacturing sector. For example, in the clothing industry, there may be a shift in demand from heavy woollen material towards light cotton clothing. Manufacturing and retail units which can predict and exploit relevant aspects of climate change will have a clear advantage.

- Impacts on agriculture and forestry, both in the UK and overseas, will have knock-on effects for industries which derive their raw materials from these sectors, affecting costs and possibly the continuity of supply. Increased transport costs at any stage in the movement of goods, from the factory supplier to the consumer, caused by strategies to limit emissions, may lead to an increase in retail prices.

- Changes in shopping habits may be expected, requiring changes in the design of shopping malls and precincts, and in opening hours. Technological advances allowing, for example, extensive adoption of electronic shopping, would make the shopping environment relatively invulnerable to the effects of climate change.

- Industry operates in a global market place. Whether or not climate change over the UK, in relation to the impacts of climate change elsewhere, would place the national industrial base at an advantage to international competitors, remains to be seen.

- Adaptation by industry is expected to be autonomous. The exception is the case of rising sea level. If the frequency and severity of surges increases, as the result of a combination of a rise in mean sea level and more frequent storm occurrence, industry located at the coast, along with other assets, will require protection. Re-location is not expected to be necessary over the time scales considered in this report.

10.1 INTRODUCTION AND BACKGROUND

For much of the developed world, the foundation of the economy lies in industry rather than agriculture. Even in the largely service-based, post-industrialized economy of the UK, 35% of output is from basic industry (construction and production in Figure 10.1), compared to only 2% from agriculture, forestry and fisheries. The specific areas of concern in this chapter are the manufacturing, retailing and service industries. Broadly, these correspond to 'Manufacturing' and to 'Distribution', hotels and catering, repairs' in official publications such as the *Monthly Digest of Statistics*. These two sectors contributed, respectively, 21% and 14% to GDP in 1993 (Central Statistical Office, 1995a).

Industry, whether the manufacturing or retail sector, is commonly perceived as 'high-tech', and relatively invulnerable to the effects of climate change. This assessment is not necessarily supported by analysis and, as noted, even a relatively small impact on the sector would have important implications for the national economies of the UK and other developed nations. However, as a result of the perception, the manufacturing and retail sectors are generally neglected in climate impacts studies, and the literature is small.

Impacts of climate change on industry must be seen in the context of the changing structure of the sector. By comparison with the 1993 figures for proportion of GDP given above, in 1963 the figures were 34% for manufacturing and 12% for the distributive trades (4% for agriculture, forestry and fisheries). In the decade of the 2050s, to which the 1996 CCIRG climate scenario relates, further changes must be expected, although the direction cannot be predicted. Important influences on present and future trends in industry include the following.

The role of the European Union. An increasing proportion of regulatory matters affecting business stem from decisions of the European Union. Issues such as monetary union, and moves to extend the free trade area to encompass eastern Europe and, more tentatively, western Asia, North Africa and North America, clearly have enormous implications for UK industry (British Chambers of Commerce, 1995a). These moves are particularly significant in the context of increasing globalization of trade, a

Figure 10.1: Contribution of sectors of business to UK output, 1990. Source: Central Statistical Office, 1995a.

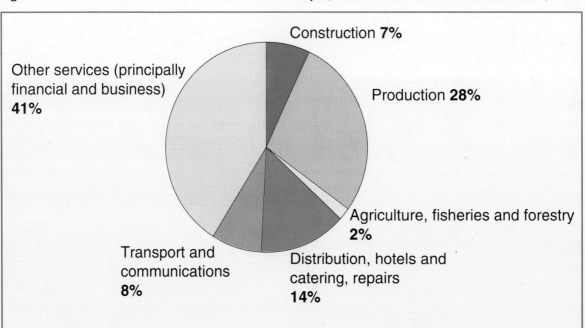

trend which is unlikely to change in the foreseeable future.

UK government policy. Within the UK, government fiscal policy is an important contributor to the environment in which industry operates. Capital projects for improvement of the infrastructure, favourable interest rates and the creation of incentives to invest are all important in creating an environment conducive to industrial development.

Environmental compliance. Small businesses alone in the UK currently spend around £1.85 billion on environmental compliance (British Chambers of Commerce, 1995b). In the future, increased pressure is to be expected from government and public opinion to limit emissions of pollutants, including greenhouse gases and ozone-depleting substances. Since energy consumption by industry and commerce (excluding transport) accounted for some 43% of UK CO_2 emissions in 1990 (HM Government, 1994), a substantial proportion of of the burden (and cost) of emissions reduction to meet targets under the Framework Convention on Climate Change is likely to fall on these sectors. Actions designed to limit emissions from transport clearly have 'knock-on' financial implications for industry at all levels. In 1990, transport accounted for around 24% of UK greenhouse-gas emissions, of which almost 30% was contributed by industrial and commercial road transport. Stated government strategies to limit transport emissions already include annual increases in fuel duties by at least 5% above the rate of inflation for the forseeable future.

Demographic and social trends. Within the workforce, the increasing presence of women, and the pressure for a shorter working week are just two factors which may require adjustment within industry. Levels of unemployment will affect the demand for goods, as indeed will changing social aspirations.

Changing technology and management practices within industry. New methods of communication, such as the Internet and electronic mail, must have important implications for the way in which industry works. These, combined with changes in management practices such as the move towards just-in-time delivery of components, suggest a transition towards smaller, more diversified, industrial units.

The manufacturing, retail and service industries are not homogeneous. A wide variety of activities are undertaken, with different sensitivities to the impacts of climate change. However, for the vast majority of individual industrial units, their sensitivities can be considered under one of three headings: supply, processing and market demand. Supply sensitivities exist through the requirement for raw materials, and reliable transport of these materials to the processing point. Direct sensitivies arise in many industries from the need for a reliable supply of electricity and water at a minimum price in order to carry out processing activities. Market demand fluctuations can be expected in some areas of consumer choice in response to climate change: for example, a shift away from woollen clothing towards light cotton materials. This model of sensitivities is adopted in the discussion which follows.

10.2 ESTIMATED EFFECTS OF CLIMATE CHANGE AND SEA LEVEL RISE

10.2.1 Sensitivity of manufacturing, retailing and services industries to weather and climate

The recent IPCC Working Group II Second Assessment Report contained a chapter on Energy, Industry and Transportation which set out the major areas of climate change sensitivity in these sectors (IPCC,1996b). This notes that the opportunities for autonomous adaptation within the sector are considerable, because of the short life time of industrial products in relation to the time scales of climate change. Industrial enterprises seen as particularly vulnerable to climate change are those reliant on agricultural and forestry products (including textiles) and the food industry.

Some idea of the potential impact of climate change on industry can be obtained from the few studies of the impact of climate variability. Palutikof (1983) showed that certain sectors of manufacturing are sensitive to the effects of extreme seasons. Various studies of the retailing sector have indicated that purchasing patterns are influenced by the weather (Weller and Prior, 1993; Ballantine, 1994; Agnew and Thornes, 1995), and this fact now forms the basis for consultancy services by, for example, the UK Meteorological Office (The Weather Initiative).

From the point of view of their impacts, there are differences between climate variability and climate change which indicate that care is required in extrapolating from the former to the latter. Under present-day conditions, for example, a rainfall deficit in a hot, dry summer may result in water shortages in industry. However, in response to climate change, water companies may be expected to invest in capital projects which will ensure a reliable supply during more frequent droughts. In this respect, extrapolation from the impacts of climate variability may be unwise. Conversely, temporary changes in consumer spending patterns during present-day hot, dry summers may become entrenched in a warmer climate, and are a good indicator of possible future trends.

The response of industry to climate change impacts cannot, in general, be inferred from responses to climate variability. An industrial enterprise may opt to absorb losses incurred due to what is seen as a 'one-off' extreme due to natural variability, whereas if it is perceived as being a result of climate change, a decision may be made to re-structure as necessary. For this reason, reliable information on the scale and rate of regional climate change is of vital importance if industry is to adapt successfully to a warmer climate (see Section 10.6.2).

Below, the small amount of evidence on climate variability impacts, and the responses to those impacts, is explored and used to extrapolate where possible to the case of climate change (see also Maunder, 1986, for a useful discussion). It is important to remember that, in the event of a climate change, variability will still occur around the new mean climate. This may be greater or less than at present, determined by the perturbation in the variance arising from the climate change. It can be assumed that industrial production will continue to respond to climate variability in the new environment.

10.2.2 Effects of climate change

Manufacturing
Vulnerabilities to climate change arise at all points along the chain from supply of raw materials through processing to demand for the finished goods.

Supply. Particularly clear vulnerabilities exist in the supply of raw materials derived from agriculture or forestry. For many crops, the increased length of the growing season and higher temperatures, combined with higher levels of atmospheric CO_2, may lead to higher yields. Factors leading to a reduction in yields, and therefore higher costs of industrial raw materials, include the possibility of an increase in the frequency and severity of water shortages, and of more losses due to disease and pest outbreaks (Harrington and Woiwod, 1995). The most obviously vulnerable industry is food processing. Timber-based manufacturing, such as the paper industry, and textiles, may also be affected. The contribution of these industries to GDP is shown in Table 10.1, and amounts in total to 7.7%.

British industry is heavily dependent on imports for its raw materials. Figures which are exactly appropriate for this study are difficult to obtain.

Table 10.1: Manufacturing industry with supply-based vulnerability to climate change, 1990

Industry	% of GDP
Food, drink and tobacco	3.1
Wood and wood products	0.7
Pulp, paper, printing and publishing	2.5
Textiles and leather	1.4
TOTAL	**7.7**

Source: Central Statistical Office, 1995c

Table 10.2 shows the value of imports of materials judged vulnerable to climate change. Of the first two categories (food and live animals, and beverages and tobacco) some unknown proportion will require no further processing and will be routed directly to the wholesaler. However, the final three categories are essentially raw materials imported as the basis for manufacture. They alone represent an expenditure of £2.7 billion, or 2% of all imports. Climate change in the country of origin can have important implications for costs of imported industrial raw materials. In some cases, there may be opportunities for replacement with home production, if the climate warms (see Chapter 5). Maize, for example, may be expected to become a reliable crop over a greater area of the UK, but this is not a viable option for crops from tropical latitudes: foodstuffs such as tea, coffee, cane sugar and tropical fruits, and industrial raw materials such as rubber. Reductions in rainfall amounts and reliability in these latitudes would lead to an increase in the cost of these raw materials (Rosenzweig and Parry, 1994).

The impacts of climate change on supply of raw materials to industry will be moderated by changes in economic and cultural factors. There is, for example, an increasing market for organically-grown fruit and vegetables. If the incidence of disease and pest outbreaks increases as a result of warmer conditions, it may not be possible for this market to survive in the face of rising prices.

Processing. Industry is a large consumer of water and electricity and, for both, requires a consistent supply at a minimum price (see Chapter 7 for impacts on water supply).

Table 10.3 shows the utilization of water supplies in the UK in 1993. By far the largest user is the electricity supply industry. However, direct abstractions by industry account for a further 8%, and some unknown additional amount will be taken from the public water supply. This table illustrates two points. First, the electricity industry must be vulnerable to water shortages, which in turn may threaten the electricity supply to industry in drought seasons. Second, because manufacturing industry is itself a large user of water, it is directly vulnerable to water shortages in drought seasons. Evidence for these effects was presented by Palutikof (1983), who studied the effect of extreme seasons on industrial production. For extreme summers (both hot and dry), and taking the example of 1975-76, a drop in production was recorded by the metal industries (both ferrous and non-ferrous), utilities and metal goods. These are large consumers of water in the production process, and it was suggested that water shortage was a contributing factor to the decline in production. Water consumption is not the only area of vulnerability: lower river flows will affect water quality, and hence the availability of discharge consents.

Table 10.2: Imports of raw materials vulnerable to climate change, 1990

Material	£m. value	% value all imports
Food and live animals	12,271	8.3
Beverages and tobacco	2235	1.5
Wood, lumber and cork	1400	0.9
Pulp and waste paper	638	0.4
Textile fibres	636	0.4
TOTAL	**17,180**	**11.5**

Source: Central Statistical Office, 1995d

Table 10.3: Actual abstractions from surface and ground water, 1993

Sector	Ml/day	%
Public water supply	16,651	32.3
Agriculture, including spray irrigation	303	0.6
Electricity supply industry	26,579	51.5
Other industry	3895	7.6
Mineral washing	198	0.4
Fish farming, cress growing, amenity ponds	3817	7.4
Private water supply	82	0.2
Other	92	0.2
Statistical error		-0.5
TOTAL	**51,618**	**100.0**

Source: Central Statistical Office, 1995b

The 1996 CCIRG scenario indicates that changes in mean summer temperature in the decade of the 2050s due to global warming will be greatest in the south-east of the UK: between 1.6 and 2°C. Mean summer precipitation is estimated to decrease in southern England, by up to 9%. These figures suggest that the incidence of hot, dry summers over southern England will increase, even in the absence of any change in variability, as a result of climate change. Manufacturing industry in the UK is overwhelmingly concentrated in the south-east of England, as shown in Table 10.4, and 31% of UK water usage by industry (excluding electricity generation) is accounted for by the Southern Water region (Central Statistical Office, 1995b). Thus, the regional patterns of climate changes over the UK indicated in Chapter 2 would tend to exacerbate the vulnerability of the sector to water shortage.

Manufacturing industry is also a large user of electricity. Table 10.5 shows energy costs as a percentage of total production costs. This is an up-date of Table 10.3 in the First CCIRG Report, and equivalent figures from that table are reproduced in the final column of Table 10.5. The table shows clearly that, although electricity supplies still formed a substantial proportion of total production costs in 1989 (the most recent year for which figures are available), the proportions are considerably lower than in 1984. This may be due to increased efficiency, or to an absolute increase in other costs such as labour and raw materials. Possible climate change impacts on the manufacturing sector are interruptions to the supply due to water shortages in the electricity industry in drought summers, and price increases caused by capital investment within the electricity industry to adapt to the new climate regime.

Table 10.4: Regional distribution of industry in the UK (measured as the percentage contribution to GDP within sector, 1993)

Region	Manufacturing	Distribution hotels and catering, repairs
North	6.3	4.3
Yorkshire & Humberside	9.3	8.6
East Midlands	9.0	6.7
East Anglia	3.6	3.6
South East	25.7	35.3
South West	6.9	8.4
West Midlands	11.2	8.4
North West	12.9	10.2
Wales	5.1	4.1
Scotland	8.0	8.3
Northern Ireland	2.0	2.1
TOTAL	**100.0**	**100.0**

Source: Central Statistical Office, 1995b

Table 10.5: Energy costs to industry in Great Britain

Industry	Total production costs (£m), 1989	Energy costs as % total production costs 1989	1984
Bricks and ceramics	1970	8.7	13.2
Cement, lime and plaster	3910	7.7	11.5
Glass and glassware	1650	7.3	13.2
Basic chemicals	5270	8.0	9.7
Paper and board	2540	8.8	10.2
Iron and steel	8260	5.6	7.9
Water supply	1450	10.8	11.4

Source: Department of Trade and Industry (1995)

Other areas where processing may be affected by climate change include, for example, days lost through sickness in the work force (see Chapter 15, Health). Uncomfortable work conditions due to excessively high temperatures will require more frequent attention.

Demand. Manufacturing industry must be responsive to changes in demand in both the home and export markets. Within the UK, changes in demand due to climate change can be expected to arise from the following:

i) Changes in the mean climate, particularly in temperature. Changes in consumer behaviour are dealt with in detail in the next section, and will clearly have implications for manufacturers. An increase in demand for air-conditioning units (see Chapter 8), refrigeration plant and refrigerated vehicles is to be expected.

ii) Changes in the occurrence of climatic extremes. Fewer very severe winters, as a result of climate change, would reduce the amount of disruption caused to industry by such events. Palutikof (1983) demonstrated a substantial drop in output from sectors such as brick and cement manufacture (linked to a lack of activity in the construction industry) in the severe winter of 1963. The reduction in the frequency of such events in a warmer climate should therefore be beneficial.

Exports amount to 30% of all UK sales from manufacturing industry (Central Statistical Office, 1995c). Any changes which affect our global competitiveness are therefore of significance to the national economy. We have already noted that industries dependent on raw materials derived from agriculture and forestry may be vulnerable to climate change impacts. However, these industries trade for their raw materials in the same global market place as their overseas competitors, and there is no reason to suppose that climate change will place UK manufacturers at an added disadvantage or advantage in this respect.

Disruption to external markets might arise as a result of extreme climate changes due to global warming. The purchasing power of these markets may be reduced by, for example, severe drought, leading in the worst case to unrest and problems of national security (Myers, 1993). Table 10.6 shows UK exports by final destination. In fact, the bulk of exports, by value, go to temperate latitude nations where the extent of

Table 10.6: UK exports by final destination, 1992

Region	% total value
European Community	55.9
Other Western Europe	7.9
North America	12.9
Other OECD	3.6
Oil-exporting countries	5.5
Eastern Europe and former USSR	1.6
Other countries	11.7
Low value trade	0.9
TOTAL	**100.0**

Source: Central Statistical Office, 1995c

climate change impacts can be expected to be no more disruptive than in the UK.

Transport and communications. The movement of raw materials and finished products to and from the factory will be subject to the impacts of climate change on transport networks (see Chapter 12). The increasing move towards just-in-time manufacturing systems will increase the vulnerability of the sector to any disruption of transport by severe weather events.

If the frequency of storms such as those experienced in October 1987 and January-February 1990 were to increase as a consequence of climate change, disruption of communications would become more frequent, with consequent negative impacts on supply lines and markets. Chapter 2 states that such an increase is tentatively suggested by the limited model data available. Conversely, a reduction in winter severity would have a positive impact on transport efficiency and reliability (see Chapter 12).

Retailing

Sales patterns. The sector 'wholesale and retail trade, repairs' contributed 11% of GDP in 1990 (Central Statistical Office, 1995c). The breakdown of sales from retail outlets in 1990 is shown in Table 10.7. Retailers of food, clothing and footwear, and household goods constitute 64% of the total. These are the areas which may be affected by changes in consumer spending patterns in response to climate change.

Table 10.7: Sales from retail outlets, 1990

Retail trades	Sales, £m.	% total
Food retailers	47517	36.7
Clothing and footwear	12752	9.9
Household goods (inc. hire and repair)	22473	17.4
Other non-food (inc. off licences, newsagents and tobacconists)	24492	18.9
Mixed retail	22090	17.1
TOTAL	129324	100.0

Source: Central Statistical Office, 1995c

Certain types of goods within a specific category such as foodstuffs or clothing appeal more to the shopper under certain weather conditions (Steele, 1951). Some at least of these swings in consumer behaviour in response to short-term variability (Agnew and Thornes, 1995; Maunder, 1986) may become entrenched under conditions of longer-term climate change. Warmer and longer summers would lead, for example, to increased consumption of 'light' foods such as salads and probably to a higher consumption of beer (Zeisel, 1950), soft drinks (Weller and Prior, 1993) and ice cream. In the clothing sector, we could predict a shift from woollen and synthetic fabrics to light cotton materials (Linden, 1962). Furnishing can be expected to shift away from fitted carpets and heavy upholstery towards tiled flooring and a greater use of exposed wood. Sales of garden furniture should increase.

The above remarks apply in the situation of 'other things being equal', in particular overall economic trends. Most probably, the situation is one where the overall volume of sales will be dictated by the economic environment, but the pattern of sales will be influenced in part by changes in the climate.

Shopping patterns. Shopping is now described as a major hobby by the British public. It is interesting to speculate whether this may change as a result of global warming. An improvement in winter weather might lead to an increase in retail activity in these months, where previously people were deterred by unpleasant weather. Summer shopping, conversely, may become less popular, or shift to air conditioned shopping malls and away from open air centres.

Services

Substantial impacts on service industries are difficult to visualize. Minor impacts, which can be easily accommodated by the sector, may be suggested. However, the greatest impact is likely to be felt in catering. This sector lies at the end of the chain linking raw agricultural materials to the finished foodstuff. Costs incurred at all points along the chain due to climate change will accumulate in the catering sector and are likely to be passed on to the customer. Other impacts include a possible increased demand for outdoor eating establishments. The hygiene standards for restaurants and cafes may require up-grading.

The profile of repairs may change. Fewer repairs will be required to equipment which is used less frequently; for example, central heating systems, and white goods such as dishwashers and washing machines, the use of which may be restricted by water shortages.

The summer of 1995

Some of the hypotheses suggested above can be tested against events in the summer of 1995. Although June was not exceptional, July was the third warmest on record, and August the hottest ever (Dukes and Eden, in press).

Statistics are not yet available for a full quantitative analysis of the economic impacts of the 1995 summer, but anecdotal reporting by the media has been widespread. This has included reports of increased sales of soft drinks and ice cream, a fall in sales and repairs of lawn mowers, and increased use of swimming pools at the expense of squash courts. Many of these reports concern a shift within sectors; for example, an increase in ice cream sales will be difficult to detect in official statistics if it is accompanied by a parallel decline in sales of chocolate, which are counted within the same sector. However, as noted above, we may expect absolute declines in some sectors in the summer season, as shopping becomes a less popular activity in hot weather. This would include purchases which can be deferred until conditions become more pleasant.

Data on retail sales is available to the end of August at the time of writing (Central Statistical Office, 1995d). Figure 10.2 shows the seasonally-adjusted volume of sales in predominantly non-food stores (where purchases may be deferred), expressed as an index relative to a

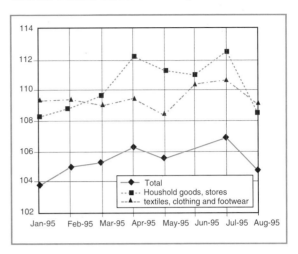

Figure 10.2: Volume of retail sales in predominantly non-food stores (1990=100). Source: Central Statistical Office, 1995d.

value of 100 in 1990. There is a clear decline in August, particularly noticeable in household goods and stores. It is probable, but by no means certain, that some part of this decline is due to the very hot weather of August 1995.

10.2.3 Effects of sea level rise

According to the first CCIRG report, Chapter 10, some 40% of UK manufacturing industry is located along coastlines and estuaries. It was stated that industry could probably absorb the costs of a rate of sea level rise of up to about 5cm per decade. The scenarios in Chapter 2 of this report indicate a rate of change of approximately this magnitude. This is unchanged over the projections in the first report.

It is important to note, however, that the figure of 5cm per decade relates only to the change in mean sea level. As discussed in Chapter 17, such an increase in the mean would be amplified in the frequency of extreme water levels caused by storm surges. Flooding of coastal installations would occur more often, and the area at risk would be more extensive. Any tendency towards a greater frequency or severity of storms (see Chapter 2) will further increase this risk.

The immediate impact on industry of this scenario would be disruption of work arising from:

- Inundation of the plant,
- Disruption of power transmission lines and water supply; and/or,
- Disruption of transport routes to and from the workplace.

Longer-term problems include structural damage to buildings and equipment caused by flooding. Salinization as a result of rising sea levels could cause contamination of the water supply of those industrial enterprises which abstract their supplies directly from the groundwater.

10.3 ASSESSMENT OF POTENTIAL ADAPTATION

We may expect adaptation by industry to climate change to be primarily autonomous; that is, without the need for specific policy intervention. The lifetimes of most assets are short in comparison to the projected timescales of climate change, allowing the time and the opportunity for adaptation within the sector. On balance, impacts on industry are likely to be beneficial. Disruption due to severe winters will be reduced. New market and sales opportunities will be created.

The largest impacts on the manufacturing and retailing sectors can be expected through changes in consumer demand. Although an absolute increase in demand may be expected for air conditioning and refrigeration equipment, in general the situation is one where the overall volume of sales will be dictated by the economic environment, but the pattern of sales will be influenced in part by changes in the climate. Expected relative changes in purchasing patterns include:

- An expansion of sales in light summer foods such as salads and cold meats, affecting the food processing industry and food retailers.
- An increase in demand for lightweight natural, particularly cotton-based, fabrics within the clothing industry, and a shift away from synthetic and heavy woollen materials.
- Within the furniture sector, a reduction in demand for carpeting and heavy upholstery fabrics, and an increase in demand for wooden furniture and tiling for floors.

Changes in buying patterns may affect the profitability of certain enterprises, and require some shift in emphasis, both by manufacturers and retailers. Changes in consumer preferences occur very rapidly, and industry must be alert and responsive. However, they present opportunities which can be fully exploited through flexible and adaptable management. Although certain shopping habits in southern Europe are culturally determined, businesses in the UK would profit by an examination of possible climate-influenced buying patterns. In summary, businesses should be aware of the possible implications of future climate change and, where possible, consider them in the decision-making process. Manufacturers and retailers who are able to predict and exploit these trends will have a clear advantage.

Within the manufacturing sector, possible impacts on supplies to agricultural and forestry-based processing industries were noted. There must be both winners and losers as a result of climate change, and it is most likely that alternative sources of supply will emerge.

Manufacturing industry relies on reliable supplies of water and electricity and, for some processes, consumption is high and forms a substantial proportion of manufacturing cost. The water industry can be expected to adjust to an increased demand, which may have to be supplied from a smaller regional resource. To maintain the current quantity and reliability of supply under a changed climate may require substantial capital investment by the water industry (see Chapter 7), leading in turn to higher prices. Manufacturers may choose to accommodate this by economies of use, by absorbing the additional cost, or by passing the added cost on to the consumer (which may affect global competitiveness). Restrictions on the discharge of pollutants, arising from low river flows, is likely to be dealt with through pollution prevention. In summary, the most likely industrial response to prolonged water shortage would probably be efficiency of use. The more extreme option of re-location to wetter areas of the country is unlikely to be required: note that the scenarios in Chapter 2 indicate wetter winters for the whole of the UK and, in summer, an increase in rainfall in the north, a decrease in the south.

Electricity generation is itself a large user of water, and it is possible that the reliability of the supply will be affected in hotter drier summers. Overall, however, consumption should decline due to warmer winters (see Chapter 8) and therefore it appears unlikely that prices will increase as a result of global warming. Table 10.5 suggests that economies of electricity use are available to the manufacturing sector.

Adaptations to the shopping environment may be required in response to climate change. Physical discomfort whilst shopping can be mitigated by installation of air conditioning. Air conditioning is increasingly common in large retail outlets, and is increasingly expected by the buying public. This prospect, combined with the reasonable expectation of warmer summers in the future, means that any retail outlet which wishes to remain competitive should regard the installation and operation of air conditioning as a high priority. Shopping malls already provide a controlled environment, but design procedures may need to take account of the need for more shade, lower densities and more efficient air circulation. A change in opening hours would take advantage of cooler evening conditions.

The exception to the likelihood of autonomous adaptation lies in the case of rising sea levels. If the frequency and severity of surges increases, as a result of a combination of a rise in mean sea level and more frequent storm occurrence, industry located at the coast, along with other assets, will require protection. Coastal protection and flood defence policy are the responsibility of central and local government (see Chapter 17). In the absence of adequate protection, the alternative may ultimately be re-location. This would be extremely expensive, involving not only the physical plant, but also the workforce. It is unlikely that the costs of such an extreme measure could be borne by the sector alone.

10.4 UNCERTAINTIES AND UNKNOWNS

The ability for autonomous adaptation within a framework of slowly-evolving climate change has been noted. However, this does not take into account the possibilities that:

- Climate change occurs more rapidly than currently estimated.
- The frequency and severity of extreme events, particularly wind storm, storm surges and hot/dry summers, increases more than is estimated from the anticipated perturbation in the mean climate.

Any impacts upon industry due to climate change take place in the context of technological, economic and social trends. These ultimately dictate the success, or otherwise, of manufacturing and retailing. Attempting to forecast future trends, and how these may affect the response of industry to climate change, is beyond the scope of this chapter. One example is the advent of electronic shopping which, if taken up widely, will render obsolete many of the comments made here about the shopping environment.

More so than any of the activities considered by CCIRG, except finance and insurance (Chapters 12 and 13), industry operates in a global context. Whether or not climate change over the UK will give the national industrial base an advantage over international competitors, who will have their own experience of climate change, remains to be seen.

10.5 PRINCIPAL IMPLICATIONS FOR OTHER SECTORS AND THE UK ECONOMY

Industry is a major contributor to GDP and to exports. Any major disruption to the sector would have wide economic implications leading, for example, to rising unemployment. However, the principal conclusion of this chapter is that, provided climate change takes place slowly, in relation to the lifetime of industrial assets, and is not accompanied by a substantial increase in extreme events, the sector should be able to accommodate the impacts and even turn them to advantage. This should have benefits for the wider economy.

10.6 RESEARCH AND POLICY ISSUES

10.6.1 Future research effort

Reliable information at the regional scale on the rate of climate change, and changes in the occurrence of extreme events due to global warming, is essential for full evaluation of the potential impacts for industry. In the near future this information is unlikely to be available, which negates any attempt to carry out a full impact assessment, and the following suggestions should be interpreted in this light.

Industry relies on supplies from sectors which are directly affected by climate change. In particular, reliable supplies of electricity and water at minimum cost are required. Assessment of the extent to which these sectors are vulnerable to climate change would be useful. Research into opportunities for industrial economies in water and power use has the advantage of generating savings even in the absence of any climate change impacts.

Industries based on agricultural and forestry produce are vulnerable to disruption in their supply base as a result of climate change. Research on the potential impacts of climate change on agro-forestry products, both home and overseas, would allow these industries to plan their future resource base.

10.6.2 Policy issues

Adaptation to climate change within industry is likely to take place without the need for policy intervention. This adaptation will proceed more smoothly where reliable information is available on the scale and rate of regional climate change. The information would permit accommodation of the predicted changes, and adaptation to take advantage of new marketing opportunities. Government has a role to play in the provision of this information, as indeed do organizations such as the Confederation of British Industry and the British Chambers of Commerce.

Some 40% of UK manufacturing industry is located along coastlines and estuaries. Sea level rise at a rate greater than that suggested in Chapter 2, and/or a substantial increase in storm surge occurrence, would threaten the security of coastal industry. Financing of coastal protection, and decisions regarding the feasibility of continuing coastal protection in a situation of rising sea levels, can only be undertaken by government. Timely planning, decision-making, and information transmission will ensure an orderly transition by industry and all other affected sectors of the economy.

11. Construction Industry

SUMMARY

- The most important climate change impacts on buildings and other types of construction are likely to arise from higher summer temperatures, increased winter rainfall, and, in the north of the UK, increased extreme winds. If sea defences are breached, the damage to construction in the affected areas will be very considerable.

- Soil moisture movements on shrinkable clays will increase, so more careful attention to foundation design will be needed in vulnerable areas.

- Thermal conditions in winter are likely to improve. The energy needed for space heating in fully heated buildings is likely to decline by around 20%. The lower mean winter indoor-outdoor temperature differential will favour greater use of passive solar energy in buildings.

- Summer conditions in large towns and cities, in the southeast of the UK especially, will become less acceptable as there will be more hot days, and the urban heat island will make conditions worse. Pressure to expand the use of air conditioning will mount. This pressure can to some extent be resisted through climatically sensitive design using natural cooling techniques. However, urban traffic noise and pollution will continue to make it difficult to find acceptable solutions in town centres.

- The life of buildings now being erected will extend beyond 2059, the end year of the climate change scenario period in this report. New construction should be designed with the probable climate of the decade 2050/59 in mind. However, clients and their designers do not presently have the appropriate information to approach the implicit problems objectively in terms of costs and benefits.

- The construction industry can respond to the expected changes by providing adaptation at the rate required. However, the actual outcome will depend on the attitudes of building owners to adaptation. They will need to be persuaded that the the expected long term benefits are greater than the costs of adaptation.

- The periodic major refurbishments of existing buildings will provide suitable opportunities to respond appropriately to climate change needs. Change will only happen if building owners understand the need for appropriate action. Such decision making needs to lie within a structure of cost benefit analysis concerning climate change and construction. This structure does not yet exist. The approach will need to include consideration of future insurance costs for construction.

11.1 INTRODUCTION AND BACKGROUND

This chapter considers the potential impacts of expected climate change on existing and future construction in the UK. The built environment is designed, erected, maintained and renovated for a wide range of owners by the construction industry. The ownership of this vast capital resource of long-life artefacts does not rest with the construction industry. It follows that the present owners of existing constructions and the future owners of new planned constructions must be the key decision makers in the process of adapting construction to climate change. The challenge is to help construction owners make appropriate decisions with the aid of sound professional advice. It is an issue of safeguarding the main capital wealth of the nation. The chapter also addresses the potential impacts of climate change on the construction industry *per se* as a service industry.

The majority of buildings are set in urban environments, but urban climates differ from rural climates (World Meteorological Organisation, 1984; Page, 1992). The climates of towns are appreciably warmer and more polluted than the climates of their surrounding countrysides. The larger the town, the bigger the climatic modifications tend to be (Oke, 1973). Climate change may therefore, impact particularly adversely on people living and working in large cities in the southeast of the UK where temperatures are currently the highest, and where temperature increases are expected to be greatest.

This chapter specifically concentrates on construction as artefacts. Most existing construction and practically all new construction will still be in use at the end of the 1996 CCIRG scenario period of 2050/59. We are therefore addressing a current issue and not a future one.

The construction industry provides buildings, the tracks for road and rail systems, transport interchanges, e.g. airports, rail and bus stations and ports, energy systems, e.g. power stations and power lines, underground tracks for information technology systems, gas systems, sea defences, river works, dams, off shore platforms, etc. The term *buildings* will be used to describe all these constructed artefacts to avoid confusion with construction per se as an industrial process. The construction process invariably involves interaction with the ground, which is often wet or even flooded. In fact the finished artefact may lie entirely below the ground surface. The constuction activity is mostly carried out in the outdoor climate. The completed product is set in the outdoor climate and needs to be durable and secure against long term climatic risk. Construction is also an industry guided by many regulations, including environmental ones, supported additionally by many standard codes of practice, many of which have important climatically based components, for example the Code of Practice on Wind Loading (British Standards Institute, 1972 [BSI]), and the Standards on Control of condensation in buildings (BSI,1989) and assessing the exposure of walls to driving rain (BSI, 1992). This regulatory environment must be properly tuned to the evolving climatic change situation.

There are five classes of problem to address in considering the direct impact of climate change on construction:

- How to manage existing constructed artefacts better to adapt their day-to-day operations to climate change.
- How to modify existing constructed artefacts to adapt them better to climate change.
- How to design new construction to make it more suitable for expected future changed climatic conditions.
- How to ensure an appropriate range of building materials and components are available to meet the emergent new design needs.

and, in the longer run:

- How to achieve more efficient production on site in the face of climate change.

There is also the strategic issue of the probable future balance of work in different construction industry sectors consequent on climate change. For example, sea level rise might demand significant civil engineering investment in production facilities.

11.2 ESTIMATED EFFECTS OF CLIMATE CHANGE AND SEA LEVEL RISE

11.2.1 Sensitivity of the construction industry to climate and weather

The sensitivity of construction to climate and weather can be assessed under three main headings:

- Impacts on people in the indoor environment.
- Impacts on the basic building fabric.
- Impacts on the construction process.

The most important sensitivity issue is how will human beings, living and working in buildings, be affected by climate change. The basic demand from people in buildings is that they should be comfortable and healthy. Of greatest importance will be the sensitivity to increased summer overheating of buildings. The frequency of this risk will increase with global warming, especially in cities. This will affect human performance, and, for more vulnerable groups, their health.

The key climatic issue concerning the structure and its cladding is that the construction should be safe when exposed to anticipated climatic extremes. In most areas of the UK, high wind is the principle risk (Buller, 1993a; Buller, 1993b), although on low lying sites near rivers, flooding may also be a critical risk. In upland areas and in Scotland snow loading can be significant. Building foundations on clay soils may be vulnerable to excessive shrinkage (Hunt, et al., 1991). This sensitivity is reflected in substantially increased insurance claims for repair and underpinning in dry summers following dry winters in sensitive areas, e.g. the summer of 1976. This issue was dealt with at some length in the first CCIRG report (1991). Sensitivity assessments of risk always must be site based.

The fabric must not only to be able to withstand the imposed extreme structural loads when first built, but must also be durable under long term exposure to sunlight, rain, snow, ice and frost and other temperature changes. Durability must be assured over very many years. There is also the risk of biological attack by pests. The biological attack path may change as the outdoor climatic environment becomes more favourable to the harbouring of specific pests, for example the House Longhorn Beetle (Building Research Establishment, 1994 [BRE]).

Deterioration is usually dependent on a combination of weather factors, e.g. driving rain (BSI 1992) involving wind and rain, frost damage involving freeze thaw cycles and sequences of rainfall. Acid rain driven onto building surfaces by the wind causes considerable building damage (Building Effects Review Group, 1989 [BERG]; CSERGE, 1993). Condensation, which is dependent on vapour pressure, air temperature, wind and rain, remains an especially serious problem in the UK, with some 30% or so of the UK housing stock still adversely affected (BSI, 1989). The sensitivity assessment of damage will usually require the study of combinations of future weather elements rather than a single weather element in isolation such as temperature or rain. Such combined assessments for construction have yet to be made in the context of climate change.

Production on building sites is affected by weather, and construction site management continually addresses this issue day by day. The greatest interruption to work in the UK tends to be due to rainfall, especially if accompanied by high winds (BSI, 1992). Wales, N. Ireland, Scotland and northwest England are most affected. Snow is less frequent in low lying areas in England and Wales, especially in the south, but can produce serious interruptions when it does occur. New building work is vulnerable to frost, and often needs initial protection. In dry weather control of dust on site can become very troublesome. The state of ground is important, especially at the start of construction, because equipment can easily get bogged down. Hard ground is less easy to work, and is more liable to produce excessive dust.

Finally, construction design is also sensitive to changes in governmental policies to control climate change through the abatement of atmospheric emissions. The debate continues as to how much should be attempted through market forces and how much through governmental regulation. The industry is affected by uncertainty about policy in these sectors.

11.2.2 Effects of climate change

Indoor thermal comfort, winter and summer

Under the 1996 CCIRG scenario mean winter temperature in the decade 2020/29 is expected to have risen by 1°C in the east of England and by 0.4°C in the northwest of Scotland. The corresponding expected increases by the decade 2050/59 are 1.8°C and 0.8°C. (See Figure 2.2). The decrease in heating degree days by the 2050s is about 20% (Table 2.4). This will give considerable financial savings in heating costs. The future winter wind speed is anticipated to be higher, especially in southern Britain. This will increase the background ventilation rate in buildings, especially in poorly weather stripped buildings on exposed sites. This will offset some of the benefits of temperature rise.

In summer in southeast England, the mean temperature is expected to rise by about 1.0°C by the 2020s and by about 1.7°C by the 2050s. This increase in the mean temperature produces a strong increase in the number of hot days above 25°C. (Table 2.4). The mean annual number of days with T_{max} above 25°C is expected to double inland in the south of England (Figure 2.10).

This increase will be associated with a small rise in solar radiation in the southern part of the UK. This will make inappropriately designed buildings unacceptably hot quite frequently in summer in the southern part of the country. Under the 1996 CCIRG scenario the number of cooling degree days also increases (Table 2.4). This implies a more extended period of operation of air conditioning plant in future with a rise in associated costs.

Figure 11.1 illustrates the temperature rise/comfort issue. The challenge is to cope with the external annual range of temperature effectively and economically. At the top, this diagram suggests that, at present, relatively few days in the UK are too hot for comfort out of doors in the shade. The box immediately below shows the typical current associated indoor temperature range. The lower pairs of boxes illustrate the combined effect of the current heat island and macroclimatic change. Future climate change will increase hot weather stress in buildings, unless appropriate off setting measures are taken.

Air conditioning demands

The demand for air conditioning will increase. In a free running naturally ventilated building without air conditioning, the mean daily indoor air temperature will typically be a few degrees above outside air temperature, especially in situations where there are considerable internal heat gains from equipment and lights. This internal rise above the ambient will make it more difficult to accept an increase in outdoor temperature without air-conditioning. While there have been recent strong pressures to reduce the amount of air conditioning in new buildings in the UK, and an encouraging increase in the number of properly designed natural ventilation systems, it seems unlikely that an expansion in the amount of air conditioning used in the UK can be resisted. The continuing deterioration of the quality of air in cities and the increasing levels of noise due to traffic growth (Royal Commission on Environmental Pollution, 1994) will

Figure 11.1: The annual range of temperature outdoors and in heated buildings without summer cooling, shown diagrammatically. *Top boxes* at present. *Middle boxes* after climate change without heat island. *Bottom Boxes* after climate change with an urban heat island. Indoor temperatures at times are set below the comfort temperature, e.g. after working hours or during the night in houses. In summer they may exceed the comfort level. The design challenge will be to reduce the maximum indoor temperatures through improved shading, better ventilation, indoor energy control and appropriate choice of building mass (Note: the Oxford scenario data are not scaled)

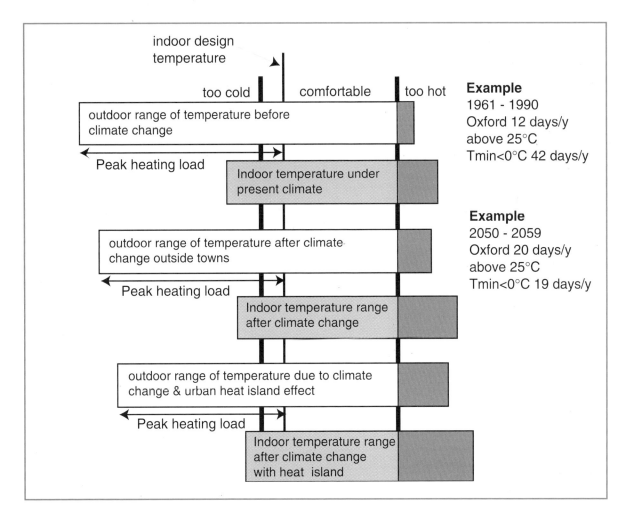

continue to make the design of acceptable natural ventilation systems in cities difficult.

Failure risks associated with wind

Wind loading is a significant cause of building failure. Such failures may result in loss of life, both within buildings and around buildings (Cook, 1985). The wind speed recurrence period information only extends out to 5 years in the 1996 CCIRG scenario, but a 50 year recurrence period is typically used in construction design. The scenario predictions are also based on mean daily wind speeds, and not on the maximum hourly or maximum gust speeds normally used in construction design. In the southern part of the UK there seems to be little difference anticipated at the 5 year return period level, but in the north an increase of some 15% is indicated in winter. Since the load on the building increases as the square of the wind speed, i.e. an increase of 34%, this must be viewed as an issue of concern to the construction industry. It could imply the revision of loading codes. Better siting and layout of development can significantly reduce wind loading risks and can be used to counter increased extreme wind speeds (BRE, 1990).

Risks associated with soil movements in shrinkable clays

As discussed in the 1991 CCIRG report, there are large areas of the UK where the drying out of clay soils in summer is responsible for serious building damage (Boden and Driscoll, 1987). These areas are well mapped. The southeast of the UK is most affected. Damage is especially widespread if a dry summer follows a dry winter. In addition to climate impacts, tree roots are often responsible for the extraction of large amounts of water from around foundations, so, in such areas, the use of trees for shading to counter warming must be carefully considered against the increased risks to foundations. The change in potential evapotranspiration (PE) has a strong bearing on the change in risk. Figure 2.13 reveals a strong gradient in summer PE from the Scottish border (no change in the 2020s) to the south coast (up 40% in the 2050s). Thus the greatest changes are likely to be experienced in the already most vulnerable southeast areas.

A much better understanding of the predicted changes in soil moisture content now exists than was available at the time of the 1991 CCIRG report (Chapter 3). The recent predictions suggest that soil shrinkage risks may become greater in southern England than was thought in 1991. An interesting development for the construction industry, since the first CCIRG report, is the fact that many insurance companies are assessing risk of subsidence damage within their premium structures on a postal code basis.

Risks associated with storms, including flash storms, soil erosion and river flooding

The issue of river flooding is discussed in Chapter 8 (Water Resources) in this report. The construction industry will need to take careful note of the statement that *'thus, a small change in flood frequency could lead to major changes in the frequency with which certain critical thresholds, or design standards, are passed'*.

The predicted increase in intensity of summer rainfall in northern England could have important implications for site water drainage. Chapter 2 states that, in that region, a 20 mm daily total could occur almost once a year by the 2050s compared to once every four years under the current climate. This has important storm water and sewerage design implications. Traditional design criteria may need revision.

Chapter 3 on soils in this report has drawn attention to the significantly increased risks of soil erosion. The foundations of buildings may be undermined if they are not protected from soil erosion. Large paved areas around buildings can produce very substantial locally erosive flows. An organisation such as the Building Research Establishment should be able to produce suitable guidelines on the handling of future soil erosion risks on building sites.

Condensation in buildings

There is a need to assess the probable impacts of climate change on condensation in buildings. The wetter, windier winters expected by the 2050s seem likely to increase the risks. Vapour pressure is the key humidity variable in making climate change assessments concerning condensation, but wind, rain and temperature changes also must be be considered. Unfortunately the 1996 CCIRG scenario only gives information about relative humidity changes. However, the winter vapour pressure must be assessed to rise in Scotland because higher temperatures combine with higher relative humidities. Scotland is already a high risk area for condensation.

Climate change and the durability of building materials

The durability of many building materials is affected by frost, especially freeze and thaw cycles in humid conditions. It seems likely there will be less deterioration due to these causes, if winter temperatures rise.

Higher temperatures in the colder parts of the UK would significantly reduce the need for the air entrainment of concrete without increasing the risk of freeze thaw damage.

Higher summer temperatures and increased global radiation will affect thermoplastic materials such as bitumen and asphalt, and specifications may need to be changed.

An assessment is needed of the probable impacts of greenhouse gas driven climate change on the effects of acid rain and low level ozone. Acid rain and acid gas deposition currently cause a great deal of economic damage to buildings, causing corrosion of metals and deterioration of protective films like paints (CSERGE, 1993). Low level ozone is a powerful damaging oxidant. The higher radiation levels expected in southeastern Britain, acting in association with the higher expected air temperatures, will enhance the photochemical formation of smog. The global climate change factors therefore seem likely to increase the amount of photochemical pollutants formed from the expected increases in vehicular pollutants, and hence to increased damage to buildings as well as presenting a rising threat to health.

Raised carbon dioxide concentrations will have a direct effect in increasing the rate of concrete carbonation and thus increase the rate of corrosion of steel reinforcement.

Future conditions on site

The greater interruption of work on site due to the greater frequency of days of heavy rainfall in the north in summer will probably be the most significant impact. The reduced incidence of snow and frost will cut down frost damage and shorten interruptions to work schedules on site. However, the predicted increase in winter wind speed in England will be less welcome, affecting, for example, crane-based operations. (Figure 2.11). The winter increase in precipitation will not be welcome on site. In summer the south of England will be drier, but in the already wet areas of the UK like Scotland, even more rain will not help construction on site, nor will the greater number of rain days in Scotland be welcome (Figure 2.7). The findings of Chapter 3 on soils are likely to prove of particular interest to the construction industry in assessing future soil conditions. The changes in the summer potential evapotranspiration (PE) (Figure 2.13) imply that the ground in southeast England will be harder to dig for a significant part of the year.

11.2.3 Effects of sea level rise

The construction risks linked to sea level rise were discussed in the previous CCIRG report (CCIRG, 1991). The greatest risk to unfavourably located construction would be inundation due to the overtopping resulting from sea level rise in combination with storm surges. Saline intrusion in coastal areas may also occur due to diffusion through the subsoil as well as through the over topping of sea defences. Such saline intrusion will have adverse effects on the underground elements of buildings in the affected areas. The salts would also tend to rise up the wall in situations with no damp proof protection (BRE, 1973).

Some local authorities have in recent years allowed additional building in coastal areas already subject to significant risks due to sea level surges overtopping sea defences under storm surge conditions. Serious but limited damage has occurred. Not only has there been the initial damage due to salt water flooding, but also the continuing damage due to the subsequent saline environment of the foundations. This is liable to cause considerable additional damage. These saline pollution problems remain long after the over topped water has disappeared.

Any national policy for enlarged and/or substantially improved sea level defences would create considerable construction opportunities. However the underlying policy issues concerning such decisions are discussed in Chapter 17.

11.3 ASSESSMENT OF POTENTIAL ADAPTATION

Buildings have always had to adapt to climate, and there are no fundamental technological obstacles to adaptation at the rate of climate change envisaged. The issues are primarily economic. Will the benefits of adaptation exceed the costs within an acceptable period of time? Building owners will not act until they are convinced of an economic benefit. They will remain unlikely to be convinced of benefits in the absence of economically based objective assessments of the impacts of climate change on construction. These are lacking at the moment.

There is an important distinction between the evolving action on adaptation needed in existing buildings, the majority of which will be still standing in the 2050s, and action needed now on new construction that will last well beyond the 2050s. There are, therefore, two climate change constituencies to address: owners of existing buildings and clients commissioning new construction. The responsibility for guiding them rests with the building design professions. They should be the key advocates for realistic adaptation, but their advice must have a sound economic foundation, and be based on competent knowledge of the assessed risks.

Adaptation of existing buildings to climate change

Climate change could undermine the soundness of the original design decisions made at the time of construction. Building owners therefore need to be made more aware of the risks they might face, if they fail to adapt appropriately to the changed levels of climatic risk. The changed risk may be environmental or structural. Significant damage risk reduction is often achievable very cheaply, once the problem is properly identified. However, building owners face difficulties in making such assessments unaided. Their assumption tends to be that the initial design was competent, because it achieved building regulation approval. However, this assumption may not be sound, even without climate change. For example, the Building Research Establishment (Buller, 1993b) has found that much climatic damage from wind is replaced by similar construction to that which existed before the damage. The same failure may reoccur again for the same reasons.

While buildings have very long lives, most buildings undergo very radical refits every 25 to 30 years. This provides a window of opportunity for more radical adaptation to meet climate change needs. Objective guidance will be needed on priorities, and on the cost and environmental effectiveness of different approaches. For example, the Building Research Establishment has developed BREEAM, a method for assessing the environmental impact of buildings, including some types of existing buildings (Birtles, et al., 1994). Many building owners, for example some property companies, have found it a useful framework within which to guide their environmental decision-making in an accredited way.

Future building and climate change

We are currently designing buildings and other new physical infrastructure to last well beyond the 2050s. The industry needs an objective view of the implicit risks and opportunities. For a start, the construction industry needs to be provided with better tools to make such objective assessments possible. Such long range climate change data must be made available in forms applicable to current construction design decision making. Methods for assessing urban microclimates need to be added to make the tools useful in the context of urbanism. (See Commission of the European Communities, 1992)

Construction designers need to make long term risk assessments concerning performance and safety. They need long range climate scenarios that relate specifically to their decision making processes. They seek professional legal protection through designing to agreed Codes of Practice and Standards. It is essential that such codes reflect the long range issues needing to be addressed as a consequence of climate change.

Building Regulations are formulated by central government but implemented locally. While they provide an important governmental tool for directing building policy, e.g. the current regulations setting minimum building insulation standards, such regulations set the minimum acceptable rather than the desirable standards. Competent design can be significantly better. Codes and regulations should be reformulated to foster realistic adaptation to climate change needs in forms that do not inhibit innovation, either in design thinking or in product development.

Computer simulation of building performance at the design stage is being more and more widely used and is totally dependent on reliable climate data inputs. It could become a key tool for assessing future adaptation effectiveness, but the data in this new tool must first be created from the Hadley Centre data bases.

Contractors work in real time weather. They will be able to adapt continuously as conditions change. However, they too face issues of long term liability for any shortcomings in meeting their original contractual obligations. So they too must assess future risk carefully.

Specific aspects of adaptation

There are two basic response strategies open:

- To make the necessary adaptations at the time of initial construction.
- To make provision for enabling adaptation to be made economically at some later date when the need for it is demonstrated in use.

For economic reasons the industry is likely to take the second route.

Three key adaptations will be needed:

- Adaptation to reduce energy demands.
- Adaptation to respond better to the greater anticipated frequency of unacceptably hot conditions in buildings.
- Adaptation to reduce construction risks due to climatic extremes.

All three have design and construction aspects. The first issue is not part of CCIRG's remit. However, reducing internal energy demands in summer helps in the second adaptation.

Hot weather adaptation may be fabric-based, i.e. shading and improved natural ventilation, or equipment-based, i.e. air conditioning, and forced ventilation. The most energy efficient adaptations are fabric-based. The effective design of natural ventilation systems was neglected for many years. Designers are now showing increasing interest in the improved design of natural ventilation systems in buildings. This approach is also being coupled /with increased use of night cooling of structures in hot weather, for example using cooling ducts cast into the structural floor (Burn, 1995).

Improved summer shading will prove an important adaptation, but it must be designed on sound thermal principles. This shading is best fitted external to the glazing. New, more effective shading equipment and shading control systems are needed. The improved use of daylight is also important to counter excessive internal heat outputs and costs from artificial light sources. The challenge is to integrate daylighting and controlled artificial lighting effectively and link it in with the shading system design. More advanced glazing materials could play an important adaptive role. These needs throw out a challenge to the construction materials industry to provide better and more economically effective sub-assembly systems to guard against overheating. Very significant work on natural cooling has been carried out under the CEC DGXII Solar Energy Research and Development Programme PASCOOL. This work should be used by the UK building industry to guide the development of improved building components to enhance summer thermal performance. There should be a considerable international market for such improved systems.

The landscaping of the ground around one and two-storied buildings can be adapted to provide effective summer shading of buildings, especially on the east and west sides (US Department of Energy, 1989). However, caution must be exercised when buildings are founded on shrinkable clays (Hunt *et al.*, 1991). Trees are ineffective shading devices on the south sides of buildings in summer, and take away some of the valuable winter sunshine.

The integration of the various overheating adaptation strategies is best achieved using the graphic capabilities of microcomputers in conjunction with their numeric capabilities. Visual aspects of design can then be interrelated with thermal aspects of design.

Successful performance simulation in the face of climate change requires suitable climatic data discs to be prepared which include, in a fully incorporated way, current climate change predictions at different stages into the future. These data discs must be sold to the industry at reasonable prices in forms suitable for use by the industry.

Adaptation at the international level

Major UK design groups and construction contractors operate internationally. They would not wish the UK governmental approach to the climate change agenda to be drawn too narrowly, as they would wish to adapt their industry in the international context.

The majority of building materials are now supplied by relatively large firms, many of them international. British Standards and International Standards, especially European CEN Standards, are very critical to their operations. The building materials sector tends to prefer to see climate issues as embedded in norms and standards for specific products, which can then be met through appropriate quality controls in the production process. The expected durability of building materials exposed to weather in different parts of the world is always a key issue for materials manufacturers. The potentially changing nature of building materials exposure internationally is thus of central interest to this group. The impacts will have to be assessed material by material, using detailed appraisal based on appropriate combinations of projected future weather data.

It is particularly important that the powerful international knowledge base emerging from the UK's Hadley Research Centre's study of global climate change should be made available in appropriate forms to aid the long term international competitiveness of the UK construction industry, as well as meeting the specific needs of the construction industry within the UK.

11.4 UNCERTAINTIES AND UNKNOWNS

The key uncertainties are:

- The lack of objective information in climate change scenarios on possible changes in the probable recurrence periods of extremes of weather, especially wind. This makes future risk assessment very uncertain.

- Lack of information about the precise impacts of climate change on buildings. Studies of these will need to be based on the use of combinations of future weather elements, for example:

    driving rain: wind in conjunction with rain,

    air conditioning: temperature in conjunction with solar radiation and dew point,

    condensation: vapour pressure in conjunction with rain, wind and temperature.

- Lack of information about the cost benefit structure for different types of construction adaptation to climatic change.

11.5 PRINCIPAL IMPLICATIONS FOR OTHER SECTORS AND THE UK ECONOMY

Buildings are central to most human activity. The built environment and the construction industry have links into most of the topics discussed in other chapters of this report. Insurance issues covered in Chapter 13 are especially important, both for the construction industry and for building owners. Insurance provides the pathway whereby risks for building owners are interpreted financially through the premium mechanisms. Increased risk premiums will provide an economic incentive to building owners to achieve more effective adaptation.

11.6 RESEARCH AND POLICY ISSUES

11.6.1 Future research effort

Building owners will primarily respond to climate change in terms of the potential economic benefits to be gained through additional investment. It is essential to create, through systematic study, a more objective understanding of the probable costs and benefits resulting from different lines of action in different parts of the country at different stages into the future. Climate change risks and benefits will be unevenly distributed geographically, as will be the associated costs, such as insurance. The expected pattern of risk and opportunity needs to be considered.

It follows that there is a need for objective research into the probable impacts of relevant weather elements acting in combination in significant impact areas such as air conditioning design, driving rain, condensation, combined snow and wind loading, ground condition and drainage. The identification of action boundaries for step changes will be important, for example, where and when is there likely to be a switch to a significant use of summer air conditioning to sustain human productivity in buildings? In what parts of the country and when will the invasion of a specific pest becomes significant? It would helpful if geographically based risk assessments could be mapped into standard GIS systems.

There is an urgent need to develop building performance simulation climatic data tapes that will allow simulation of building performance in the climate change mode. These must be formatted to suit current needs in the different professions involved. They would provide a key tool for research as well as for practice. The creation of an international climate change data base to assist UK designers, manufacturers and contractors operating internationally is also needed.

As insurance premiums, more and more, are being linked to local assessments of climatic risks rather than national assessments of risk, there is a need to study feedback loops between the possible future policies for the insurance of construction and the responses in construction practice to such policies. The construction aim should be to reduce climatic risks. This should be reflected in reduced insurance costs, justifying the investment.

There should be a drive to translate basic climate change scenario data into standard construction design guides such as the CIBSE Guide and Specification.

11.6.2 Policy issues

A study should be made in conjunction with one or two environmentally aware local authorities of how best the urban climate change agenda could be worked into the local strategic planning process, possibly inter-linked with the Agenda 21 process.

The key policy issues in this area are:

- How to achieve hot weather mitigation through improved urban layout and landscape policies taking account of urban heat island effects.

- How to promote the greater use of natural ventilation in urban situations in the face of growing air and noise pollution from traffic.

- How to improve the air flow characteristics of towns to mitigate other adverse effects of climate change

- How to evolve urban layouts that foster the use of new and renewable sources of energy, especially the effective passive use of solar energy, and simultaneously ameliorate the summer climate and reduce pollution in the urban environment.

12. Transport

SUMMARY

- The extent to which the predicted levels of climate change will be important for transport is very much dependent upon the extent of other predicted changes in the transport system, such as the growing need to incorporate changes which are environmentally sustainable. It is expected that changes resulting from these other factors could be significantly more important than those arising from climate change.

- There are a few areas where transport is particularly sensitive to changes in climate and where policy measures will need to be devised in order to enable appropriate adaptations to take place. The aspects of transportation likely to be more significantly affected by climate changes include: transport links close to areas of coastline and estuaries which may be subject to increased likelihood of flooding as a result of sea level rise; an increased likelihood of disruption of all modes of transport resulting from an increase in the frequency of strong winds; and the possibility of greater levels of damage to infrastructure as a result of higher summer temperatures.

- There are likely to be some positive effects of climate change for transportation, for example, a reduction in the numbers of days with frost and lying snow could lead to less disruption of air, road and rail transport modes.

- The lifetimes of most transport artifacts are quite short and many, such as motor vehicles, can be expected to be replaced several times over the next half century, hence giving a high level of adaptability to potential changes. Some more long lived pieces of transport infrastructure, such as roads and bridges, are less easily adaptable in the short term. While the potential for adaptation is good, it is dependant upon a willingness to change and upon the projected climate changes occurring sufficiently slowly.

- Transport is a derived demand and hence is highly dependent upon changes in other sectors documented elsewhere in this report. The linkages between transport and other sectors of the economy are not clearly known. Changes in these sectors as a result of climate change (or otherwise) will have considerable secondary impacts on transport. These may exceed any direct impacts of climate change on transport itself. For example, changes in patterns of tourist activity in the UK resulting from climate change could generate considerable changes in the overall levels of demand for transport or could alter the existing balance of demand between various areas of the country.

- Research is needed on the changes in demand for travel that may result from climate change, looking specifically at the willingness and capability of individuals and corporate enterprise to adapt. More information is also required on the likely frequency of weather events which have the potential to severely disrupt transportation services.

12.1 INTRODUCTION AND BACKGROUND

This chapter considers the possible implications of climate change on the transport system of the UK including land, water and air based transportation modes. As far as can be ascertained all of the main conclusions from the transport chapter of the First CCIRG Report (CCIRG, 1991) are very much relevant today. This chapter will, however, aim to cover some of the more important findings in detail and, where possible, make reference to new research findings.

It is important to note that the transport system is currently one in which many changes, some very significant, are likely to occur over the period up to 2050 and hence the predicted extent of climate change may have relatively little impact compared to these. For example it has been suggested (IPCC, 1996b) that in the long term, around 2025 to 2050 and beyond, changes in travel culture and lifestyle could combine with and contribute to changes in urban layout, leading to reductions in motorised travel of up to 50% in some North American and Australian cities. The potential reduction for Europe is predicted to be smaller, nearer 25%, as a higher proportion of access needs are already met by walking and cycling. Within the transport field there is much emphasis being placed on the development of more sustainable and environmentally friendly transportation. Considerable research effort is being made on ways to improve transport technology to meet these aims, on ways to reduce the use of modes which impact badly on the environment, on ways to reduce the overall demand for transport and on measures to promote less environmentally damaging modes of transport.

Figures for 1993 for Britain (Department of Transport, 1994) indicate the importance of transport within the economy and show that transport and travel accounts for 16% of total household expenditure, amounting to a total of approximately £1 billion per annum in 1993. A further £10 billion was spent by local and central government in 1993. Ninety four percent of passenger kilometres travelled (excluding walking) are undertaken on the road, compared to only 5% for rail and 1% for air. Road also accounts for 63% of goods moved (tonne kilometres) and 82% of goods lifted (tonnes) compared to rail (6% and 5% respectively) and water (25% and 7% respectively). The Department of Transport predictions for the future in terms of road transport indicate a 83% - 142% growth by the year 2025 compared to a baseline of 1989 (Department of Transport, 1989). Means of coping with this increase in demand will need to be developed.

There have been few studies to date which have explicitly aimed to investigate the effects of various climate change scenarios on transportation. Perhaps the major recent exception to this has been the IPCC Second Assessment Report (1996b) which considers the effects of climate change on various sectors of the economy including transport.

12.2 ESTIMATED EFFECTS OF CLIMATE CHANGE AND SEA LEVEL RISE

The IPCC Second Assessment Report concluded that the lifetimes of most transport artifacts are quite short and many, such as motor vehicles, can be expected to be replaced several times over the next half century, hence giving a high level of adaptability to potential changes. Some more long lived pieces of transport infrastructure such as roads and bridges are less easily adaptable in the short term. While the potential for adaptation is good, it is dependant upon willingness to change and upon the changes in climate occurring sufficiently slowly.

Table 12.1 (derived from IPCC, 1996b) shows the main climate sensitivities in the transportation sector.

Transport is becoming increasingly important and this is accelerated with increasing economic activity (IPCC, 1996b). Given a growing reliance

Table 12.1. Summary of climate sensitivities in the transportation sector

	Energy transport/ transmission	Transportation infrastructure	Transportation operations
Temperature	permafrost and pipelines; capacity of power lines	effect of freeze-thaw cycles on roads	road maintenance costs; air conditioning in cars; ice and coastal shipping in high latitudes
Precipitation	icing of power lines		impacts of snow and ice on road and air transport
Frequency of extreme events	effects on power lines	effects on roads, railways and bridges	safety and reliability of operations e.g. airports
Water availability			inland navigation during drought
Sea level rise		effects on coastal infrastructure, migration of coastal activity	
Other		changes in demand due to changes in movements of agricultural products; changes in settlement patterns	effects of fog, snow, rain, ice on operations and safety

Source: IPCC, 1996b

of other sectors of the economy on transportation for continuing development it is essential that planners and decision makers are made aware of the potential impacts of climate change.

12.2.1 Sensitivity of the transport sector to weather and climate

The conclusion of the 1991 CCIRG Report was that transport was likely to be very sensitive to changes in weather and climate, while some modes of transportation and some geographical areas could be expected to be more sensitive than others. Table 12.2 shows a rank order of impacts on UK transport operating procedures for different modes of transport.

The table shows that there are very different levels of sensitivity to weather conditions for each of the modes of transport considered. Specific types of geographical locations, for example, those close to rivers and the sea, or high mountain passes are just some of the areas that may be expected to be particularly sensitive to weather and climate changes. More recent communications with transport experts indicate that some of the rankings in this table should be updated. In particular under the heading of rail the effect of wind should be ranked quite highly, perhaps even as high as '1' given the extensive disruption and potential for injuries which occur when overhead wires or trees are blown onto rail lines. Also, the effect of low cloud on the road system should be given a ranking in respect of its importance for hill and mountain roads.

Table 12.2. Ranked first order magnitude of impact on UK transport: operating procedures

Variable Transport system	Low cloud	Snow	Rain	Ice	Fog	Wind
Rail	n.a.	4	2	5	1	n.a.
Road	n.a.	5	3	4	3	2
Air	4	5	2	3	4	2-4
Sea	n.a.	n.a.	n.a.	2	4	4

1=low; 5=high; n.a.=not affected
Source: Parry and Read (1988)

Experience shows that more extreme weather events do have obvious effects on transportation, for example the wind storms of October 1987 and January 1990 led to major disruption (though mostly short term), while extremes of rainfall and snow can have similar effects. In the Department of Transport's 'Calendar' of important events over the last 10 years there are several occasions where the impacts of severe weather are considered sufficiently important for inclusion in the list. These include December 1984 (multiple crash on the M25 in fog, 10 people killed), January 1987 (severe weather conditions: record breaking low temperatures and heavy snowfalls disrupted all forms of transport), December 1990 (severe weather disrupted British Rail services) and February 1991 (severe weather disrupted rail services). Aside from these national events, there have, been many more examples of weather related impacts on transport at a much more local scale. Evans (1995) in an article on major British road accidents (10 or more fatalities) also includes an incident on the M4 near Hungerford in March 1991 which was caused, at least in part, by the foggy conditions prevailing at the time.

Department of Transport (1995) statistics for road accidents show that in 1994, of the 3650 fatalities on British roads, 561 (15.4%) took place when it was raining, snowing or foggy. When the road was wet or covered in snow or ice, 1323 (36.2%) took place. Unfortunately, there is no way of relating these events to the relative frequency of travelling in such conditions and hence it is not possible to comment on the relative riskiness. The same document also provides similar statistics for each county, though it is not possible to determine a relationship between counties where the frequency of snow and ice are relatively high and the incidence of accidents involving these conditions. This also depends upon the level of service of winter maintenance.

Certain recent developments in transport practices could result in greater sensitivity to climate and weather effects, for example the increasing use of 'just-in-time' distribution. This is a reliance on the capability of being able to despatch goods at short notice, with a high degree of certainty that they will arrive on time. Any factor, such as changes in weather or climate, which acts to upset this reliability, will necessitate a rethinking of such a distribution system.

12.2.2 Effects of climate change

The 1996 CCIRG scenario suggests that changes in average temperatures and rainfall will not be rapid and hence the likelihood is that both the transport system as a whole and individuals within that system will be able to adapt at a rate consistent with the change. Indeed, it seems likely that many aspects of the transport system will not need to change. If any change occurs it is likely to be as a result of other pressures on the system, for example, that arising from increasing levels of concern about the environmental implications of increased demand for certain highly polluting modes of transport. The recent Royal Commission Report on Transport and the Environment (Royal Commission, 1994) highlights the need for major changes in the UK transport system as a result of increasing concern over the environmental implications of rapid growth in demand for transport, in particular road and air traffic.

One of the problems of predicting the impacts of climate change on the transportation system is lack of knowledge of the combined effects due to changes in all of the climate parameters (e.g. rainfall, temperature, storm frequency etc). It is generally easier to make an assessment of the individual parameters. Some of these individual effects may work counter to each other, while others may be additive. The following outlines some of the major effects which may result from changes in individual cli-mate parameters.

Frequency of frost
The 1996 CCIRG scenario of future climate change indicates a substantial reduction in the

number of frost days throughout the country. For the most part this change is likely to have positive effects, in particular through a reduction in the annual disruption to ground and air transportation, though it should be noted that these changes are likely to be felt more positively in the south of the country than in the north. This will lead to economic benefits and a reduction in individual stress levels caused by uncertainty over whether journeys should be undertaken particularly in the winter months. Many of these changes may turn out to be small when compared with the effects of other changes in the transportation sector. For example, while rail journeys may become more reliable as a result of changes in climate, the effects of ageing rolling stock on certain lines and rail privatisation may act counter to this, and have a much greater impact on the system.

A reduction in numbers of frost days could be expected to lead to an increase in the number of journeys made (journeys which would have been cancelled as a result of adverse weather conditions). This will result in an increase in the levels of air pollution. Such a change could, of course, help to increase the rate of climate change.

Less snow and ice will mean a reduction in the application of salt on roads and hence a reduction in levels of corrosion, both to vehicles and to infrastructure. Such a change should result in an increased lifespan of vehicles (although this is dependent upon other factors such as the buoyancy of the second-hand car market, decisions with regard to car purchase and the costs of motoring), which could lead to increased pollution levels due to older, more polluting vehicles remaining in use. Work by the Royal Automobile Club (RAC) has suggested that 10% of vehicles currently account for 50% of carbon monoxide emissions (House of Commons Transport Committee, 1994).

A reduction in the frequency of icy road conditions could be expected to lead to a decrease in the number of road accidents. However, experience shows that accident numbers, particularly the more serious accidents, are lower on days with ice and snow as a result of a reduction in the number of journeys made. There is some evidence, however, that the numbers of accidents involving only minor damage increases on such days. Generally people will drive more slowly and safely to compensate for the increased risk that icy roads impose. If there is to be an increase in the level of more serious accidents then this will result in an increase in the level of pain, grief and suffering associated with such events as well as the imposition of the economic costs involved. Department of Transport costs for an average road accident are currently £39,810, whilst a fatal road accident is valued at £913,140 (Department of Transport, 1995).

Changes in Precipitation

Under the 1996 CCIRG scenario there is an increase in winter precipitation and the intensity of rainfall over the whole country, plus an increase in rainy days throughout the UK except in the far north. In summer the picture is more diverse with increases in rainfall intensity throughout the country, but a north-south divide in terms of mean precipitation: the north becoming wetter and the south drier.

Such changes, particularly in the intensity of individual rainstorms, could be expected to lead to an increase in flooding episodes and hence localised disruption to transportation services. However, rainfall, unless very extreme, is unlikely to have such a large impact on transport as snow and ice.

Increases in rainfall can be expected to lead to higher levels of corrosion of various transport infrastructure and vehicles. This would result in shorter lifespans and increased maintenance costs. This will need to be considered when planning budgets and designing future infrastructure. Whether this will be sufficient to counteract the reduced levels of corrosion expected as a result of less salt and grit use, as discussed above, is unknown.

There is likely to be a reduction in the number of days with snow lying, although the number of snowfalls may not be less. Such changes will generally have effects on transport similar to those for frost.

Temperature

Changes in temperature are likely to have more subtle effects than changes in precipitation. These will depend upon individual behaviour and responses to the changes, as well as on other factors. For example, a general increase in temperature could result in an increase in the number of tourists visiting the UK (although this depends on relative amounts of sunshine and cloudiness as well) and possibly an increase in the numbers of UK residents who choose to stay in the UK for their holidays. However, the tourist market is likely to be very dependant upon many other factors (not least the relative value of the pound), any one of which could swamp any effects of temperature changes. Changes in the level of tourism in the UK will affect the demand for transportation.

Whether increases in temperature will result in a greater level of outdoor activity and an associated increase in demand for transport is almost impossible to predict. Certainly the number of leisure journeys is generally higher in the summer than in the winter. However, there may be factors operating to reduce the levels of outdoor activity, e.g. an increase in the frequency of rainy days, public awareness of the dangers of skin cancers from exposure to UVA/UVB radiation as a result of damage to the ozone layer. Generally hotter weather coupled with increasing traffic levels could result in a higher frequency of air pollution episodes and hence increased impacts on human health.

In order to assess fully the possible effects of increased temperatures, more information is needed on how the temperature will be distributed throughout the year. If there are longer periods of hot, dry days, there will be more rapid evaporation from lakes, rivers and canals. Canal transport, would be directly affected, particularly the canal tourist industry. Reductions in water levels may possibly affect the incidence of algal blooms and thus have a further negative impact on the tourist industry.

Sustained high temperatures could result in physical damage to infrastructure, for example, in the hot summer of 1995 there were several instances of rails buckling and numerous instances of tar melting on roads. The increased likelihood of such events will need to be allowed for in the design standards of infrastructure. Changes in temperature distribution may also lead to altered timings of annual events which regularly disrupt transport, such as 'leaves on the line'. Temperature also affects transport operations; for example, rising temperatures may result in greater reliance on air conditioning in vehicles. This has implications for increased fuel use (see Titus, 1992).

One considerable effect of changes in temperature on transport would result from changes in the patterns of agricultural production in the country. Such changes would affect the distribution patterns of various agricultural foodstuffs throughout the country. Differences in the types of crops grown could also lead to changes in the import/export markets and thus the requirements for international transportation of freight.

Wind

Mean seasonal windspeeds are estimated to increase throughout the country in both winter and summer, but with the increase greater in winter, especially over southern England. Such changes could have substantial impacts on transport, especially on road and air transport. For example, the 25th January 1990 wind storm resulted in 46 deaths, most being transport related events. Motor insurance weather damage claims totalled 55,000 in January and February 1990 (Thornes, 1991).

Disruption to transport can also be increased due to windiness. Extreme storm events can result in many road closures, mainly due to trees or vehicles blown on to roads. However, certain exposed sections of road are particularly vulnerable to lesser winds. Records of lane closures on the Severn Bridge indicate that in the 1980s lanes were closed because of high winds on average some 130 hours per annum on 20 days annually (Perry and Symons, 1994). In addition, high sided vehicles are barred from the bridge on about 20-25 hours annually. Effects on shipping and air travel may also be significant as well as effects on overhead rail power cables.

Other changes

There are a number of more general impacts which are much harder to define, but which mainly involve changes in individual travel behaviour. This is very difficult to model with any level of accuracy even when major changes in transportation are proposed (e.g. major infrastructural or scheduling changes). Existing models are unlikely to be sensitive enough to be able to cope with many of the small, subtle changes that climate change may cause. This is particularly the case when compared to other predicted changes which are expected to occur over the next 50 years or so. It is quite conceivable that the transport system, in particular road based transport, will be very different to that currently the norm. Increasing concerns about the environmental impacts of current transport practices could be very important in leading to such changes.

There may be some minor impacts of climate change on inland waterway transportation. Arnell *et al.*, (1994) indicate two main impacts of climate change on navigation on inland waterways: firstly, the water available for the navigation system may change, leading to problems in maintaining water level; and secondly, changes in river flow regimes will affect sedimentation in navigable river channels and hence maintenance. The expected first order effect of lower flows is toward less sedimentation. Given the importance of certain waterways for the movement of goods it will be important to monitor water flow and levels carefully in order to identify any climate induced changes.

12.2.3 Effects of sea level rise

Effects of sea level rise are likely to be relatively serious where they occur, but not very widespread. There will be some damage to coastal transport links which may necessitate limited rerouting or strengthening of existing sea barriers. Low lying areas of the country such as East Anglia are most likely to be adversely affected. Some urban areas may be affected badly, for example parts of Central London as a result of increased likelihood of Thames flooding. In some cases transport routes could be combined with sea defences such as barrages. Certain aspects of port infrastructure and jetties may need to be adapted to cope with increased water levels.

12.3 ASSESSMENT OF POTENTIAL ADAPTATION

There is considerable scope for adaptation of the transport sector to possible climate changes particularly in view of the relatively long duration over which the changes can be expected to take place. Adaptability is dependant upon the willingness and ability of individuals and organisations to change. The IPCC Second Assessment Report (IPCC, 1996b) outlines four groups who will need to adapt:

- Individuals in their role as consumers or citizens;

- Business enterprises, operating over a wide range of scale and technological sophistication, which may be privately owned or run as state enterprises;

- Policymakers concerned with land use planning and the development of transportation, energy or industrial infrastructures which may be affected by climate change; and

- Policymakers making high level decisions about the adequacy of policies to deal with climate change and the appropriate balance between adaptation and migration strategies.'

Adaptation to some of the undesirable impacts of climate change could include the following:

- Rerouting elements of the transport network away from low lying areas of the country, or in some cases development and strengthening of sea and river defences.
- Better planning of new infrastructure to meet the changing demands of temperature and rainfall.
- Increasing individual and organisational awareness of changes that may occur and their own roles in the process of adaptation.
- More accurate prediction and advance warning of severe weather events.

Most difficult to adapt to are changes in patterns of storm events in particular if there is a higher frequency of more severe storms.

12.4 UNCERTAINTIES AND UNKNOWNS

Given the uncertainties over the projections of climate change expressed in the 1996 CCIRG scenario this implies a further level of uncertainty over the ways in which the transportation sector might change.

It seems likely that factors other than climate change will play an important role in shaping the development of the transportation sector into the 21st Century. Crucially important are factors such as the cost and continuing availability of oil, progress in the development of vehicles powered by alternative energy sources, the strength of the growing environmental movement in the UK, government policy towards transport, and changes in individual motivations and needs for transportation.

Transport is implicitly linked with many of the other sectors in this report such as the financial sector, tourism, agriculture and construction and hence is very much dependent upon changes in those sectors. The ways in which such changes might affect transport demand is very difficult to predict.

12.5 PRINCIPAL IMPLICATIONS FOR OTHER SECTORS AND THE UK ECONOMY

Should changes occur in the nature and amount of transportation as a result of climate change this will have financial implications. Some of these have been discussed above, such as the possible change in the numbers of serious road accidents and hence a reduction in the associated costs to society as a whole. Other changes may have positive financial effects such as a reduction in the number of working days lost as a result of inclement weather conditions. However, the situation may not be so simple. If more journeys are made because of a lower frequency of icy roads then, all things being equal, the total amount of emissions will rise, other social costs of transportation will rise, and there will be a greater demand for an increase in transportation links or improvements to those already existing. The difficulty is to determine whether the benefits in terms of financial gains of such scenarios will outweigh the disbenefits. It should be noted that this is very much dependent upon the methods which are used to calculate the costs and benefits of different strategies.

12.6 RESEARCH AND POLICY ISSUES

12.6.1 Future research effort

To date there have been few comprehensive attempts to determine the impacts of climate change on transport. In particular, survey work is needed on the ways in which individuals and corporate bodies might react to the predicted changes in climate. Such data could enable the inclusion of climate change considerations into various transport modelling packages which would allow the more detailed impacts to be assessed. Few individuals or companies are aware of the potential impacts of climate change on transportation, and no surveys have been undertaken to examine the willingness of people to change. Such changes that do occur as a result of climate change will need careful management. More information is also required on the likely frequency of weather events which have the potential to severely disrupt transportation services. Because of the general lack of public knowledge about the precise impacts of climate change, it would be worthwhile developing publicity materials which specify the nature of expected changes in climate and outline the changes to transportation systems which may be required.

Although unlikely to be readily obtainable, more information on the ways in which the predicted changes in climate will affect day to day weather patterns would be useful.

12.6.2 Policy issues

Transport should on the whole be able to adapt to any direct effects of climate change without the need for the development of long term policy. However, there are a few areas of transportation which are especially sensitive to climate change and where it is crucial that policy decisions are taken in the near future. These include: decisions on **i)** how to protect or move transport links and facilities which are at increasing risk of flooding as a result of sea level rise; **ii)** how to avoid increased disruption to all modes of transport resulting from an increased frequency of strong winds; and **iii)** how to design and strengthen transport infrastructure to reduce susceptibility to damage as a result of high temperatures (for example melting tar and buckled rails).

The importance of climate change on transport should not be exaggerated. With the exception of the points mentioned above, the likely impacts will be small, especially when compared to the magnitude of changes expected to occur for other reasons.

The major policy issue facing transport planners at present is how to develop a transport system which is sustainable over the period up to the mid 21st Century and beyond. Changes to the system as a result of perceived climate change over this period will have to be developed within this overall goal of achieving sustainability.

13. Insurance

SUMMARY

- Property insurance would be immediately affected by a shift in the risk of extreme weather events. It is anticipated that changes in their frequency and/or severity will occur, which would necessitate pricing or product changes, particularly for the most vulnerable areas or activities covered. At present a detailed description of future weather is not available; thus it is not possible to make specific planned responses.

- Since the risk of coastal flooding will increase due to sea level rise, this will expose the property insurance industry to the greatest potential losses in the UK (e.g. the insured value of the property protected by the Thames barrier is estimated to be £10-20 billion).

- Because of the international nature of UK insurance institutions, the impacts from climate change or sea level rise abroad are important. UK insurers incurred substantial claims in Hurricanes Gilbert (1988), Hugo (1989), Andrew (1992) and Opal (1995) directly through foreign subsidiaries, and also through reinsurance activities in London, including Lloyds. Hurricanes and other tropical storms may increase as a result of global climate change.

- Property insurers have begun to adapt to the run of increased UK weather losses from storm and drought-induced subsidence. More efficient methods of handling claims have been introduced. Building insurance rating now recognises susceptibility to subsidence. Major exercises are or have been conducted individually and collectively to assess the approximate pattern of risk from coastal flooding, and the potential impact of storms. New taxation arrangements are being negotiated with government to facilitate the creation of funds to meet the irregular but high cost catastrophes.

- UK insurers are actively involved in exercises to reduce the vulnerability of buildings to wind damage. The insurance rating of weather hazards is continually reviewed together with assessing the financial impact from potential extreme events in order to control exposure.

- Apart from property insurance, there is little literature on the impact of weather. There will be however, impacts on human health and longevity in the UK, on the transport sector (including international marine business) and on energy exploitation. The insurers of such activities have not yet addressed the issue of climate change specifically.

- For the property insurance sector more detailed information is required on the likely pattern of extreme events in the UK and abroad in areas with a significant exposure to property damage for UK insurers. Improved computer models to translate the impact of such potential events into economic terms are required. The life insurance and pensions industry require to establish the likely implications for longevity.

- The insurance industry should co-operate with other stakeholders in the property market to reduce society's vulnerability to extreme events, by assembling and providing information on damage and best practice in other countries to assist adaptive measures such as improved construction and land development. Property owners should be encouraged to mitigate their risks through appropriate pricing, deductibles and other insurance policy conditions. Insurers should regularly review their capabilities to handle extreme events, and promote the adoption of robust disaster plans by clients and suppliers. The industry's professional bodies must ensure that internal education and training gives greater attention to climate risks. Legislation to allow the creation of tax-allowable catastrophe reserves should be concluded, and insurers should integrate these into their overall plans to manage their exposure to climate risks.

13.1 INTRODUCTION AND BACKGROUND

This analysis will concentrate on the general, or non-life, underwriting industry which is much larger than the UK reinsurance sector and is more sensitive to weather than broking or life insurance. However, it should be recognised that the reinsurance sector plays a major part in coping with weather catastrophes, by spreading the costs more widely. Historically climatic impacts have been viewed as random. However, the repercussions of climate change could include severe property damage. Within insurance, only the property insurance industry is beginning to recognise climate change as a strategic variable, at the UK and international level.

The insurance sector of the UK economy generates huge turnover in the UK and in invisible exports. Statistics are fragmentary, but Table 13.1 gives an indication of the sales volume. The other measure of scale, capital employed, is not readily available. In 1993, it is estimated that the sector contributed about £4 billion surplus to UK balance of trade (London Business School, 1995; Tillinghast, 1995). The UK is peculiar in that it has a large sector (the London Market) dealing with overseas risks (Home Foreign), reinsurance, and international transport. Insurance is essentially concerned with re-distributing the financial risk of human activities. Therefore, although it is useful to consider the direct effect of climate on the insurance industry, another benefit of this focus is that it provides information about the cost of climate in other sectors, especially domestic consumption. The ancillary aspects of insurance such as risk management and settling claims give insights into the practical problems of damage prevention or recovery.

Most insurance in the UK is provided by free enterprise limited companies, or Lloyd's (the insurers). Many of these insurers are now foreign owned, but continue to operate in the UK because of the large domestic market and specialist 'London Market'. Those buying insurance (the policyholders) can be subdivided into 'Commercial' (business) and 'Personal' (domestic) policyholders. Although there is a variety of types of insurance cover, including Motor, Marine and Liability, it is Property policies which are particularly affected by weather. Usually Personal policies are 'comprehensive' i.e. cover a wide range of risks, including weather damage, while Commercial policies are more variable; the policyholder may choose to bear part or all the risk and indeed the insurer may exclude or limit certain risks. Businesses can also insure against the consequential loss due to property damage, though this cover is less commonly purchased.

Since the first CCIRG report (CCIRG, 1991) there have been no major weather disasters in the UK, although there have been some localised floods (e.g. Perth 1993, Chichester 1994, Glasgow 1994, Aberdeen 1995) and a period of drought in 1995. Overseas the trend towards increasing property damage from storms, wildfire and flood has continued, but flood is often not insured. In the UK, the Chartered Insurance Institute has published a report which identified several areas of concern to UK insurers in climate change, such as their exposure to flood in the UK (Dlugolecki et al., 1994). Global climate models (GCMs) are still not capable of providing detailed information on extreme

Table 13.1: Size of UK Insurance Industry (1992)

Type of Business	Source of Business		
	UK	Overseas	Total
	(net written premium £ billion)		
Life & Pensions	43.3	9.2	52.5
General	21.6 (est)	6.4 (est)	28.0
London Market	n.a.	n.a.	12.0
Total	n.a.	n.a.	92.5

Source: Tillinghast (1995), London Business School (1995), Carter and Falush (1995).

events in future, particularly storms. Policymakers within the industry are reluctant to act until scientists are more specific. The formation of the United Nations Environment Programme (UNEP) Initiative on Sustainable Development and the Environment for the Insurance Industry may trigger more strategic activity on adaptations. Its objectives include defining key environmental issues such as climate change and trying through research to identify a position which the sector can support internationally.

The IPCC's Second Assessment Report (1996b) is now complete, and for the first time includes a chapter on the impacts of climate change on Financial Services, including Insurance. The IPCC report identifies that poor land development practice, inadequate construction and greater wealth and population have considerably increased society's exposure to risk from weather hazards and flooding. Also insurance is considered as a policy option by the working group on policy issues. Since problems will become more serious in some areas due to climate change, closer co-ordination between policymakers and industries such as insurance will help the process of adaptation. This is particularly so, because insurance in some form is an implicit pre-requisite for wealth creation and possession, particularly property. Finally the IPCC report notes that policies to mitigate climate change (e.g. emission controls) would have implications for investment activities in the primary (capital - raising) and secondary (stock exchange) markets.

13.2 ESTIMATED EFFECTS OF CLIMATE CHANGE AND SEA LEVEL RISE

13.2.1 Sensitivity of the insurance sector to weather and climate

Property insurance

Figure 13.1 gives an illustration of the impact on UK-based property of severe winter weather since 1960, at 1987 values. The figures present an increasing trend, but they are affected by i) increasing real exposure during the period, due to growth in population, wealth, wider insur-

Figure 13.1: Cost of major UK winter weather incidents, 1960-1995. (Excluding £275m freeze in December 1995, 1987 value)

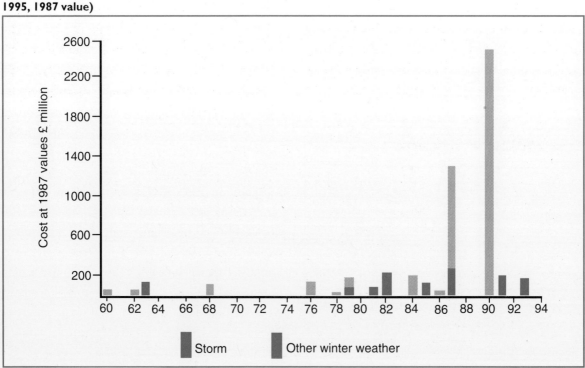

Source: Association of British Insurers

ance cover etc; ii) greater propensity to lodge claims; and iii) a return to 'normal' variability following the very uneventful 1960's. From a study of non-catastrophe costs, the effects of i) and ii) during the period 1980-89 were to double the impact of weather damage (Dlugolecki et al., 1994). It can be seen that the costs during this period have in real terms escalated much faster than that. Recent storms have proved far more costly than the cold winters of 1981/2 and 1987. Figure 13.2 isolates the effect of weather for domestic buildings insurance. Here again the impact doubled over 25 years. Disasters are included, but no major floods occurred in this period. The high cost in the most recent period is due to subsidence following drought, and the 1990 storms.

The UK insurance industry is also exposed to losses through its international activities in the London Market (reinsurance) and from overseas subsidiaries. Prior to the 1987 hurricane there were no 'billion-dollar' storms. Since then there has been one almost every year (see Table 13.2), and this generated considerable stress on the profitability of Lloyd's due to the poor understanding of the exposure at that time (compounded by other problems such as asbestos-related claims). There have been many severe inland floods overseas in recent years (e.g. Mississippi 1993, Rhine 1994, Rhine 1995) but such events are still not commonly insured. Individual storms have been large enough to bankrupt some smaller insurance companies. The real threat is probably from a cluster-in-time of extreme events which might exhaust the reinsurance protection and channel back the bulk of later events to the primary insurance market. Despite the upward trend in weather losses, it is still not possible to say categorically that this is due to a shift in weather patterns arising from man-made climate change (IPCC, 1996b).

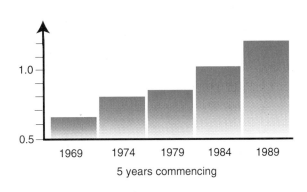

Figure 13.2: Weather costs vs. time for UK domestic buildings. (Source: General Accident, using industry and internal data).

Non-property insurance and climate

Motor claims are sensitive to fog and road-ice (as well as precipitation). Heat, atmospheric pollution and increased levels of ultra-violet radiation may be important for health and life insurance. Clearly, large-scale loss of life in a natural catastrophe would have implications for life-insurers.

Table 13.2: 'Billion Dollar' Storms

Year	Event	Insured Cost
1987	'Hurricane' in SE England/NW France	$ 2.5 B
1988	Hurricane Gilbert, Jamaica/Mexico	$ 0.8 B
1989	Hurricane Hugo, Puerto Rico/South Carolina	$ 5.8 B
1990	European Storms - (four)	$10.4 B
1991	Typhoon Mireille, Japan	$ 4.8 B
1992	Hurricane Andrew, Florida	$16.5 B
1993	'Storm of the Century', Eastern USA	$ 1.7 B
1995	Hailstorms, Texas	$ 1.1 B
1995	Hurricane Opal, Florida	$ 2.1 B

Source: IPCC, 1996b

Liabilities insurance may be affected due to the risk of pollution from waste sites following large-scale inundation. The question of public authorities' liability to protect the community from natural hazards might arise if protections or warnings are inadequate. Liability for damage caused by tree roots during drought is already recoverable in law.

The London Market 'marine' sector, (which includes aviation, cargo, transportation and much energy-exploitation, e.g. oil rigs), is vulnerable to severe weather, but in practice most losses have occurred from operator error, e.g. crashes, spillages, explosions. The literature is virtually silent on weather hazards. However, a recent example was the loss of the tanker Braer on the Shetland Isles in 1993 in a series of storms. The resultant oil pollution claims seem likely to exhaust the International Oil Pollution Compensation Fund (Shand, 1995). In general, however, modern equipment with well-trained operators is not the source of many sinkings.

13.2.2 Effects of climate change

Chapter 2 in this report outlines a scenario for future climate in the UK. As noted in this chapter, there is considerable uncertainty for four reasons: the sensitivity of climate to atmospheric change is imprecisely known; other GCM formulations appear equally valid - the UKTR model yields relatively cooler and wetter conditions than other scenarios of the future; GCM findings require considerable interpretation for regional use; and the model ignores sulfates, which are recognised as an important modifier of climate change. The most likely way in which the potential changes will affect the insurance industry are through shifts in the pattern of extreme weather.

Extreme weather causes several types of claim:

Drought	- subsidence of buildings
Freshwater flood	- water damage from rivers/thunderstorms.
Cold	- burst pipes leading to water damage
Heat	- minor effects
Storm	- wind damage, associated with water damage.

Drought

Long periods with reduced precipitation lead to soil shrinkage, particularly in clay with high shrink/swell potential which in the UK is located in the Midlands and southeast England. In turn this can lead to costly claims for *subsidence* to the buildings, or *heave* when heavy rain causes the soil to swell subsequently. Generally only dwellings are covered for this eventuality, not commercial property. This cover was introduced in 1971, and 'subsidence' claims have amounted to about 10% of the cost of all household claims since 1979. The 1996 CCIRG scenario envisages less summer rainfall in the south, which combined with higher temperatures is certain to cause subsidence. If the current housing stock were still present, this could result in a cost of perhaps £200m at 1990 values in years of severe drought. These may occur twice per decade by the 2050s.

Freshwater flood

Such events arise from local thunderstorms, or from prolonged periods of precipitation, possibly exacerbated by snow-melt and coastal tides. In the UK this is a relatively minor hazard. If mean annual precipitation were to increase by about 10% this could result in noticeable additional cost. Nevertheless, the effect is not seen as critical, provided land development is managed. For example, if the number of local floods doubled, this would imply an additional annual cost of £50m at 1990 values.

Low temperatures in winter

The main threat to property from low temperatures is the bursting of waterpipes. The major cost arises from spoilage of stock, fittings and furniture. The rise in consumer wealth has made this a particularly costly hazard in domestic property insurance. The potential rises in winter temperature will not eliminate 'winter' burst pipe claims, but could substantially reduce them: perhaps by as much as 30% for the temperature increase projected for the 2050s (+1.4°C). The annual cost was estimated at around £125m for the 1980s (CCIRG, 1991). Since the minimum (night) temperatures are likely to rise faster still, this means that winter freeze may disappear as a significant UK hazard. One factor which could invalidate this would be any strong weakening of the Gulf Stream.

High temperatures in summer

Property is relatively insensitive to higher temperatures (within UK parameters), thus increased temperature is not likely to increase insurance claims. There is a possibility of heath or woodland fires, but this is not seen as a serious threat.

Windstorm

This hazard is a major component of insurance costs. It also causes operational problems due to the high volumes of claims. Storms accompanied by rain are more serious due to the water damage to exposed interiors. Windstorm can also cause sea flooding by increasing wave height and generating a storm surge.

The 1996 CCIRG scenario envisages a potential 6% increase in mean winter windspeeds in southern Britain by 2050. This could possibly lead to a **doubling** in storm costs because of three factors: the damage curve is highly non-linear; there is likely to be more precipitation; and sea levels will also be higher (and major storms in the UK always originate off-shore). This might conservatively suggest an increase in average annual cost of £200m, although the cost in a single year might be £3 to £4 billion. The infrequency of major storms makes it difficult to calculate an impact. Over the 25 years 1960-94, the annual cost at 1990 values was around £200m. This was very heavily skewed towards the end of the period, but it has been argued that this is purely chance, not significant (Christofides *et al.*, 1992).

Overseas

IPCC scenarios envisage an increase in global precipitation and temperature levels. They incline to an increase in the variability of rainfall, with spells of drought and intensive precipitation. The scenarios are not very specific on the regional level, nor do they give details for convective events (tornadoes etc., which can be very destructive), or for cyclonic windstorms. Regional phenomena like El Nino are not modelled. In the absence of detail, it is impossible to give any quantitative assessment for overseas risks. Recent trends would suggest caution.

13.2.3 Effects of sea level rise

There is no doubt that property could be severely affected by sea inundation. Since 1953 when flood cover was relatively uncommon there has not been a major incident. Unpublished estimates give a figure of perhaps £4 billion for a recurrence of the 1953 incursion with today's exposures. The values at risk in London are higher (perhaps £10 - £20 billion), but protected by the Thames Barrier. A rise in eustatic water level could be exacerbated by severe storms, with associated higher wave height and wind surge. Spurred on by this, the Association of British Insurers (ABI) has carried out a study of vulnerable areas in England and Wales, in collaboration with the National Rivers Authority (NRA). However, it is clear that detailed topographical data will be required to make proper risk assessments. Internationally, the potential damage from sea level rise is very great (IPCC, 1996b). However, this is an area which is generally not commercially insured, and so the impact on insurers would be relatively small.

13.3 ASSESMENT OF POTENTIAL ADAPTATION

There has been little in the way of strategic adaptation to climate change, i.e. planned with long-term objectives. The industry believes it has time to react. Two initiatives are the imminent introduction of tax-allowable catastrophe reserves (needed also for international competitive reasons) and the ABI initiative on coastal vulnerability, resulting in a high level assessment of England and Wales NRA defences. Abroad, insurers in vulnerable areas like Florida and Fiji are co-operating with the authorities on initiatives to improve building design and construction.

Technically, there has been autonomous reaction through a shrinking of capacity and increased rates following extreme weather events in the UK in 1987 and 1990 and overseas. However, the international catastrophe market has been easing. Insurance has coped with the past rate of change by using pricing, minimum damage limits (deductibles or excesses), and risk-spreading through reinsurance to alleviate the financial effects. Risk management has been tackled through distribution of leaflets and publicity, and physical inspection of individual premises followed by recommendations for physical protection, often with some adaptation of the cover given. The operational effects of catastrophic storms have been alleviated by easier and faster guidance to claimants on what to do. This has enabled the industry to cope with a doubling of costs in a decade. The 1990 events represented a further doubling. The industry is coping, particularly through price increases. In particular, differential rating by postcodes for subsidence has led to a range of 600% in domestic building rates where once there was a uniform national rate. A similar approach to flood risk seems likely. However, competitive pressures may limit the application of price increases. Also, reductions in the labour force by underwriters and loss adjusters may make it difficult to cope with a future disaster.

Any financial strain is likely to be exacerbated by multiple incidents occurring either in the UK or internationally, placing a great need for substantial capital and reserves. This will encourage the trend towards consolidation in what is a very fragmented industry.

13.4 UNCERTAINTIES AND UNKNOWNS

Better information on future precipitation patterns will allow a firmer assessment of likely changes in subsidence and freshwater flood claims. Winter warming will reduce burst pipe claims. A major unknown is the future wind-circulation pattern, which has serious possibilities for property damage. Better information is required on the sensitivity of property to weather damage -historical statistics are hard to assemble. With regard to sea level rise, one of the major unknowns is the value and location of property most at risk. A number of independent attempts are being made by technical consultants, brokers and public bodies to develop computer models particularly to explore exposure to storm, flood and subsidence. Similar work is well advanced abroad on windstorm. Transport, life and health insurance are all likely to be affected by climate change, but no estimates can be given.

An additional concern is the chance combination of events, which can yield exceptional extremes, as happened at Towyn where the storm surge, wave height and water level were individually not exceptional, but when combined had a return period of 1000 years (Roe, 1990).

Perhaps the largest unknown is the nature of future economic activity. This could dramatically alter the sensitivity of the sector if activities/assets become more or less vulnerable to interruption/damage from climate hazards. Recent trends have exacerbated society's exposure, through poorly planned development for example. Electronic systems are sensitive to water damage. Finally, many business activities are so interconnected, that interruption to one stops a whole chain. An increasing proportion of weather-related costs is due to business interruption, not simply property damage.

13.5 PRINCIPAL IMPLICATIONS FOR OTHER SECTORS AND THE UK ECONOMY

Property insurance allows consumers to purchase and enjoy major assets, business to invest and expand, and other financial industries (banking) to lend funds secured by property. Therefore, if the amount of uninsured property damage were to rise significantly due to the absence of insurance for weather hazards like flood and subsidence, that could seriously affect those activities. Firstly, there would be hardship requiring public assistance (e.g. Glasgow 1994), and business failures or underperformance. Secondly, confidence would be eroded, leading to economic blight in vulnerable areas.

If catastrophes become more frequent, this could lead to a growth in the demand for insurance. If catastrophes can be better understood, this could also lead to a growth in supply. As long as other countries are proportionately more affected, and the UK tax regime is favourable (*pace* off-shore locations) this could boost the London Market, including Lloyd's, through reinsurance inwards. There has been a move in some countries towards government intervention, but this will always be limited by the squeeze on public finances, and by administrative problems.

Other countries are experimenting with alternatives to reinsurance (Catastrophe Futures, Catastrophe Bonds). So far there is little success, since there are more natural buyers than sellers, and the instruments are too generalised for particular buyers. However, the possibility of substitution through financial innovation cannot be ignored.

13.6 RESEARCH AND POLICY ISSUES

13.6.1 Future research effort

- Predictions of future probability distributions of temperature, rainfall and windspeeds are required, to allow assessment of the frequency and severity of different losses, including the possibility of combinations of events.

- More precise quantification of the links between different levels of meteorological event and physical and financial damage is needed.

- Information on the exposed risk must be improved (particularly the locations of individual buildings insured under 'block' building society accounts) and, in general, on vulnerability to flood. Detailed topographical maps are lacking.

- Effective ways to encourage risk management by policyholders need to be established.

- Resources should be allocated to explore climate change as a strategic economic variable, at UK and international level.

13.6.2 Policy issues

- It is vital that insurers have adequate reserves to meet catastrophe payments. At present UK taxation policy discourages 'catastrophe funds', contrary to practice in other countries, e.g. Germany. Currently legislation is in hand.

- Owners should physically protect and maintain their assets to bring the risk of damage within insurable parameters. Expert advice should be made available, and insurance terms (e.g. deductibles, rates and policy conditions) should be adapted to individual risk circumstances.

- Emergency procedures in all commercial and public services should be improved to ensure that operations can continue with minimum disruption following natural catastrophes. Public emergency procedures should be improved to alleviate losses when severe weather is imminent, e.g. more effective public warning systems.

- The 'recovery' phase should incorporate prompt advice and reliable repairs properly controlled. Such services are increasingly provided by insurers, and remove a considerable burden from the public authority.

- Insurers should provide information on damage patterns to technical design centres and others to ensure adequate attention to climatic factors and inundation, and to assist cost-benefit studies, when setting standards for the design and maintenance of buildings or coastal defences.

- Personal and business assets should be able to get insurance cover during their lifetime in order to provide a solid basis for credit. However, it is important that the market should operate as freely as possible, to avoid cross-subsidies and financial crisis, as described in a recent study of Florida windstorm insurance. (Academic Task Force, 1995). Particularly vulnerable property may require other financial mechanisms to spread losses and protect poorer members of society from financial hardship. (In the 1990 Towyn floods 6% of the affected buildings and 38% of contents were uninsured).

- Historically, natural perils have been relatively ignored by insurers, compared to the risks of fire, crime and trade processes. The various professional bodies, especially the Chartered Insurance Institute, should give them more attention, e.g. in technical and management qualifications and training.

- The International Decade of Natural Disaster Reduction initiative offers the UK insurance industry the opportunity to capitalise on its expertise through collaboration in other countries on such issues as how to reduce vulnerability to natural hazards like storm and flood.

14. Financial Sector

SUMMARY

- The greatest implications for the sector are primarily indirect, being felt through pressures upon customers, both commercial and personal. If the risk of flooding increases, the sector will face indirect pressures as business customers lose income through disruption, while homeowners and businesses face direct financial pressures through higher insurance premiums.

- The availability of insurance coverage has reduced the indirect effects upon the sector; as a result little information is available from which to project figures. The price of capital to different sectors/geographic areas may increase if lenders/investors are faced with higher risks as a result of the withdrawal of adequate insurance coverage.

- The financial sector is of great importance to the UK economy and in view of the international nature of much of the sector's business, the consequences of climate change outside the UK are likely to be of major significance. The financial sector tends to focus on the short term and discount the longer term, but can react relatively quickly to changes in its market, thereby off-setting some of the longer term issues experienced by other sectors which have more fixed investments.

- Climate change could affect the geographical distribution and balance between business sectors within the UK, impacting differently upon individual institutions as a result of their own particular portfolios. Financial operations are vulnerable to short term disruption due to failure of communications or denial of physical access, such as resulted in October 1987 in the London area.

- If climate change seems likely to have an effect on socio-economic activities, this will affect the appraisal of existing and new debt and equity investment opportunities, particularly in agriculture and for coastal regions.

14.1 INTRODUCTION AND BACKGROUND

The UK's financial sector generates huge turnover and contributes substantially to the economy through invisible exports. For example, in 1993 the financial sector's contribution to UK GNP was 24.5 %. Despite significant reductions in the numbers of staff employed within the sector over recent years, financial services remain a major employer with an estimated 150,000 people involved in international wholesale activities, generating between £10 - £15 bn in the City of London alone (London Business School, 1995).

The sector consists of a wide range of businesses from large, multinational financial services institutions to small, niche players. These include banks, building societies, venture capital houses, fund managers, investment institutions and leasing houses. Within this diverse grouping there are further separations into UK and overseas (foreign) owned banks and investment houses, each operating within a range of activities which include international and domestic financing (for example, the sterling money market) and management of international funds. This wide range of institutions and diversity of activity means that it is difficult to present an accurate picture in terms of employees, impacts and future risks at local, national and international levels.

Within the banking sector alone, the range of services offered to clients includes the provision of funds to meet short term needs (often by way of overdraft and in the form of working capital for businesses), specialised and trade finance, investment banking (corporate advice, mergers and acquisitions, etc.), investment management (fund management, venture capital, etc.), treasury (money markets, foreign exchange, derivatives. etc.), securities (capital markets, equities. etc.), mortgage, savings, life assurance, pensions and insurance broking services.

The full range of financial services is used by a relatively small number of businesses, primarily multinational companies (MNCs) of which there are approximately 38,000 worldwide. Indeed, it is important to remember that the vast majority of businesses are more accurately described as small and medium sized enterprises (SMEs), being geographically local and employing under 50 people. For the UK (National Westminster Bank, 1994) the figures are :

< 100 employees   99.0 %
< 50 employees   98.5 %
< 20 employees   96.0 %

There are approximately 3.4 million small businesses in the UK, with about one third of them borrowing money at any one time (on average £20,000), and on a relatively short term basis (British Bankers Association, Personal Communication). Much of this borrowing is secured by a legal charge (or mortgage) over the assets, most frequently in the form of property. For larger organisations which do not provide tangible security, the norm is to offer various forms of 'comfort', often by way of covenants or warranties. In such circumstances the onus is on the financial institution to ensure that the performance of the business matches that which was envisaged.

Additionally, lenders will generally require adequate insurance coverage over the assets and against damage caused to the assets (whether this be buildings, stock, work in progress, etc.). The availability of insurance coverage may be crucial to the granting of funding to a business, and demonstrates the close inter-relationship between the two sectors.

Many larger businesses, particularly MNCs, choose to spread their financial sector 'purchases' across a range of institutions and this has the effect of complicating, and perhaps blurring, any potential impact an environmental issue may have upon any one member of that sector. Additionally, MNCs rarely provide security to support their borrowing, the underlying strength of their balance sheets determining the pricing and access to funding.

The personal customer market is mainly split between the banks and building societies, both providing a range of services to the individual which involve the assurance and insurance sectors. In a deregulated and highly competitive market, many individuals choose to segregate their borrowing and savings requirements, spreading these across a range of institutions depending upon their short term and long term needs (for example, access to borrowing facilities, such as an overdraft or mortgage and life and pensions arrangements). Again, where security is required, the availability of adequate and appropriate insurance coverage is crucial, and for the domestic housing market this is particularly the case.

While the relationship between climate change and property coverage provided by the insurance sector is perhaps becoming clearer, for the reminder of the financial sector this is not the case; indeed, the issues have only recently started to be considered. This includes not only the banks but also the investment community whose funds are invested in the present with payment to third parties taking place many years in the future. Most of the analysis in this chapter will, however, be limited to the banking sector.

It is important to bear in mind that the financial sector, its products, services, customer base and operational practices, have been subject to rapid and ongoing change over the last ten years. This change has included deregulation of the markets, a reduction in the differentiation between players, such as building societies and banks, and merchant and clearing banks. The changes have led to a reduction in the number of organisations operating in the UK market, greater competition, a broadening in the range of products and services offered and a blurring of the customer base.

This trend is not only likely to continue but to increase and the industry will be substantially changed by the turn of the century and significantly different by the 2020s and 2050s. The trend towards mergers is likely to continue leading to fewer, but larger, players in the sector and the move towards international diversification and expansion will increase, thereby spreading operational (including climatic) risks.

14.2 ESTIMATED EFFECTS OF CLIMATE CHANGE AND SEA LEVEL RISE

The financial services industry currently has a strong tendency to work on a short term basis, the decisions being taken with a view to repayment of any funding, generally within a time horizon of 20 years. For the majority of lending propositions the time scale is far shorter, frequently between one and five years, although recent trends in terms of refinancing indicate that this is moving towards a 15 - 20 year time horizon. This should be beneficial to business and encourage the financial sector to take a longer term view of the impact of externalities, such as types and real costs of energy usage on the underlying viability of a business or sector.

However, the absence of 'hard' data in respect of risks associated with environmental issues and climate change in particular may result in inaction until either significant losses are experienced or a clear trend emerges. While there is some data for the insurance sector, this is not the case for the financial sector. Indeed, the absence of data relating to the financial sector generally and geographically confirms the need for greater efforts now if the implications are to be adequately understood and appropriate action taken.

The availability of insurance coverage for the domestic and commercial markets has further reduced the indirect effects on the banking/ building society sector as claims for damage to property and loss of income to business have generally been met by the insurance sector. As a result, there has been no real need to analyse the situation and little information on incidents to date is available from which to project figures. Indeed, for the insurance sector too the information and interest primarily lies within the relatively narrow area of property; their corresponding investment strategies do not cover climatic concerns. The absence of data may well compound the ambivalence expressed by other areas within the financial sector.

14.2.1 Sensitivity of financial sector to weather and climate

Direct implications

As greater reliance is placed upon the use of rapid communications media, both within and outside the UK, to complete business transactions within highly competitive markets, any changes of climate conditions which put this at risk (e.g. damage to power supplies) will become increasingly significant. The storm in October 1987 caused severe disruption to both the London and international financial markets as banks were unable to complete transactions in the majority of cases because of power failure. Although some banks did have sophisticated back up systems and could maintain 'normal' business operations, the absence of such systems by others rendered this unusable at the time.

Subsequently, most major financial institutions now have back up and emergency systems to cover such disruption although these have not yet been tested for a significant period of time on a large scale and internationally. Evidence of increased probability of disruption might result in the relocation by larger institutions of sensitive parts of their business to regions which are less sensitive to climate disruption (either on a permanent or seasonal basis) or to developing international networks which can be activated when necessary. This could, in time, reduce the overall direct (invisible) importance of the London market but allow for the development of overseas business, thereby allowing for the continuation of lower risk income streams fed back to individual parent organisations. There would, however, be potential socio-economic implications in respect of such action.

Indirect implications

As the UK's financial sector is predominantly international in nature, the impact of climate change outside of the UK is likely to have a greater impact than changes within the UK for the majority of banks. The situation for the building societies will be somewhat different as their business interests are limited almost exclusively to the UK. Therefore, the individual exposure of the different sectors and institutions within them will vary. For example, there may be greater impact upon mortgage lenders than commercial banks as damage to property and increased costs to individuals, either through increased insurance costs/excesses or absence of cover puts further pressure on their available funds to repair any damage and correspondingly upon their ability to repay any borrowing. Those whose business has a greater international diversity should be able to spread their risks more effectively.

14.2.2 Effects of climate change

Apart from extreme circumstances, such as major windstorms disrupting power supplies, and preventing business operations, changes to weather patterns in the UK are likely to have relatively few direct implications for the financial sector; in these circumstances the majority of the impact would tend to be indirect. As both domestic and commercial customer relationships tend to be spread throughout the country, it is unlikely on present projections, that single institutions will be impacted significantly, unless the changes take place rapidly and across a wide range of areas and industries.

There may, however, be longer term implications if the balance of industries within the economy changes. Additionally, as the UK has a predominance of small businesses, broadly spread in geographic terms, (Department of Trade and Industry, 1994) there should be an opportunity for change to take place in advance of any significant harm to the financial sector's own viability. This is unlikely, of course, to be the case for the private individual or businesses concerned and there may be knock on effects for certain geographic areas as a result and a corresponding demographic and socio-economic impact.

Taking the three issues of precipitation, temperature and wind speed (storm), only major storms are likely to impact directly and potentially significantly on the banking sector within the foreseeable future. The remainder of the impact will be indirect, affecting the financing of certain sectors, such as agriculture, and also involving those holding investments and venture capital.

In this context, it is critical that the financial sector continue to develop effective and timely risk assessment techniques. These should include aspects of climate change, where appropriate, allowing for the information to be incorporated into the risk management and decision making processes so that necessary checks and balances can be put in place. In the absence of relevant and meaningful data, however, this will prove difficult, if not impossible for individual banks to do.

While general large-scale trends may indicate that action is needed, the current information is so general that it cannot be applied to individual businesses or sectors. For example, it could be argued that there may be a need to consider the long term viability of the UK's agricultural sector in terms of its existing products but how can this information be adequately translated for an individual farm and against what timescale? The financial sector operates on the basis of a multiplicity of 'contracts' and relationships and in the absence of this type of data, financial institutions are unlikely to change their current debt or equity practices for no apparent or clear business reasons.

As mentioned earlier, the banking sector tends to operate in the short term when set against the time horizons for climate change. Although other areas within the financial sector, such as the pensions funds, in theory operate over a longer term, insofar that they are investing today to make payments to individuals in perhaps forty years time, in practice their purchasing and sales decisions for investment purposes are far shorter (and often much shorter than those of the banks and building societies).

Recent stock market fluctuations indicate that there is a trend towards the more rapid turnover of stocks and on a larger scale to that previously seen. Climatic effects which lead to a disruption of business in an important economic area, such as that illustrated in the non-climatic field by the Kobe earthquake in 1995, may trigger significant market movements. The adoption of twenty four hour dealing on an international basis coupled with technological enhancements and the ongoing globalisation of investments may exacerbate this change further, resulting in a tendency towards shorter term investment exposure to any one sector. The pace of change may become still faster as the emerging markets begin to develop their investment capabilities and encourage a greater intensity of stock market activity.

Precipitation

Both prolonged drought and freshwater flooding would be of relatively minor direct importance in terms of the location of financial sector premises, with perhaps some potentially vulnerable operations being relocated. As enhanced technological capability makes it less important to have 'on site' financial services, this problem will decline but there could be possible implications for some individuals and businesses, primarily in terms of inconvenience.

However, changes in precipitation patterns could impact on the financial sector as the effects (on both the domestic and commercial sectors) limited their ability to meet their financial obligations. This would be likely to be significant if the insurance sector withdrew property and loss of earnings/disruption coverage in the affected areas. Such a change would be likely to have cost implications for the financial sector which in turn would affect their customers.

Freshwater flooding would affect the agricultural sector in particular, as well as the distribution of materials to third parties, impacting upon the profitability of a business and its financial robustness. Apart from possible difficulties arising from additional precipitation, a reduction in rainfall could clearly also affect agriculture, requiring a shift to alternative crops/products. Assuming that this sector could make the transition gradually, any difficulties should be relatively easy to manage, the financial sector being able to arrange adequate refinancing of viable businesses.

For the domestic sector there could be implications in terms of the value of property if mortgagees decided en masse that certain vulnerable areas of the country, such as coastal lowlands, were no longer suitable as security in support of borrowing, either for a home or as collateral for business. This in turn would affect employment as well as other socio-economic areas. There would be greater impact upon the building societies in view of their particular portfolios, as well as on the construction industry.

Temperature change

There would be limited direct impact upon the financial sector, other than possible increases in energy consumption to counteract the changes of hotter summers and cooler winters (northern UK). Indirectly, temperature rises could alter the distribution and balance between business sectors within the UK, such as agriculture and tourism, affecting individual financial institutions differently depending on their own particular portfolios. The greater the planning period and advance knowledge of change and its likelihood, the lower the likely impact as both business and financiers planned for the change. For example, changes to the types and location of farming/agricultural practices would be less profound if an orderly change was effected. This in turn could allow for food distributors/retailers to reassess their sourcing requirements and practices.

Wind speed

The direct implications for the financial sector in terms of its information technology requirements and capabilities and the need for continuous energy resources to enable it to match international competitiveness issues have already been discussed. Other than this, the majority of impacts caused by storm damage would be indirect, as they affect both commercial and domestic properties. Again, the availability of competitively priced insurance coverage is crucial, allowing the banking sector to provide supporting products and services for customer and commercial needs (Advisory Committee on Business and the Environment, 1994.)

Damage to buildings caused by storms may require amendments to their design and the use of construction materials if they were to be part of an investment portfolio, but are likely to be driven by requirements determined by the insurance industry for coverage purposes. This would, of course, necessitate ongoing co-operation between insurers, financiers, research institutions (such as the Buildings Research Establishment) and the construction industry to minimise the impacts.

14.2.3 Effects of sea level rise

Once again, there would be relatively little direct impact upon the financial sector as a result of an increase in sea level, apart from a need to vacate vulnerable areas to protect communication needs. Indirect consequences could be significant, primarily in terms of disruption to business and the ability of both the domestic and commercial markets to repay their financial obligations. As with areas susceptible to freshwater flooding, vulnerable geographic areas susceptible to sea water encroachment may find it difficult to attract finance resulting in local economic and social consequences. These businesses could, however, be transferred in many cases to different areas, resulting in a possible increase in prosperity in those locations.

If the risk of flooding increases substantially due to sea level rise, the sector will face indirect pressures as business customers lose income through disruption, while homeowners and businesses face direct financial pressures through higher insurance premiums. Within the domestic market, mortgage lenders could be affected as the values of properties in these areas declined or individual dwellings were abandoned. This in turn could impact upon the investment houses, many of which hold substantial tranches of land for investment purposes. A move away from investing in certain areas would be likely to exacerbate the socio-economic difficulties.

In view of the importance of London as a financial centre, both in terms of its value to the UK economy and the global transfer of capital, it is critical that there are flood defence systems which ensure that any change in sea level rise does not result in significant flooding of the Thames basin. Although many large, financial institutions maintain back-up (shadow) locations and information technology system outside of the areas of likely flooding, the resultant disruption to the transport systems could cause severe difficulties as employees failed to reach their offices. The move towards teleworking by many of the larger institutions may help to reduce this risk, but it will still be important to ensure that the current security value of the Thames Barrier is maintained.

14.3 ASSESSMENT OF POTENTIAL ADAPTATION

The financial sector is adaptable in terms of its speed of reaction to information, in the creation of new products and services for its diverse customer base and in respect of the flexibility of location. This latter aspect may increase as information technology develops and allows for greater decentralisation of operations. Against this, however, is the trend towards out-sourcing of services which could lead to some vulnerability, such as in service/communication continuity, if not effectively managed.

There is, nevertheless, a close working relationship between the banking and insurance sectors, the former tending to provide finance at the outset with repayment being made at an agreed future date, while the latter collects premia at the start of an agreement and releases funds in the event of an unforeseen circumstance, provided that this is not caused wilfully or through negligence.

It is, therefore, important for the banking sector that insurance products, designed and priced as part of a broader risk management system, are available for the wide range of personal and commercial clients and across the whole of the UK, not just in selected areas and for selected sectors.

14.4 UNCERTAINTIES AND UNKNOWNS

In order for the two sections of the financial sector to develop their products and services, it is likely that new and improved sources of information will be required, both from the potential clients and from the scientific community at large. The provision of pertinent, accurate and timely information would enable both markets to respond to the issues associated with climate change, thereby allowing for minimal competitive (in terms of the UK overall), economic and social disruption.

As mentioned above the availability of insurance coverage for the domestic and commercial markets has reduced the indirect effects on the sector; as a result little information on incidents to date is available from which to project figures. Withdrawal of insurance cover, however, thereby making the costs more transparent to the financial sector, would be likely to result in a requirement for more information in order to allow for more accurate risk pricing.

From an economic perspective, the financial sector in its broadest context is highly competitive and climate changes, if not handled in a timely manner, could result in greater damage than apparent at first sight due to relocation of parts of the industry to other geographic areas as well as withdrawal of products and services.

14.5 PRINCIPAL IMPLICATIONS FOR OTHER SECTORS AND THE UK ECONOMY

The insurance sector's decision as to whether or not to continue to provide cover may result in higher risks to lenders and, as a consequence, result in the higher price of capital. Additionally, investment capital might be moved from the UK to other jurisdictions, further damaging the wealth of the nation.

All sectors could be affected depending upon their size and location, giving rise to serious socio-economic consequences. However, there are likely to be differences in timing, with sectors such as construction and water being affected at an earlier stage. Additionally, with over 70 % of the UK's land currently being used for agricultural purposes, changes within this sector could be significant, although greater production efficiency may counteract this impact if a timely transition in terms of products and their location takes place.

14.6 RESEARCH AND POLICY ISSUES

The absence of sufficiently detailed information/data on the potential effects of climate change within the financial sector, coupled with the current operating mechanisms within the banking sector, means that there is limited likelihood of action being taken in the foreseeable future.

This situation would markedly be altered if clear and material changes take place on the potential effects of climate change and are seen to have an effect on profitability or if insurance coverage is withdrawn and banking clients are forced to carry the costs themselves, resulting in a greater need for finance or a higher level of defaults from those affected.

14.6.1 Future research effort

In order to assist the financial sector there needs to be :

- Greater disclosure of information in respect both of business activities which may cause changes to climate and in respect of disruption to business activity by climate change so as to allow the sector to track the changes/trends over time.

- Improved communication between the providers of debt and equity finance and the insurance sector (including underwriting and loss assessors).

- Better tracking of incidence, that is climate changes and their financial implications by sector and location, (for example, the development of area maps by sector/location/vulnerability over time).

- Enhanced co-operation between all interested parties, including the regular exchange of information.

- Increased research into the economic implications of climate change for trade, transport and distribution systems in the UK.

14.6.2 Policy issues

- Improved co-operation is needed between all interested parties (bankers, investors, insurers, government departments) in the assessment of effects of climate change on the financial sector. In particular, clearer reporting of incidents of weather impacts is needed so that adequate analysis of trends can be made.

- Those institutions within the financial sector which have not developed contingency plans to take into account the possible effects of climate change should be encouraged to do so.
- Businesses and individuals need to participate fully in the assessment of potential effects and not consider insurers to be providing total assurance nor bankers to be the lenders of last resort.
- Government departments and agencies should work closely with the financial sector, particularly with the aim of providing the most appropriate and timely information that science allows.
- Financial institutions, particularly those providing longer term finance and venture capital should reassess their investment strategies and business practices in the light of new information on climate change.

15. Health

SUMMARY

- Climate change would have both direct health effects (e.g., heatwave-induced deaths) and indirect health effects (e.g., altered rates of infectious diseases and altered exposure to certain air pollutants). While some health benefits would result from climate change, the net impact upon health is likely to be adverse. In the longer term, the indirect health effects (especially from infectious diseases, allergic disorders due to pollens and spores, reduced supplies of freshwater) may become dominant.

- Most of the data that allow quantitative forecasting of health impacts refer to the direct effects of thermal stress and extreme weather events. US studies suggest that an increase in heatwaves would cause several thousand extra deaths annually, especially in large urban settings. UK research indicates that extra summertime deaths may be largely offset by reductions in winter-related deaths.

- Despite recent observations of short term climatic influences upon vector-borne infectious diseases in some tropical regions, and modelled predictions that suggest a wider geographical spread of various such infectious diseases (e.g. malaria, dengue and leishmaniasis in response to climate change), the UK should not be much affected. Public health defences appear to be sufficient to cope with climate-related increases in potential transmission 'pressures' from these diseases.

- Other food-borne and water-borne infective agents, as causes of diarrhoeal and dysenteric infections, are likely to spread more readily in warmer and wetter conditions.

- The existing problems of urban air pollution would be exacerbated by climate change, by enhancing the production of photochemical pollutants (e.g. ozone) and, perhaps, by amplifying the biological impacts of certain pollutants. The geographical distribution and the seasonality of aeroallergic disorders (hayfever, asthma, etc.) would also change.

- The overall burden of additional illness or premature mortality cannot yet be quantified, in particular because of the complexities of forecasting the various indirect impacts upon human health. Simplistic calculations based solely on, for example, heatwaves and weather disasters would therefore be incomplete and misleading.

15.1 INTRODUCTION AND BACKGROUND

Climatic variations affect many of the processes and systems that, in turn, affect human health. The impact of longer-term climate change upon human health would occur predominantly via mechanisms that would impinge on whole communities or regional populations. Thus, the scale of concern is at the population, not individual, level. More broadly, there is a growing awareness that the sustained health of human populations, across generations, depends on the continued stability and productivity of the earth's natural systems and its human-managed ecosystems. Each of these systems is, in various ways, influenced by climate.

Our insights into this complex potential hazard to human population health are limited and incomplete. It is an environmental hazard with a different scale, timeframe and complexity from those that we typically address as 'environmental health hazards' (McMichael, 1993a). Scientists have therefore only recently begun to pay formal attention to this topic.

Within any one country, such as the UK, the particular balance of positive and negative health impacts of climate change would reflect local environmental and socio-economic circumstances. In general terms, some of the anticipated health effects would result directly from altered exposures to physical agents (e.g., heatwaves, floods, storms) or to familiar infective agents. However, many effects would arise indirectly, via disturbances of natural systems. These would affect, for example, the range and activity of various infectious diseases and availability of fresh-water supplies. It is a reasonable expectation that, in general, disturbances of ecological relationships and biogeochemical systems (as distinct from mere shifts in mean temperatures) would pose risks to human health. Meanwhile, some benefits to health would also result from climate change.

The main categories of potential effects of climate change upon human population health (without particular reference to the UK population) are summarised in Table 15.1.

Table 15.1: Main types of potential health impact of climate change
(Note: this list does not refer specifically or exclusively to the UK)

Direct impacts
1. Deaths, illness and injury due to increased exposure to heat-waves. Reductions in cold-related diseases/disorders.
2. Altered rates of death, illness and injury due to changes in frequency or intensity of climate-related disasters (droughts, floods, forest fires, etc.)

Indirect impacts
3. Altered distribution and transmission of vector-borne infectious diseases (viral infections, malaria, leishmaniasis, etc.)
4. Altered distribution and transmission of certain communicable diseases (water-borne and food-borne infections, some respiratory infections, etc.)
5. Impacts upon agriculture and other food production. Adverse effects are most likely in tropical and subtropical regions; beneficial effects may occur in some temperate zones.
6. Effects upon respiratory system, via increased exposure to pollens, spores and certain air pollutants[1]
7. Consequences of sea-level rise-via flooding, disrupted sanitation, soil and water salination, and altered breeding sites for infectious disease vectors
8. Impacts on health caused by demographic disruption, displacement, and a decline in socioeconomic circumstances, as might result from impacts of climate change on natural and managed ecosystems

[1] Acute mortality relating to respiratory distress accounts for approximately 10% of heat-related mortality (US Environmental Protection Agency, in press). In addition, immediate marked increases in asthma-related mortality can occur during stressful weather (Goldstein, 1980).

15.2 ESTIMATED EFFECTS OF CLIMATE CHANGE AND SEA LEVEL RISE

15.2.1 Sensitivity of health to weather and climate

There are, as yet, few relevant data that enable the quantitative forecasting of most categories of anticipated health impact. This reflects both the nature of the impacts and the newness of this field of research. For some of the effects that would result from a direct impact of climate change upon human health, quantitative extrapolation from prior human experience is possible. For other effects, arising via indirect pathways, there is more complexity, non-linearity and uncertainty; therefore, there is little current prospect of their quantification.

An increase in the frequency and severity of heatwaves would cause an increase in (predominantly cardio-pulmonary) mortality and morbidity, particularly in the very young, the frail-ill, and the elderly. Physiological acclimatisation to higher mean temperatures would lessen the impacts. Studies in large urban populations in several continents indicate that the number of summer-time heat-related deaths would be increased several-fold by the typical 2050 climate-change scenarios (Kalkstein 1993; 1995). In very large cities this would represent several thousand extra deaths annually. On the other hand, there would be fewer cold-related deaths. However, insufficient research has been done to allow estimation of the balance between extra and averted deaths.

The organisms that cause and spread infectious diseases are very sensitive to climatic variables, especially temperature, relative humidity and surface water. In many parts of the world, relatively early indirect health effects would therefore be anticipated from changes in the abundance and range of insects and other vectors that spread vector-borne infectious diseases such as malaria, dengue, yellow fever and schistosomiasis. Temperate zones would become more hospitable to certain insect vectors and to the infective pathogens they carry (Shope, 1991; Sutherst, 1993). At a global level, it is predicted that there would be net increases in the potential for transmission of many vector-borne infectious diseases (IPCC, 1996b).

Some increases in non-vector-borne infectious diseases, such as salmonellosis and other food-related and water-related infections, could also occur because of climatic impacts on water distribution and temperature and hence on micro-organism proliferation.

Other climate change-induced health impacts would include changes in risks of death, injury and psychological disorders from altered patterns of extreme weather events (e.g., floods, storms, cyclones). A warmer climate, with bouts of hot weather and higher levels of incident ultraviolet radiation, may increase and prolong the exposure to certain air pollutants in urban environments. The effects of climate change upon agricultural, animal and fisheries productivity, while still very uncertain, could cause some regions of the world to experience gains, and others losses, in food production. Regional increases in the prevalence of malnutrition and impaired childhood development would result. The physical, social and demographic disruptions caused by rising sea levels, and by climate-related shortages in fresh water supplies, would also impinge adversely on health.

15.2.2 Effects of climate change

Heat-related and cold-related mortality

The most readily anticipated 'direct' health effect is that due to increased thermal stress. As is evident from Chapter 2, the frequency of very hot days in the UK is likely to double by the 2050s as a result of climate change. There have been consistent reports, in temperate countries, of increases in morbidity and mortality during heatwaves and of overall higher death rates in winter than in summer (Kilbourne, 1992). Data from London for 1987-1992 show the widely reported j-shaped relationship of daily mortality to daily temperature, with a stronger upturn at high than at low temperatures (Ponce de Leon and Anderson, Personal Communication).

Model-based forecasts (for CO_2-doubled climate scenarios) of the mortality impacts of an increased frequency of heatwaves in certain cities in the USA indicate that, by mid-21st century, there would be an increase of several percentage points in the number of deaths occurring over a summer (Kalkstein, 1993; IPCC, 1996b). Data from recent US heatwaves show that the increased mortality impinges most on vulnerable segments of the population, especially those living in poor and crowded housing (Kalkstein, 1995). By extrapolation, in a large city such as London this may represent several thousand extra deaths annually. However, analysis of daily time-series data indicates that many (20-40%) of these heatwave-associated deaths entail a forwards-displacement of the time of death by just several weeks. This accords with the fact that a large proportion of the deaths occur in frail and susceptible persons - the very old and those with pre-existing chronic diseases (Kalkstein, 1993; IPCC, 1996b).

This potential increase in heat-related deaths would be offset, partially if not fully, by a decrease in winter-related mortality. As in other temperate zone countries, there is a seasonal excess of deaths in winter in the UK (Curwen, 1991). This excess predominantly comprises additional deaths from cardiovascular disease (due to an increased propensity of blood to form clots, for reasons that include temperature-related physiology, seasonal variation in the intake of particular micronutrients, and, perhaps, reduced physical activity) and from infectious diseases.

However, no single study has yet been published that enables a direct comparison of the anticipated winter gains and summer losses that would accompany a background warming. Indeed, a quantitative estimation of the losses and gains in person-years of life will require new research into the phenomenon of displacement of time-of-death. One recent UK study has estimated that approximately 9000 fewer winter-related deaths (representing a 2-3% reduction) would occur annually by the year 2050 in the UK, under climate change scenarios that entail 2-2.5°C winter temperature increases (Langford and Bentham, 1995). Just over half of those avoided deaths would have been from ischaemic heart disease and stroke, with chronic bronchitis and pneumonia each contributing 5-10%. That study, however, only examined winter deaths, and therefore could yield no corollary estimate of the impact of increased summer temperatures.

Extreme weather events

Climate change scenarios for the UK indicate that there may possibly be changes in the frequency and/or intensity of storms or high winds. It is thought likely that the risk of river flooding will increase due to changes in rainfall regimes (see Chapters 2 and 7). The risk of coastal flooding is likely to increase due to sea level rise and possible changes in the frequency of storm surge events. It is uncertain whether storm activity may also increase in parts of the UK.

Changes in the frequency or severity of extreme weather events (floods, storms, etc.) would have both direct and indirect health impacts. This would encompass an increase in deaths, injuries and post-traumatic psychological disturbances, as well as the more diffuse health consequences of loss of employment and enforced relocation.

The public health impact of floods includes damage or destruction to homes and displacement of the occupants that may, in turn, facilitate the spread of some infectious diseases because of crowded living conditions and poor personal hygiene. Flooding may also cause contamination of water sources with faecal material due to the disruption of water or sewer systems and prevention of solid-waste collection and disposal. The psychological effects of natural disasters may have a considerable impact on the health care system. For example, following floods in Bristol in 1968, in which 3,000 homes were flooded and one individual died, there was an increase of 53% in visits to doctors for the year following the flood (Bennet, 1970).

Indirect Effect on Health

Vector-borne and other infectious diseases

The incidence and geographical distribution of vector-borne diseases are likely to be affected by climate change. The life-cycle and viability of both the vector (typically a cold-blooded organism) and infective pathogen (or 'parasite') are responsive to climatic variables. Recent unexpected experiences with infectious diseases, such as cholera (rapid spread in South America), dengue fever (ongoing northern spread through Mexico, into southern fringe of USA), and hantavirus (new source of respiratory infection, in response to environmental stresses in southwest USA), are a reminder that suprises occur frequently with infectious disease. We cannot expect to foresee clearly the impacts of climate change upon all infectious diseases.

Malaria is an important example, with around 350 million new cases worldwide annually, including approximately two million deaths. Malaria existed in the UK for several thousand years, and it is only within the past half-century that it has effectively been eradicated (largely via drainage of wetlands and use of insecticides, particularly in southeast England). Several recently-published studies predict that, under the IPCC's scenarios of global climate change, the *potential* geographical range for malaria transmission would expand (Martens *et al.*, 1995; Martin and Lefebvre, 1995). The *potential* for malaria transmission already impinges upon approximately 45% of the world population, and these highly-aggregated (and simplified) models forecast that this would increase to around 60% by later next century. Although this includes an increased potential for transmission in much of Europe, the existing environmental and public health defences should be sufficient to prevent its reintroduction. Elsewhere, actual increases are likely to occur in populations currently at the margins (in terms of latitude or altitude) of established endemic areas, particularly in tropical and subtropical regions.

The 1996 CCIRG UK climate change scenario of increased mean summer and winter temperatures and wetter winters would affect the proliferation of arthropod vectors (usually an insect, tick or mite) since these are more sensitive to temperature than to the availability of food (see Chapter 4). An increase in the abundance, activity, and geographical range of many insects species is likely to occur. Sea level rise may increase flooding and the presence of brackish water which may also encourage some insect vector species.

The seasonal activity of various insect vector species may be altered. Species may become active throughout the winter, instead of hibernating, or may breed earlier in the spring (see Chapter 4). For example, Lyme disease (caused by the spirochaete *Borrelia burgdorferi*) occurs occasionally in the UK in regions where the tick vector, *Ixodes ricinus*, is prevalent. Some such areas are of high recreational value, e.g. the Lake district, New Forest, and Richmond Park (Guy and Farquhar, 1991). The tick's seasonal behaviour is temperature-dependent, with bimodal peaks in spring and autumn in warm regions and a unimodal pattern in colder regions (Craine *et al.*, 1995). Hence, the transmission cycle of the spirochaete in the UK (Nuttall *et al.*, 1994) would be sensitive to changes in climate.

Climate change could result in various vector-borne diseases becoming established in the UK. The risk would be further increased by an increase in the number of infected travellers arriving in, or returning to, the UK. Vectors that have been historically limited by climate, but which are transportable in planes, ships, trains and trucks, could be more easily distributed beyond their present range. In addition to the possible climate-affected occurrence of previously unsuspected or unknown infectious diseases (as with Hantavirus Pulmonary Syndrome in the southwestern USA in recent years), the following are three examples of vector-borne disease types that could occur within the UK in the wake of climate change.

Arboviral infections: There are various arthropod-borne viral infections that cause haemorrhagic fevers and encephalitis (brain inflammation), the transmission of which is climate-sensitive. For example, tick-borne viral encephalitis, present in central Europe, is transmitted by ticks which are already present in the UK.

Malaria: Five of the Anopheles mosquito species found within the UK can transmit some strains of *Plasmodium vivax* (one of the several main forms of the malaria parasite). Locally-transmitted malaria was last present in southeastern England around the middle of this century. (During the 1960s and 1970s, coordinated mosquito control strategies in many European countries essentially eliminated malaria from Europe, after its presence for at least several thousand years.)

Although the likelihood of transmission of vivax malaria would increase in response to warmer temperatures, public health surveillance and control measures should prevent its becoming re-established in the UK. While there is doubt about the ability of (some) native mosquito species to transmit the more life-threatening *Plasmodium falciparum* (Bradley, 1989), efficient vectors of falciparum malaria from Eastern Europe or the Mediterranean could, in theory, become established in the UK if mean summer temperatures were to increase by several degrees centigrade. (falciparum malaria also requires a higher temperature than vivax malaria for the development of the parasite within the mosquito.) However, as in the rest of Europe at present, reasonable public health measures should suffice to prevent its transmission.

Local transmission of falciparum malaria has recently occurred in New York City, USA, when daily temperatures exceeded 25°C for at least 11-14 days (Layton *et al.*, 1995). Such episodes indicate that any climate-related world-wide increase in malaria may affect the UK – at least by increasing the number of cases of imported malaria (and 'airport malaria'). This has surveillance implications for blood transfusion services.

Leishmaniasis: This vector-borne infectious disease is endemic, at a low level, in rural areas of the Mediterranean region of Europe (including most of Portugal, Spain, Italy and Greece). Climate change would make more probable the establishment of the sandfly vector of leishmaniasis in the UK, and hence the possible introduction of Leishmaniasis.

Food-borne and water-related illness

Climate change would also affect the ecology and population dynamics of bacteria and other pathogens that can contaminate food or drinking-water supplies. Warmer temperatures would generally encourage the proliferation of disease-causing micro-organisms.

There are seasonal patterns in the incidence of food-borne illness in the UK (see Figure 15.1: Cowden *et al.*, 1995). Food-borne illness in England and Wales during 1982-1991 occurred much more frequently in the late summer months; further, a particularly strong relationship was observed between the incidence of food-borne illness and temperature in the preceding month, suggesting that high ambient temperatures have their most significant impact at points in the food system prior to the food reaching the consumer (Bentham and Langford, 1994). Assuming continuation of the current structure of the agri-food system, the UK could expect monthly increases in food-borne illness of between 5% and 20% by the 2050s in response to the IPCC's anticipated temperature changes, with the highest proportional increases occurring in spring and autumn (Bentham and Langford, 1994).

**Figure 15.1: Seasonal distribution of outbreaks of infectious intestinal disease, 1992 and 1993 (n = 458), England and Wales.
Source: Cowden et al., 1995.**

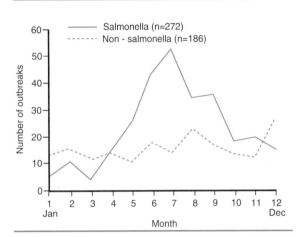

The UK has experienced periodic outbreaks of water-borne diseases through contamination of drinking water supplies, often associated with the inadequate treatment of river water (Anon, 1990). Climate change may alter the availability of subterranean freshwater in some parts of the UK (see Chapter 7) and thus lead to the use of more easily contaminated sources of drinking water. Recent outbreaks of cryptosporidiosis (Public Health Laboratory Service, 1995), such as in Devon, which may have been associated with consumption of contaminated mains water during water shortages, illustrate the nature of this problem. The contamination of recreational water courses may increase because of increased water temperatures, milder winters, increased agricultural run-off (which increases the level of phosphate-nitrate nutrients), and reduced riverflow which concentrates pathogens.

Conditions which favour the occurrence of blooms of blue-green algae (cyanobacteria) include hot summers and nutrient-rich waters (caused by changes in agricultural run-off or reduced riverflow) and, therefore, the incidence of freshwater toxic algal blooms may increase with climate change (see Chapters 4 and 7; Arnell et al., 1994). Some species of cyanobacteria are toxic but there is little correlation with the composition of a particular bloom and the concentration of its toxin. Further, very little epidemiological research has been done on the risk to human health from algal blooms, although contact with cyanobacteria has been associated with skin reactions, rhinitis, gastrointestinal complaints and atypical pneumonia (Elder et al., 1993).

The contamination of shellfish with marine biotoxins occurs seasonally on the northeast coast of England and the east coast of Scotland, during April to July, often when blooms of marine phytoplankton occur. The various environmental factors that influence the occurrence of marine biotoxins are poorly understood. The production of biotoxins associated with temperate waters (such those which cause paralytic shellfish poisoning and diarrheic shellfish poisoning in the UK) is temperature-sensitive, and may be expected to increase at higher water temperatures (Hallegraeff, 1993).

Food Production

At the global level, it is anticipated that climate change would affect crop (and other food) production (Parry and Rosenzweig, 1993). To the extent that this reduces food supplies available to any particular population, this would affect nutritional status, hunger, child growth, and health. Such adverse effects would be most likely in tropical and sub-tropical countries, especially in poor rural areas. In many temperate zones, including much of the UK, agricultural production may increase (at least in the early decades of climate change) (see also Chapter 5).

Warming might actually facilitate further 'mediterraneanisation' of the British diet, in particular by extending both the growing season for salad vegetables (closer to their point of final consumption) and the period of the year in which the consumption of such foods seems appropriate. This would have favourable effects on risks of various non-communicable diseases, especially cardiovascular disease. Further, any adverse impacts on aspects of food production in the UK should not affect population health since the UK would continue to purchase food imports as necessary. Climate change may, however, cause some economic and social dislocation of farmers and fishermen, and this, in diverse and indirect ways (associated with socio-economic dislocation, psychological stress, and internal migration) would tend to affect health adversely.

Respiratory disorders

An increase in the frequency and stability of anticyclonic conditions in summer (hot weather) with longer sunshine hours would enhance the formation of photochemical oxidants such as ozone in the breathing-zone lower atmosphere; it may also increase the biological impact of air pollutants (Ponce de Leon et al., in press; IPCC, 1996b). Changes in temperature and rainfall would also affect the production and dispersion of pollens and spores. These changes would, in turn, influence risks of various respiratory disorders and diseases.

Seasonal allergic disorders would be affected by changes in the production of pollen and other biotic allergens; plant aeroallergens are very sensitive to climate (Emberlin, 1994). Changes in pollen production would principally reflect changes in the natural and agriculturally-managed distribution of many plant species (e.g. birch trees, grasses, various crops such as oilseed rape, sunflowers, and ragweed species) along with their annual pattern of pollen release. Hay fever (seasonal allergic rhinitis) occurs directly in response to pollen release. The increase in the frequency of hay fever in Europe over the last few decades, including in the UK, has been greater in urban than in rural areas (Emberlin et al., 1993), and this may reflect an additional influence of urban air pollutants. While the causation of asthma remains uncertain, some particular urban air pollutants and spring-time pollens clearly contribute (e.g., Goldstein, 1980). The seasonal distribution of asthma in temperate regions is complex, with a peak in the pollen season and a further increase later in the year; in the tropics asthma increases in the wet season (IPCC, 1996b).

15.2.3 Effects of sea level rise

Rising sea levels would have a range of health impacts around the world, especially on vulnerable coastal populations (e.g., Bangladesh and small island states). UK coastal communities would be affected by changes in the frequency and severity of storm surges and subsequent flooding (see Chapter 17). A particularly vulnerable region is East Anglia. Indirect public health impacts may occur via economic, social and physical dislocations (e.g., salinisation of coastal farm-land).

Globally, climate-related processes such as sea level rise, the drying and erosion of agricultural lands, and exacerbated water shortages in arid and semi-arid countries will all tend to increase the migratory pressures upon the wealthier temperate countries, such as the UK (Myers and Kent, 1995). The dilemmas of balancing socio-economic, public health, political and moral considerations would thus intensify.

15.3 ASSESSMENT OF POTENTIAL ADAPTATION

There is a wide range of adaptive options to minimise adverse health impacts from climate change. These will involve several sectors, especially environment, meteorology, health-care, and education (and public media).

Major adaptive options include the following: environmental management (e.g., control of waste-water engineering to deal with heavier downpours, and control of surface water in relation to mosquito breeding sites); disaster preparedness (e.g., weather watch/warning systems for heatwaves and storm surges); protective technology (insulated housing, energy-efficient air conditioning, water purification, etc.); enhanced vaccination programmes (where vaccines exist for those infectious diseases that may increase in incidence); improved and extended medical and hospital care services (especially for the impacts of heatwaves, other extreme weather events, and the rapid treatment of persons with infectious diseases); public education directed at personal behaviours (e.g., heat exposure during heatwaves); and appropriate professional and research training and re-training.

It will also be important to assess in advance any risks to population health from proposed technological adaptations (such as alternative energy sources).

15.4 UNCERTAINTIES AND UNKNOWNS

As a category of environmental health hazard, climate change differs markedly from the types of hazard with which human society has been predominantly concerned. We are well familiar with the nature of the hazard posed by *local* environmental pollutants. However, climate change and its environmental consequences would be the result of a *global* overloading of the earth's natural systems. Because of the resultant changes and disruptions to various ecological relationships and biogeochemical systems, it would pose health hazards on a different scale and of a somewhat unfamiliar content (McMichael, 1993b). In light of this wider 'ecological' dimension, it is a prudent assumption that climate change would, on balance, constitute a serious long-term hazard to human health.

Since future climate change and its environmental consequences can only be addressed as a scenario, rather than as an observable reality, the forecasting of consequent health impacts necessarily entails uncertainties (McMichael and Martens, 1995). A 2-3°C increase in average global temperature within a century would fall outside the range of recorded human experience, and would differ qualitatively from past experience of temperature fluctuations around a long-standing stable-state mean temperature. In these respects, the past can provide only limited guidance to the future.

It is also well recognised by climate scientists that, as yet, there is limited capacity to model and forecast climate change at the regional and local levels. Relatedly, as consideration of potential health impacts becomes focused on local populations, such as the UK population and its member subpopulations, so the uncertainties compound in relation to the many specific circumstances (including local adaptive responses) that, ideally, should be taken into account.

Prediction of the health effects of climate change must (for the moment) rest largely on extrapolation and reasonable conjecture. Comparative studies of disease patterns in relation to existing climatic differences within Europe would facilitate this. As more information accrues, the modelling of population health impacts will improve. There may also be early clues, such as shifts in infectious disease patterns that are reasonably attributable to defined climatic fluctuations or changes. Unlike the health hazards posed by localised pollutants, however, this category of health hazard does not readily allow empirical observation of health impact in one population as a basis for estimating the risk in other populations. Climate change will impinge widely, affecting vulnerable populations approximately simultaneously; these circumstances of scale and timing would generally preclude the useful accrual of empirical epidemiological observations. That is, the usual sequence of observation, causal inference, and preventive intervention will be of little relevance. Rather, the assessment of future health impact will require a new capacity for systems-based thinking, predictive modelling, and dealing with uncertainty.

15.5 PRINCIPAL IMPLICATIONS FOR OTHER SECTORS AND THE UK ECONOMY

The burden of additional illness or premature mortality cannot yet be calculated in any comprehensive fashion. Hence, the estimation of economic costs remains elusive, in particular because of the various indirect and diffuse impacts upon human health. For example, there would be health consequences associated with climate-related disruption of employment (e.g., local fisheries, energy generation) or with increased immigration from environmentally-degraded tropical countries.

Simplistic calculations based solely on (for example) extra deaths related to heatwaves and weather disasters are misleading. Indeed, there is a problem here similar to the debate about 'internalities' versus 'externalities' in economic accounting. Certain health impacts are readily foreseen and counted (e.g. number of heat-related deaths) and thus become ready substrate for conventional 'costing'. However, the much wider penumbra of less direct, less immediate and less obvious health impacts must also be incorporated, in some fashion, in the quantitative assessment of social and economic costs.

15.6 RESEARCH AND POLICY ISSUES

15.6.1 Future research effort

There is an urgent need for local research:

- To identify the range of potential health impacts of climate change within the UK.
- To carry out comparative studies of disease and mortality patterns in relation to recent and present climatic variations, within the UK and within Europe. (In general, there is need for more extensive knowledge about climate-health relationships, as a basis for predicting the impacts of climate change.)
- To contribute to the development of predictive modelling (for national, regional and global application). This will require collaboration with, among others, biologists, ecologists, mathematicians, and modellers.
- To produce indicative estimations of the net health impact on the population, and the associated social and economic costs. This will require collaboration with social scientists.

15.6.2 Policy issues

Climate change and its environmental and ecological consequences, along with various other incipient global environmental changes, pose a hazard to the long-term health of human populations. In the first instance, therefore, it is important to increase the use of UK data for the monitoring of changes in population health status or in hazards to population health. Simple opportunities already exist in relation to monitoring changes in the seasonality of diseases and in the health impacts of heatwaves, warmer winters, and altered patterns of air pollution.

More generally, population health status is an important integrating outcome that reflects a range of other environmental impacts. Therefore, awareness of the potential health impacts of climate change should have substantive implications for policy-making in the various 'upstream' sectors (environment, industry, construction, agriculture, etc.) that would mediate some of the effects of climate change.

16. Recreation and Tourism

SUMMARY

- Climate forms an important part of the environmental context for recreation and tourism. Climate change is likely to provide new opportunities for the industry if a trend towards warmer, drier and sunnier summers stimulates an overall increase in tourism in the UK.

- Any climate-led changes will take place on both a domestic and an international scale but may be over-ridden by other factors, such as the availability of leisure time and the amount of disposable income.

- An enhanced uptake of many outdoor pursuits ranging from gardening to more strenuous sports, particularly those which are water-based is likely to occur.

- Any increase in either the intensity of use of leisure facilities, or the length of operating seasons, is likely to have a positive economic impact through better employment opportunities and the enhanced viability of commercial enterprises.

- Given an overall rise in tourist activity, there will be a need for better management of visitor pressures at peak periods in order to maintain quality. Greater problems of environmental protection could arise in key settings including crowded municipal parks, eroding beaches and moors and heaths threatened with an increased fire risk.

- In the north of the UK, any significant increases in rainfall, windspeed or cloud cover are likely to offset the more general advantages associated with higher temperatures. The viability of the Scottish ski industry will decline if snow confidence becomes less secure than at present.

- Potential rises in sea level will adversely affect fixed waterfront facilities, such as marinas and piers. For beaches backed by sea walls, increased erosion could lead to a loss of beach area. Other recreational locations along the coast, such as sand dunes, shingle banks, marshlands and soft earth cliffs, may also be affected.

16.1 INTRODUCTION AND BACKGROUND

Tourism may be defined as all the activity involved with visits away from home, lasting for up to one year, which are undertaken for leisure, business and other purposes. Within the UK, tourism is estimated to employ more than 1.5 million workers and has a regional breakdown in value as illustrated in Figure 16.1. The whole UK leisure market is worth £40 billion (at 1992 values) of which some £19 billion, mainly outdoor activities, is weather-related. With a long-term trend to more leisure time, partly created by a rising proportion of retired people, and more disposable income amongst the population as a whole, the importance of tourism is unlikely to decline in the future.

UK tourism relies heavily on the attraction of townscapes and landscapes. 'Cultural tourism', incorporating largely indoor visits to historic cities, museums and art galleries, is important, as is business travel and family visits. Outdoor recreation is based on the variety of coastal and rural landscapes which exist over short distances, giving rise to leisure travel for the appreciation of scenery and heritage. In addition there is growing participation in outdoor sporting activities. Many of these commitments are undertaken in spite of, rather than because of, the climate.

It is not possible to assess UK tourism in isolation. In terms of relative competitiveness, tourism in the Mediterranean region guarantees sunshine and higher temperatures for short summer visits, together with a climate which is warmer and drier at other seasons attracting longer-stay visitors. This combination of 'push-pull' factors in tourist travel decisions is evident in the relationship that exists between summer rainfall in Britain and the number of people who take Mediterranean holidays the following summer (Smith, 1990). Climate change is likely to alter these competitive international balances as well as influence domestic leisure activity.

The success of recreation and much of tourism in the UK is highly dependent on the quality of the leisure environment. This means in part the renewable biophysical resources on which rural tourism and outdoor activities depends, such as forests, lakes or beaches. But quality considerations also apply to the built environment and any trends towards more congestion, litter, traffic and noise pollution will reduce the attractiveness of historic cities. Already some key sites are beyond their effective carrying capacity at peak times and many facilities require better visitor management.

16.2 ESTIMATED EFFECTS OF CLIMATE CHANGE AND SEA LEVEL RISE

16.2.1 Sensitivity of recreation and tourism to weather and climate

A distinction can be made between weather-sensitive and climate-dependent tourism (Smith, 1993). On their own, UK climatic conditions are insufficient to generate either mass travel from overseas or guarantee planned outdoor leisure at any season. Indeed, there has been a sustained erosion of the domestic summer market for many years by holidays taken abroad in 'sunshine' destinations. Mintel (1991) claimed that 73% of respondents to a survey cited 'good weather' as the main reason to go abroad. Such 'import substitution' has been only partially offset by the emergence of short-holiday markets and the growth of the day-visitor trade, stimulated by factors such as progressive improvements in road transport.

Despite this, UK tourism displays marked seasonality and weather sensitivity. In Scotland ferry services to the west coast and the islands carry over half of all passengers during 12 peak summer weeks and 40% of all bed-nights booked into youth hostels are registered in July. Within seasons, the prevailing weather conditions become progressively more influential in affect-

Figure 16.1: The regional value of tourism in the UK in 1992. Source: British Tourist Authority and Northern Ireland Tourist Board.

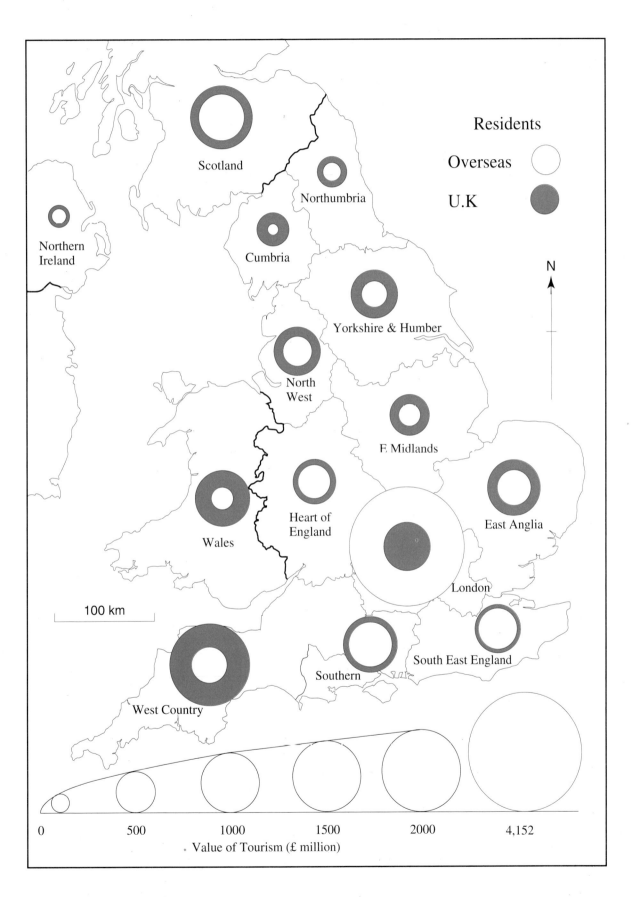

ing tourism as the amount of atmospheric contact necessary for each activity increases (Wall, 1992). This is especially true as tourists adapt to weather conditions by short-term, opportunistic decision-making. For example, during wet weather conditions unfavourable for walking or camping, these activities may be exchanged for visits to nearby towns. The switch may generate additional income for the local economy through visits to shops, restaurants and other indoor facilities. This demonstrates that 'poor weather' is a relative term and that weather-sensitive 'winners' and 'losers' already exist.

16.2.2 Effects of climate change

Summer

Any trend in southern Britain towards increased temperatures and lower summer rainfall, in association with reduced cloudiness and humidity, is likely to produce an increase in the amount of leisure time spent outdoors. Along eastern and southern coasts, for example, higher air and sea surface temperatures will permit more bathing, and could even lead to the restoration of the traditional family beach holiday. These positive changes assume that related atmospheric problems, such as poor air quality and the threat of skin cancer, do not keep people indoors.

Even then, it is possible that most UK resorts will thrive, as long as prices for accommodation and other services are competitive. For example, the Mediterranean is likely to become less attractive for health reasons in the summer. Apart from the dangers increasingly associated with skin cancer, many Mediterranean beach resorts may simply be too hot for comfort in the peak season, with a much higher frequency of severe heat waves (Perry, 1987; Giles and Balafoutis, 1990). A few destinations (e.g. Cyprus and Corsica) offer the potential to commute from hot beaches to cooler mountains but the Mediterranean area is likely to face other climate-related problems, such as marine water pollution and the scarcity of fresh water supplies.

Longer-haul flights to tropical destinations may not offer a solution. Several destinations in the Caribbean, which presently attract charter flights predominantly in the summer (Perry and Ashton, 1994), and the Far East are already prone to severe storms and hurricanes. As marine temperatures are likely to rise further, there is a possibility that the existing hurricane zones may experience greater activity in the future and that such storms may spread to other coastal areas. Even the perception of increased hazards might damage developing tourism in countries such as Malaysia or northern Australia which have few facilities to offer the holiday-maker apart from climate, as well as more established tourist areas, (e.g. Florida). In general, destinations with a great reliance on their natural resource base to attract tourists may be at most risk, although some high latitude destinations may become more attractive.

Some summer recreation already takes place along the shorelines of freshwater lakes and reservoirs, but these water spaces are often under-used in the UK. Increases in air and water temperatures may enhance this leisure potential, as long as changes in the thermal and hydrological regimes do not combine to produce algal blooms or otherwise impair the quality of the water space. Studies have shown that the demand for water-based recreation is partly determined by quality (Burrows and House, 1989). Any increase in use can lead to problems of shoreline erosion and conflicts between recreational and conservation interests. In addition, higher lake temperatures in the UK would threaten the survival of some fish (whitefish and charr) and higher river temperatures would slow the growth rate of brown trout (Arnell *et al.*, 1994). This might lower the options for game fishing.

The drier, warmer summer weather envisaged for parts of England will lead to more opportunities for certain outdoor pursuits. For example, increased garden use (already one of the most popular leisure activities) seems inevitable. Such a trend will be further encouraged by the demographic shift to people over 30 years of age together with more home ownership, more houses with gardens and growing environmental concerns. Current predictions, which ignore climate change, suggest that this market will grow at an annual compound rate of 1.8% to reach £27 million (at 1993 values) by 1998 (Data Monitor, 1994). The garden market is likely to react in different ways to climate change, with more demand for drought-resistant plants (as water supplies for hedonistic purposes become less available), and with improved retailing opportunities for accessories such as garden furniture and barbecues.

Over much of England and Wales climate changes are likely to benefit a whole range of activity holidays, which already account for about one-quarter of all UK holiday expenditure. These include walking, climbing, cycling, golf, riding, game fishing, sailing, canoeing or conservation work. The scope for some more specialised weather-dependent sports, like hot air ballooning, will also increase (Thornes, 1983). Given an expected increase in the frequency and strength of sea breezes, gliding opportunities are likely to increase along coasts.

Periods of summer heat stress and associated poor air quality in urban areas may make large cities less desirable places in which to spend leisure time. This could lead to a weekend flight to the countryside and to the coast, where land temperatures will be moderated by sea breezes during the day. Such a pattern could stimulate a rise in the ownership of second homes along the coast and in rural areas by urban residents.

Further north, the changes in climate are likely to be less favourable for outside activities. Although it is often claimed that the weather does not deter overseas visitors, around 40% of all holiday visitors to Scotland cite 'poor weather' as the major dislike. There is also criticism of the lack of wet weather facilities. Despite the fact that part of the attraction of theme parks is their all-weather facilities, Moutinho (1988) found that a 'good climate/environment' was mentioned by half the visitors sampled in a Scottish survey and it was suggested that a series of poor summers could prove disastrous for a park that relies heavily on open-air rides and similar features. There is also evidence from outside the UK that complexes such as EuroDisney suffer from lower attendances in wet weather (Perry, 1986). If it remains wet, or becomes even wetter in summer in Scotland, with greater humidity and more biting insects, the already high proportion of Scots taking summer sunshine holidays elsewhere may rise even further and there may be increased demand for indoor visitor facilities.

Other seasons

An increased level of UK tourism will not necessarily be confined to the summer. Improved climatic conditions, perhaps with more reliable medium-term weather forecasts, are likely to lead to more spontaneous day visits and short-break holidays. Domestic travel patterns are likely to be more stable than international travel in the face of climate change because the former often take place in relatively short periods of free time and these time limitations place constraints on destination choice. More short-break holidays taken at the shoulder seasons will reinforce marketing efforts within the industry to extend the season. During May, June and September the weather is often congenial with fairly low rainfall, good visibility, long hours of daylight compared with more southerly locations and temperatures still high enough for many outdoor pursuits. Activity holidays taken in spring and autumn already account for a large amount of the shoulder-season occupancy. With most age groups showing an increased interest in health and fitness, certain outdoor activities with high-spending profiles per trip, such as golf, tennis and fishing, as well as walking in the countryside, are likely to become more popular during the shoulder months.

Winter holidays are now an important part of tourism. Harlfinger (1991) showed that 86% of Germans visiting Majorca during the winter found the climatic experience to be positive and over half regarded the return to Germany as a potential burden on their health. Milder winter conditions could provide less cold stress to older people in Britain and mean that vacations currently taken in the Mediterranean, or wider afield, by British residents could become less compelling. In part, this will depend on the ability of the UK industry to develop and market good accommodation and rival attractions at home. It is doubtful that this will be a factor in the north of Britain where the climate scenarios suggest a continued exodus to winter sun.

Milder winters will lead to less disruption and cancellation of outdoor sporting fixtures, such as football, hockey and racing events. Reduced frequencies of frost and snow will provide a financial opportunity for sports managers because of the increased playability of outdoor pitches, including less need for the costly under-surface heating of turf at major venues, and also because spectators will be able to travel to such events more reliably. However, heavier winter rainfall could also lead to the need for better drainage of playing pitches, golf courses and athletics tracks.

Generally milder conditions, especially in the winter and shoulder months, could entice more people into the uplands for hill walking, creating opportunities and threats for agriculture and nature conservation. Mountaineering may provide some partial compensation for reduced skiing opportunities in Scotland but there may be a greater risk of snow avalanches in the warmer conditions. In the 1994/1995 winter season, an estimated 25 climbers died on the Scottish hills, including 10 in avalanches. Although such deaths represent less than 0.1% of those taking part in the sport, more initiatives are likely to be needed to educate people about upland hazards. Any trend to wetter conditions in northern Britain, together with more upland use, means a much greater likelihood of increased footpath erosion. This may well increase the existing management problems on key long-distance routes, such as the Pennine Way.

Warmer winter conditions will bring less snow and snow confidence (the probability of snow lying at particular times during the season) will decline. Higher windspeeds will lead to more snow drifting and less access to the slopes as a result of the closure of ski lifts and tows in conditions of gale-force winds, low cloud and reduced visibility. At present, the ski season in Scotland is from January to April. As the season contracts, suitable snow conditions at New Year and Easter will become less likely and a complete snow cover will become less certain at the critical February mid-term school holiday period. Estimates for the facility at Glenshee suggest that 60,000 skier days per season are required to maintain viability, compared to the 180,000 skier days achievable in a good year. As an indication of the effect of warmer winters, in 1991/92 the number of skier days fell to 12,500.

Poor snow conditions have several consequences, including the inability to use expensive facilities like piste grooming machines. It is important to note that, in Switzerland, the economic effects of recent snow-deficient winters have been greatest on the resorts at low altitudes, impacting adversely on both transport and accommodation (Abegg and Froesch, 1994). Given the prospect of an aperiodic series of snow-deficient winters in Scotland, it is likely that ski clients will either fail to book or at least delay booking until a few days before arrival in order to be certain of snow. Reduced sales and hire of ski equipment will follow, together with the laying-off of staff in hotels and other service sectors. More ski accidents, followed by insurance claims for medical expenses, due to the presence of thin, patchy snow cover with more exposed rocks and skiers concentrated onto congested slopes, may occur. If the recent trend toward more westerly winds continues, the locus of the ski industry may transfer from certain areas, such as Glenshee and the Lecht in the east, to more westerly facilities at Aonach Mor and Glencoe.

16.2.3 Effects of sea level rise

Rising sea levels will produce increased inundation and erosion along the coast. These processes will affect all the natural and cultural features along marine shorelines, many of which (beaches, golf courses, promenades, swimming pools, piers, marinas, sea-front hotels) are tourism-related (Boorman et al., 1989). For example, there are an estimated 450 designated 'bathing beaches' in the UK together with 370 coastal marinas, many in the south and east of England. Some of the most important links golf courses (Muirfield, St Andrews, Royal Lytham St Annes) are at risk from a combination of rising sea levels and coastal erosion. Studies of consumer attitudes suggest that the public values the coastal environment highly as a recreational resource and that there is a willingness to pay for its protection. For example, when asked to value their enjoyment of a beach visit in monetary terms, visitors to sites on the east coast of England produced an average figure exceeding £7.00 (Coker et al., 1989).

For a beach backed by a sea wall, as in many resort towns, the likelihood is that increased erosion would lead to a loss of beach material just at the time when the demand for beach-based holidays may be growing. Depending on the supply of sand, the beach facility could be totally lost with subsequent undermining of the stability of the sea wall. Other UK coastal habitats used for recreation such as sand dunes, shingle banks, marshlands and soft earth cliffs would also be affected. Attempts to develop a coastal vulnerability index for Wales suggest that some currently attractive dunes and bays (e.g. Borth Bay), could be the first type of 'beauty spot' to disappear. More flooding of estuaries could reduce the scope for specialist pursuits, such as bird watching, and lead to ecological changes at Sites of Special Scientific Interest.

The costs of shoreline protection for recreational facilities built along the shore, plus any beach preservation and replenishment activities, will be high. For example, Barkham et al., (1992) highlighted the £165,000 spent in 1990 on improving sea defences to protect the Royal West Norfolk golf course while at Sidmouth, Devon, the lowering of beach levels by over 4 metres has necessitated the expenditure of over £6 million to provide new breakwaters and enhanced beach nourishment. Once again, these effects may not be worse than impacts elsewhere. The low gradient beaches characteristic of much of the Atlantic and Gulf Coasts of the USA are very vulnerable to erosion (Leatherman, 1989). In other countries important historic sites are already under threat from rising sea levels and storm surge.

16.3 ASSESSMENT OF POTENTIAL ADAPTATION

Tourism is a continuously adapting industry, responding to changing demographic and economic conditions as well as to new demands and technologies. In view of the fragmented structure of the industry, climate change adaptation is likely to be gradual with new investment in tune with other strategic decisions.

The most severe economic problems are likely to exist in Scotland. The ski industry has less scope for adapting to climate change than exists in some other regions. For example, there is not the option, available in the Alps, to concentrate more on higher-level glaciers or to transfer clients to other resorts in order to compensate for a lack of snow. Some of the losses may be offset by an extended summer recreational season although there are relatively few alternative tourist facilities in the Highlands.

Any future investment in winter sports activities should be taken in the knowledge that the outdoor season is likely to become shorter

(McBoyle et al., 1986). Even if direct adaptation takes place, such as the development of indoor ski facilities, the investment may not be confined to Scotland. For example, the recently built Tamworth Snow Dome is near large population centres in the Midlands and can open 15 hours per day, thereby attracting some 1,500 indoor skiers per week. Within Scotland the only options appear to be a diversification into other activities, perhaps combined with a campaign to persuade people that climate should have a lower weighting in leisure decisions. Obstacles to growth can only be overcome by raising product quality and increasing value for money. At present there is a relatively poor provision of all-weather facilities and perhaps scope exists for major investment in such a facility in the more populous central belt. Strong indoor attractions have the advantage of year-long season viability, which benefits accommodation operators and provides more permanent employment opportunities.

Most indoor attractions may benefit from the general increase in tourism. For example, heritage centres increasingly offer safety for children and the opportunity to acquire knowledge. The present trend is to enhance these centres to provide more 'hands on experience'. The current health and fitness boom is also likely to lead to more indoor tourism. Sophisticated weather-proofed facilities, such as 'tropical' pools with other sporting and refreshment facilities, also offer scope for quasi-serviced accommodation and year-round operation. These centres are likely to be complemented with more community sports halls and swimming pools containing squash, indoor bowls and ten pin bowling. There is also a need for more hotels to provide indoor pools, squash courts and gymnasia which will not only provide more business use but will also attract short-break holiday makers.

Further south, theme parks are likely to attract even more than the present 75 million visitors per year, mostly day-trippers. Some parks, such as Thorpe Park, Surrey, already concentrate on water rides. Given significant changes in summer weather conditions, aqua parks (modelled on facilities in southern Spain or Tenerife) may become more popular. At Thorpe Park some 60% of visitors currently arrive between mid-July and the end of August; any extension of the season, perhaps partly as a result of non-climate factors (e.g. the semesterisation of the school year), would have important implications for the profitability of such ventures.

Along the coast, the risks of an eroding shoreline for fixed resort facilities, such as marinas and piers, must be taken into account. Some of the coastal resort facilities and services in the UK are still of indifferent functional quality by modern standards, despite recent investment. Climate change might provide an opportunity for these traditional resorts to modernise more quickly, to recover business and achieve further growth. Each resort has its own strengths and weaknesses and must formulate its own strategy for better niche marketing, whether this be for retired people or in catering for young people, with activity holidays and 'white knuckle' attractions. Ideally, each resort should have at least one key attraction with strong market impact.

In rural areas, outdoor recreation will continue to build on natural resources, such as forests and water. This must be done while maintaining a balance with conservation and local agricultural interests. For example, there may be an increase in leisure motoring, using more air-conditioned or 'open-top' cars and off-road recreational vehicles. To develop water space facilities effectively, visitor accessibility and the availability of local accommodation are key factors. The imaginative use of water can attract a high level of spending per visit and could prove successful within day-travel reach of major population centres. An alternative development, catering to a different market, could be the promotion of good-quality, rural self-catering cottages designed to meet the needs of urban dwellers fleeing poor atmospheric conditions in the cities. Such cottages can achieve a high rate of occupancy in areas which offer visitors wet weather activities and could be modelled on the successful French gite system.

16.4 UNCERTAINTIES AND UNKNOWNS

One weakness of some of the existing climate impact studies is the assumption that 'everything else will remain equal' over the next few decades. Recreation and tourism in the UK will react especially strongly to future economic, social and technological changes. For example, the English Tourist Board has claimed that the reduction of VAT on tourism services could create almost 90,000 additional jobs and increase foreign exchange earnings by £1.2 billion.

The future domestic demand for tourism in the UK will depend on both the availability of increased leisure time and disposable income. Given high disposable incomes, technology can help provide increasingly imaginative indoor facilities with high entry prices. But if the climatic conditions for outdoor recreation improve, there may be less demand for these facilities and the unit market price may not attract the necessary investment.

More extended and more accurate weather forecast information from agencies such as the Meteorological Office will be an important element in allowing the tourist to optimise on future weather sensitivity. The issue of accurate 5-day to 30-day forecasts could stimulate further growth in 'spontaneous' short-break holidays, especially in the shoulder months, or for skiing in Scotland. In turn, this is likely to increase congestion and management problems for the industry at 'honey pot' sites. For many people the search in recreation is for peace and quiet and the maintenance of the overall quality of the tourist experience will be very important.

Access to leisure facilities is often dependent on the private car, especially in those rural areas lacking an effective public transport system. Any future restrictions on the use of private cars, perhaps to combat increases in atmospheric pollution or to relieve traffic congestion, would have an adverse influence on the development of tourism in such areas.

16.5 PRINCIPAL IMPLICATIONS FOR OTHER SECTORS AND UK ECONOMY

Recreation and tourism impinges on many other sectors. For example, it is estimated that, globally, tourism accounts for at least 60% of all present-day air travel. Within the UK, transport improvements will be particularly important for ensuring accessibility to popular, short-term destinations without more road congestion and atmospheric pollution. Domestic tourism will flourish only if the travel experience compares well with the delays and stresses of air travel overseas.

Changes of climate are likely to lead to greater tourism-related pressure on the water industry, especially in summer. First, many of the main beaches designated under the European Economic Community Directive on the quality of bathing waters fail to meet the required standards. This can give rise to damaging publicity for some resorts and needs faster action for improvement if the potential use of bathing waters is to be achieved. Second, domestic water supply demands can more than double in coastal resorts due to the influx of summer visitors and this may create problems for some resorts in southern England. Peak water demands will be further increased by more garden irrigation and the watering of golf courses, grass tennis courts and public parks

The perception of improved weather conditions in the UK may attract more overseas visitors, as well as stimulate more domestic tourism. The high scenic quality of key rural areas must be preserved, especially in pockets of great pressure, e.g. Areas of Outstanding Natural Beauty and Heritage Coasts. Much better methods of visitor management will be needed if these areas are to absorb more tourism and still remain attractive environments. There may be new problems over larger areas in the National Parks. For example, more forest and heathland fires are likely during the summer. Since it is known that most rural fires are started by people, this could mean the closure of large areas of countryside at precisely the times when such areas are in demand for recreational access. In all cases, care must be taken in the location of new tourism development to ensure that its character and standard is compatible with the local environment and conservation requirements. For example, the increasingly heavy use of the coast is likely to present planners with problems, especially in zones requiring defence against sea level rise.

If tourist activity increases, leisure-related retail spending is also likely to grow, for example at garden centres and sports/leisure wear outfitters. More generally, tourism has a role to play in the economic and social regeneration of certain areas through investment and new facilities. Greater spending on tourism in the UK as a result of climate change could lead to a positive balance of payments for this sector.

Relatively few investigations have been made into the relationships between climate and tourism. Moreover, many of the studies of the weather sensitivity of outdoor recreation are now over 20 years old. Wall and Badka (1994) surveyed several national tourist agencies and found that, while the majority of respondents felt that climate was important to their country's tourism industry, very few were aware of climate change research related to tourism. This is surprising given the economic importance of the industry and the significance of potential climate-induced changes.

16.6 RESEARCH AND POLICY ISSUES

16.6.1 Future research effort

New research initiatives are urgently needed into the effects of climate, observed and perceived, on tourism. This will require some collaboration between meteorologists and tourist specialists, hopefully sponsored by the industry which, by itself, has demonstrated little capability for undertaking basic research.

The challenge will be to draw direct links between weather and climate conditions and the behaviour of tourists, especially the leisure consumer. For example, de Freitas (1990) has drawn attention to the need for alternative coastal recreation opportunities when conditions are unsuitable for beach recreation.

This requires a better knowledge of beach recreation climate, including the potential for microclimatic modifications as provided by wind breaks or shading structures. The optimum climatic conditions for other types of outdoor recreation also requires investigation.

Climatologists need to develop better techniques for portraying holiday climates in a way meaningful to tourists attempting to choose between competing destinations, as suggested by Perry (1993).

16.6.2 Policy issues

Climate change should feature more prominently in the minds of those planners and policy makers who have strategic responsibilities for recreation and tourism. Above all, the tourist industry should take a more pro-active stance in preparing for a scenario of greater tourist activity in the coming decades.

A greater awareness of current climatic variability, as well as some knowledge of future scenarios, is needed not only for long-term planning but also to deal with existing conditions. Suitable indicators are required by decision-makers planning for tourism and recreation, such as those relating to climate and pollution stress.

Within the UK, the potential upsurge in overall tourist activity will require careful management to ensure the maximum economic and social benefits.

Specifically, any increase in demand for access to the countryside will bring threats and opportunities for nature conservation, agriculture and road transport. Along the coasts, policies will also need to integrate recreational pressure with the planning of basic infrastructure, such as for water supplies and coastal flood defence.

17. Coastal Regions

SUMMARY

- A number of uncertainties still surround the prediction of future sea levels and storm intensity/frequency at the regional level. According to the 1996 CCIRG climate change scenario global mean sea level may be +19cm higher than the average of the 1961-90 period by the 2020s and +37cm higher by the 2050s. The global rise in sea level is likely to be exacerbated in southern and eastern UK by sinking land. Coastal areas could also face a significant risk of increased flooding, inundation and erosion with consequent assets damage if the frequency or severity of storm surge events also increased. Currently it is not possible to quantify this risk. Coastal lowlands around the Wash, stretches of the Norfolk and Suffolk coasts and, to a lesser extent, areas on Teeside and in south west Lancashire seem particularly vulnerable.

- Assets in the UK coastal zone are important environmentally, economically and politically. A significant proportion of the UK's social and economic welfare depends directly or indirectly on the availability of environmental goods and services provided by coastal and marine systems.

- Many coastal zone resources are sensitive to change, including human intervention, and are having to accommodate a relatively high rate of change and subsequent environmental pressure.

- Several stretches of the UK's coastline and related coastal zones are vulnerable to a combination of the effects of natural variability, climate change and human interventions. Protecting the capital (human and natural) assets, or in some cases relocating assets, will be costly if the effects of climate change are sudden rather than gradual. Any increase in the frequency with which sea and coastal defences are over-topped or breached would have major financial implications; for example, damages from the extreme flood event at Towyn in North Wales in 1990 totalled £35 million.

- Once information on the physical effects of potential or actual climate change is available, social and economic systems will begin a complex process of adaptation, which then introduces feedback that may alter the future state and change the potential damage.

- The limited number of economic studies undertaken for UK coastal areas suggest that the protect response strategy is generally economical. However there may be local areas where a strategy of either 'managed retreat' or 'do nothing' is best. A 'safety first' approach encompassed within an evolving integrated management strategy, moving beyond shoreline issues to a coastal zone-wide basis, has much to recommend it.

- The potential impacts of climate change and sea level rise are likely to be most severely felt, if defences fail, in terms of local direct and indirect human health effects and economic asset damage. There may also be losses to tourism and recreation and natural systems such as wetlands and mudflats and related flora and fauna. The national effects on other sectors such as forestry, energy, transport, industrial production, agriculture, water resources, soil resources and financial services are likely to be small.

17.1 INTRODUCTION AND BACKGROUND

A coastal zone is defined in terms of both a marine and landward boundary. The marine boundary of the UK coastal zone can be taken to be the continental shelf edge or the 'exclusive economic zone' limits, and the landward boundary to be fixed either in terms of existing local government administrative areas or the more extensive natural drainage basins. However, these boundaries coincide with only some of the areas from which demands are imposed on the resources of the coastal zone. These boundaries are also not likely to de-limit the influences of coastal processes on the designated area, such as sediment transfers and the atmospheric deposition of materials.

The 1996 CCIRG climate change scenario provides some possible sea level rise outcomes for the UK, together with non-quantitative indications of strong winds and storm frequency and severity. Global mean sea level is predicted to rise by +37cm by the 2050s, or by +23cm if the effect of sulphate aerosols are included; and by about +63cm or +42cm respectively by the 2090s. At the regional level, East Anglia can expect a net effect (after allowing for vertical land movement) of +50cm by the 2050s and the north of Scotland a net rise of 25cm by the same decade. However, the most significant impacts of sea level rise are connected to possible changes in the frequency or severity of storm surge events. Global climate models are not capable of simulating such local storm surge episodes.

The coastline of the United Kingdom has a length of 12,429 km and an exclusive economic zone of 1,785 square km. Its coastal morphology is quite varied and includes most temperate coastal formations from rock cliffs and headlands to barrier islands, sand beaches, shingle ridges, dunes, salt marshes and intertidal sand and mud zones. A significant percentage of the coastline is developed (in industrial, commercial, residential and recreational terms). Around 31% of the coast of England and Wales is developed when considering coastal frontage occupied by built environment and recreation facilities such as caravan parks, camping sites, car parks and golf courses. Economic pressure for further expansion of these facilities is likely to intensify in the future placing increasing strain on the development planning and guidance process. About 40% of UK manufacturing industry is situated on or near the coast (see Chapter 10).

The population in coastal urban agglomerations is over 26 million and may reach nearly 28 million by the year 2000. As well as these densely populated areas the coastal zone contains, among other assets, rural and agricultural land, smaller coastal towns, major infrastructure, inshore fisheries and offshore oil and gas resources (see Chapters 8, 9, 11, 12, 13 and 14). The annual volume of goods loaded and unloaded through UK ports represents the second largest European country trading volume after the Netherlands.

Some 57% of all agricultural Grade 1 land is to be found below the 5m contour line and potentially at risk from salt water damage which will limit its use (see Chapters 3 and 5). Agricultural land which most requires the protection of sea defences is located around the Wash and parts of the Norfolk and Suffolk coastal zone.

Tourism based on coastal recreation and amenity resources is a significant contributor to local and regional economies in the UK (see Chapter 16). Many stretches of coastline also provide unique combinations of assets of such cultural significance that they have been designated heritage coasts and given extra protection from development pressures.

Coastal areas support a highly varied set of ecosystems, including internationally significant wetlands and sites for a number of major bird populations. Some sites provide unique habitats for plant and animal species which are protected because of their rarity, or because these UK sites support all or most of the world population. Some 10% of the UK's notified nature reserves are located near the coast. An increase in sea level of more than 20cm will adversely affect mudflats and some salt marshes, including nature reserves that are important for birds (see Chapter 4). The coast is also important as a geological resource and many sections of the UK coast are internationally recognised as agreed geological reference points. The coastal zone is therefore environmentally, economically and politically important.

The resources of the UK coastal zone continue to be sensitive to, and modified by, human activities. The scale of modification ranges from dredging operations that directly affect local water bodies and fisheries, through sea defence and coastal protection measures (sea walls, beach groynes, beach nourishment and offshore artificial reefs, barriers and barrages etc.), to regional-scale changes in land use, such as drainage and conversion of wetlands for agriculture (now much reduced in extent), industrial and residential development. These latter impacts can affect the dynamics of runoff from catchments and offshore water quality. There can be no doubt, therefore, that the coastal zone is having to accommodate a relatively high rate of change and subsequent environmental pressure.

The process of change in the coastal zone involves pervasive interaction between the coastal environment and coastal populations and their activities. There is a sense in which economic and environmental systems are now co-evolving and that the scale of economic activity and intervention is significant. Coastal zone resources, including waste disposal capacity, have come under pressure from competing usage demands. Conflicts and trade-offs are inevitable and may be exacerbated by the variety of different stakeholders present at any given location in the coastal zone.

Given the interrelated nature of the economy-environment systems, it is not easy to disentangle the impacts of natural variability in coastal zone processes from impacts that arise from human activities, including, for example, potential global warming and related sea level rise, or practices that interfere with the movement of sediment along a coastline. Erosion of beaches is a widespread problem in the UK and has multiple causes. For many stretches of the UK coastline, storms present a significant but uncertain hazard, because of the natural variability in the strength and directions of wind and waves, in tide and surge levels and timing, and on mean sea level. Climate change, however caused, can change the severity and frequency of storms and therefore increase the hazards to the coast. Furthermore the vulnerability of these stretches of coast can be exacerbated by human activities, both within and outside the local coastal area.

The valuation and vulnerability assessment of zone-wide coastal systems can be undertaken via a 'GDP at risk' method. This approach relies on national/regional income statistics and seeks to identify the financial asset values under threat from a rise in sea level. The method calculates the proportion of the national gross domestic product (GDP) represented by the assets within the hazard zone. The result is an indicator of what is at risk in current circumstances rather than a measure of the economic damage cost or lost social value due to sea level rise (Turner *et al.*, 1995a). This is because 'lost' GDP can be recreated in an alternative location. As an example, the results of the GDP-at-risk analysis for a stretch of East Anglian coastline (Hunstanton to Felixstowe) are shown in Table 17.1. The sensitivity of the results is presented for a range of sea level rises, GDP growth rates and discount rates.

Table 17.1: An Illustration of the GDP-at-Risk Method

The total present value of real GDP at risk on the East Anglian coast between Hunstanton and Felixstowe from permanent flooding and erosion, 1990-2050 (£ million)

GDP growth rate pa (%)	RSL scenario							
	3% discount rate				6% discount rate			
	20cm	40cm	60cm	80cm	20cm	40cm	60cm	80cm
(a) Permanent Flooding								
1	3872	4003	4139	4409	1558	1610	4665	7773
3	8359	8641	8936	9517	2955	3045	3159	3364
5	19178	19824	20502	21835	6039	6243	6457	6876
(b) Erosion								
1	295	448	1142	1214	119	180	459	488
3	637	967	2466	2620	225	342	872	926
5	1462	2217	5657	6012	460	698	1781	1893

Source: Turner et al. (1995a)

Note: The losses in the table are not equivalent to actual economic damage costs but rather are potential losses only. The majority of these valuues are at risk both with and without climate change.

17.2 ESTIMATED EFFECTS OF CLIMATE CHANGE AND SEA LEVEL RISE

17.2.1 Effects of climate change and sea level rise

New estimates of expected global sea level rise have been made since the First CCIRG Report in 1991. Global mean sea level by the 2020s may be +19cm higher than the average of the 1961-90 period and +37cm higher by the 2050s (see Chapter 2). If the effect of sulphate aerosols are included the levels are +12cm and +23cm respectively. The global rise in sea level is likely to be exacerbated in southern and eastern UK by sinking land and mitigated in northern areas by rising land. The net effect in the southeast of England is an expected sea level rise of +50cm by the 2050s and only a +25cm rise in the north of Scotland. Coastal areas would face a significant asset damage risk if the frequency or severity of storm surge events also increased. The 1996 CCIRG climate scenario has the frequency of gales over the UK increasing by up to 30% by the 2050s. However, global climate models are not currently able to predict changes in the frequency or magnitude of storm surges at the regional level around the coasts of the UK. Given an increase in mean sea level of 37cm by the 2050s, the 1996 CCIRG scenario assumes that the probability of a storm surge exceeding a given threshold is likely to increase, but is not able to quantify this risk.

The distribution of UK coastal lowlands at most potential risk is shown in Figure 17.1. They range in extent from less than a hectare to over 400,000 ha in the case of the Fenlands, which is the most extensive coastal lowland in Britain. Areas in the Fenlands, Teeside and the southwest Lancashire coastal lowlands lie well below the present day high tide levels and are already at risk if current defence structures are breached (Tooley and Turner, 1995).

There is a considerable range of variability of tidal conditions around Britain. Adjacent estuaries can display markedly different tidal characteristics, and hence different altitudes at which the spring tide intersects the natural or artificial sea defence. Extreme water levels occur periodically as the result of high astronomical tides, and aperiodically as the result of storm surges. High astronomical tides affecting the

Figure 17.1: Map of the UK Showing the Distribution of Blown Sand, Marine Alluvium and Lowland Peat in the Coastal Lowlands.
Note: the shaded areas effectively indicate land at or below OD. Based on: International Quaternary Map of Europe, Sheet 6 København, 1:2,500,000. Hannover, Germany 1970; Quaternary Map of the United Kingdom, North and South, 1:625,000, Institute of Geological Sciences; and The Atlas of Britain and Northern Ireland, p.18 superficial deposits, 1:2,000,000.

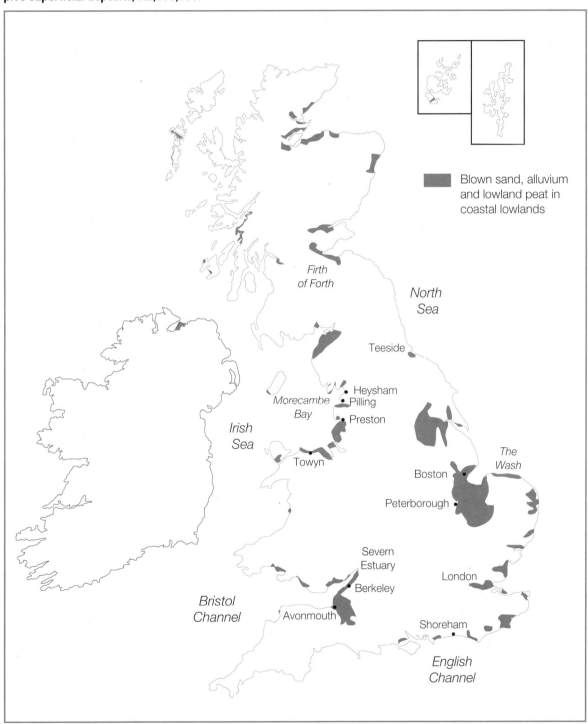

coasts of Britain range in altitude from +1.3m at Lowestoft to +8.0m OD at Avonmouth. Extreme water levels that have occurred in the past on the east coast of England have influenced the calculation of danger-level altitudes as part of the Storm Tide Warning Service. These levels have been exceeded or equalled on 112 occasions for the storm surge seasons 1972-73 to 1988-89 (Coker et al., 1989). Coastal lowlands will become more at risk if sea level rises (due to global warming), if ground altitudes fall (which is the case in the south east of England, for example, because of post-glacial adjustment) or there is an increase in storminess.

The coastal lowlands of Britain are affected by a combination of land drainage activities and long-term earth movements resulting in part from ice loading and delevelling during deglaciation, and in part from subsidence adjacent to the North Sea sedimentary basin. The combined result is that in southeast Britain the relative rise of sea level is greater than the global average. Conversely, uplift north of a line joining the Humber and Mersey, means that the rate of relative sea level rise and the consequent risks are not increasing as fast there. Any rise of sea level caused by climate change will increase the risks that already exist all round the coast.

Changes to mean sea level are much smaller than the short term changes due to tides, surges and waves. However, only a slight rise in mean sea level will amplify the frequency of extreme water levels caused by storm surges (see Box 17.1).

Box 17.1

The figure below illustrates how a hypothetical change of 0.2m in mean sea level affects return periods for different sites. Where long tide gauge records have been analysed for extremes of tide plus surge the return periods for any elevation above a datum can be plotted; by using the non-linear scale as here these plot as straight lines such as A and B, for two different sites. The lines are different because of the different local exposure and coastal morphology. The 3.2m level (solid lines A, B) has a 'return period' of 100 years at both sites at present. However, for a 0.2m sea level rise the return period falls to 25 years and 5 years at B and A respectively, since the data replot, as the broken lines A', and B'. This hypothetical amplification effect would be further magnified, as the slope of the lines would increase, if the storm frequency and severity increase. The 1996, CCIRG, climate change scenario allows for such an increase in mean wind speed and storm frequency.

Note that this figure plots data from which waves have been averaged out. To include them requires a more complicated joint probability analysis.

Source; CCIRG (1991)

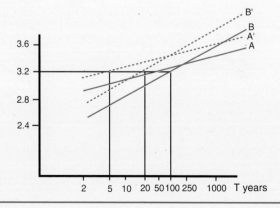

17.2.2 Sensitivity of the coastal zone to weather and climate

Vulnerability to impacts is a multi-dimensional concept encompassing biophysical, socio-economic and political factors. Coastal zones are characterised by highly diverse and unique ecosystems, with the result that a large number of functions are performed over a relatively small area. This concentration of functions, together with their spatial location, makes these zones attractive areas for people to live and work in. A significant proportion of the UK's social and economic welfare depends directly or indirectly on the availability of environmental goods and services provided by coastal and marine systems.

The susceptibility of an area is conditioned by the resilience of the coastal system, which is itself greatly influenced by population and settlement patterns, and rates of socio-economic change. Climate change could generally affect water flow and flooding in coastal catchments, with implications for wetland functions such as sediment trapping, flood control, aquifer recharging, waste processing, wildlife habitat, recreation and amenity provision, fish nurseries and biogeochemical cycling (see Table 17.2). There may also be a consequent damage cost impact on tourism, freshwater supplies, fisheries, exposed infrastructure, agricultural areas and other drylands, residential, commercial and industrial properties and cultural assets.

Coastal ecosystems themselves are sensitive to changes in water level relative to land, as well as loss of wetlands and lowlands, erosion of shorelines and habitat, increased salinity of estuaries and freshwater aquifers, new tidal ranges in rivers and bays, changes in sediment and nutrient transport and changes in the amount of light penetrating into aquatic ecosystems. The most vulnerable stretches of coast are those experiencing rapid erosion rates and with low geographic relief.

But human intervention in coastal environments, particularly in the form of hard engineering flood defence and coastal protection, has also reduced the capacity of ecosystems such as salt marshes to retreat inland in the face of sea level rise. To a certain extent the UK has made itself more vulnerable to natural climatic variability and extreme flood and storm events by increasing the population density in many of its coastal areas. A lot of capital is now invested in some coastal locations potentially at risk from sea level rise and storm events. Protecting this capital investment, or in some cases re-locating as-

Table 17.2: Climate Change Events and Impact Categories

Impact Categories	Climate-Related Events				
	Beach/ cliff erosion	Storms	Salt-H_2O infusion	Siltation	Flooding/ inundation
Tourism	£	'£'			£
Fresh water supplies & quality			'£'	'£'	'£'
Fishing	'£'	'£'		'£'	£
Coastal residences	£	£			£
Commercial buildings and infrastructure	£	£		'£'	£
Wetlands	£, nm	£, nm	£, nm	£, nm	£, nm
Agricultural and drylands	'£'	£	£	'£'	£
Life expectancy	£, nm	£, nm			£, nm
Culture and heritage sites	nm	nm			nm

nm = non-market impacts
£ = market priced major impacts
'£' = minor impacts

Source: Adapted from Turner et al. (1995b)

sets, will be costly if the effects of climate change are not slow and gradual. A high degree of foresight and integrated management and planning will also be required. Equally, the problem of vulnerability can be more readily mitigated the greater the technical, institutional and economic capabilities of the national economy (Turner *et al.*, 1995b).

Any increase in the frequency with which sea and coastal defences are over-topped or breached would have major financial implications. Flood losses during the 1953 East coast floods amounted to some £30 million in 1953 prices (equivalent to over £350 million in 1991 prices) (Arnell *et al.*, 1994). A great deal of new economic development has taken place in coastal flood risk areas since 1953.

The financial value of the coastal land along the 'central' south coast of England has been estimated to be up to £5745 million (Ball *et al.*, 1991).

The 1990 flooding of Towyn in North Wales produced a total cost figure (damage and response costs) of £35 million. That extreme event affected some 1200 properties and caused the evacuation of 3,500 residents. The return period of the water level is estimated to be between 150 and 500 years, but there are no data from which to estimate the probability of a similar engineering failure that allowed the sea wall breach.

A 1992 report to the Commission of European Communities made a very broad economic assessment (based partly on assets at risk expressed in terms of gross domestic product (GDP) equivalent) of the impact of sea level rise and other climatic factors on Member countries. The UK faced expected annual costs for storm surge damage amounting to an additional ECU 13,000m, with a 0.5m rise in sea level; and an overall cost impact of ECU 26,000m per year under a 4°C global warming scenario (Climatic Research Unit/Environmental Resources Ltd., 1992). This analysis, however, did not take into account any adaptation responses.

17.3 ASSESSMENT OF POTENTIAL ADAPTATION

Coastal protection and flood defence policy is the responsibility of the Ministry of Agriculture, Fisheries and Food in England. The National Rivers Authority (NRA) and local authorities are mainly responsible for the planning, execution and maintenance of works on the coast, with the latter having a specific duty to mitigate the effects of erosion. Capital and maintenance costs are largely financed by central government. The Scottish Office, Welsh Office and Northern Ireland Office cover these functions for Scotland, Wales and Northern Ireland.

The primary aim of the sea defence policy is to reduce risks to people and the developed and natural environment from flooding and coastal erosion by encouraging the provision of technically, environmentally and economically sound and sustainable defence measures. The enabling measures are stated to be: the provision of adequate and cost effective flood warning systems; the provision of technically, environmentally and economically sound and sustainable hard and soft engineered defence measures; and the discouragement of inappropriate development in areas at risk from flooding and coastal erosion (Ministry of Agriculture, Fisheries and Food, 1993a, b).

For coastal zones IPCC (IPCC, 1992b) recommends the assessment of three broad types of policy response:

Managed retreat - abandon the land and structures in vulnerable areas and manage the population resettlement sensitivity; this option may also include restoring or creating desirable habitat, landscape, or amenity features in an ecologically sustainable way;

Accommodate - continue occupancy and use of vulnerable areas via a range of adaptive adjustments;

Protect - continue full defence of vulnerable areas (especially population centres), economic activities, and natural resources.

The UK's strategy contains elements of all three responses but at its core it is a protect policy.

Once coastal social and economic systems are given access to information on potential and/or actual 'physical' climate change effects a complex process of adaptation responses and feedback effects become possible that may alter the future state of the coastal zone. In other words a co-evolutionary process is set in motion encompassing both natural and social systems. The socio-economic impacts of climate change will be a combination of the consequences of physical changes in, for example, sea level/ storminess or general weather conditions and adaptation responses with their related effects. The consequences of both forms of change can be expressed in terms of their human welfare (wellbeing) costs and benefits.

The IPCC Climate Change Impacts Assessment Guidelines (Carter et al., 1994) distinguish between endogenous adaptation and exogenous adaptation, although a continuum of measures is also recognised. Endogenous adjustments include biophysical responses in natural systems and consequent routine adjustments in socio-economic systems. As more relevant information passes through socio-economic systems so tactical adjustments are stimulated (e.g. flood proofing of some properties and storm warning systems). The Ministry of Agriculture, Fisheries and Food (MAFF) funds the sea level monitoring programme, for example, at the Proudman Laboratory. There are also exogenous adjustments overtly and externally imposed on the system by decision makers (e.g. hard and soft engineering flood defence measures; central government planning guidance; managed retreat planning zones). 'Counter-attack' measures may also be deployed such as the use of dredging spoil and solid waste to create new islands or shoals (for salt marsh replacement), or to form beaches, dunes and marsh terraces in front of hard engineering structures.

The information transfer process in the UK coastal zone is complex. We have already emphasised that the coastal zone has been the subject of long-term human intervention. This intervention process has conditioned human perceptions of vulnerability. Asset owners and markets have probably underestimated the risk of surprise storms, flooding and inundation. They have implicitly assumed or expected that the relevant authorities could guarantee them adequate current and future protection. The market value of a range of assets potentially at risk may not therefore correctly reflect the actual risk that exists. In erosion prone areas, the property and land markets have adjusted better to the historical risk situation, because of the nature of the erosion process and its visibility.

The NRA, which has the responsibility for sea defences over much of the coastline of England and Wales, is currently operating on regional estimates of sea level rise of 4-6mm per annum (Table 17.3). This policy is in line with the scenarios presented in Chapter 2. In 1991/92 the NRA spent over £91 million on sea-defence related capital works to protect people, property and some natural systems (Arnell et al., 1994). Major urban/industrial centres on estuarine

Table 17.3: MAFF Advisory Guidelines for Regional Sea Level Rise

Region	Rise (mm/year)	Total Rise by 2030 (over 1990 levels)(cm)
Anglian, Thames and Southern	6	24
North West, Northumbria	4	16
Severn Trent, South West, Welsh, Wessex and Yorkshire	5	20

Note: Northumbria and Yorkshire; and South West and Wessex, were merged in 1993
Source: MAFF advice on allowances for sea level rise issued November 1991.

sites are heavily defended by hard engineering structures (see Thames Barrier Box 17.2). More generally over 33% of the coastline of England and Wales is protected by artificial (hard engineering) defences. There is a concentration of such defences on the East Anglian coastline. In 1992 the NRA published the results of a survey of existing sea defences and their condition. Some 16 percent were showing some signs of wear. Maintenance and replacement costs for sea defences and coastal protection works, under a protect strategy, may therefore be substantial in the future. Bolder structures may in some cases be more acceptable than solid walls and beach nourishment (as at Sidmouth) can combine protection with the maintenance of a 'natural' beach.

Managed retreat or managed setback is a policy option which has more recently been given a higher profile. The NRA, MAFF and English Nature are currently undertaking some research and development work at sites on the East Anglian coast which do not involve population resettlement issues, but do offer opportunities for habitat restoration/creation. The Department of the Environment has published (in 1993) a Coastal planning guidance note which sets out the statutory and voluntary planning procedures and a vision of future strategy.

A recent economic study has predicted that for Britain the optimal degree of protection varies between 92% and 98% (sea level rise 20cm up to 1m over the next 100 years) for open coasts and beaches, and between 98% and 99% for urban/industrial/port areas (Fankhauser, 1995). The policy message of these tentative results is that it will probably be economically optimal for Britain to protect most of its vulnerable stretches of coastline. Provided that the cost-efficient solution is implemented, the costs of sea level rise for Britain will not be catastrophically high (Fankhauser, 1995).

Nevertheless, in some areas the main effect of recent sea level rise seems to have been the steepening of beach profiles and the consequent loss of areas of inter-tidal habitat. Some 2750 ha of salt marsh and up to 10,000 ha of inter-tidal flats, which represent a significant environmental resource, may also be under threat from sea level rise (Pye and French, 1993). The presence of hard engineering defences is probably the cause of this loss because they prevent the natural adjustment of the coastline (e.g. landward retreat of salt-marshes) as sea level increases. The solution to this is perhaps a policy of selective managed retreat where existing and/or future land uses permit (Department of the Environment, 1993; Burd, 1995).

A regional economic cost-benefit analysis of a vulnerable stretch of coastline, the East Anglian

---

Box 17.2 Thames Barrier

The Thames Tidal Flood Defence system was completed in 1982 at a cost of £1000m of which the Thames Barrier represented some £600m (1982 costs). Barrier operation and maintenance costs (including forecasting and monitoring) amount to £4.0m pa. There have been no significant economic damage costs arising from storm/flood/erosion events since 1982. The protected 'house equivalents' within the tidal flood zone and potentially at risk from a breach or overtopping amount to 1,166,834 (all land use Band A).

The design of the Barrier (operating over a notional 50 year period) includes provision for mean sea level rise based on the extrapolation of previous trends (Gilbert and Horner, 1984). Analysis of records of high water level on the Thames Estuary suggested that the level was rising at a rate of about 0.8m/century. Therefore the CCIRG 1996 scenario estimates of sea level rise are unlikely to compromise the effectiveness of the barrier before 2030.

coast, has also been completed (Turner et al., 1995a). The study area was a stretch of coast bordering the North Sea from Hunstanton to Felixstowe. The vulnerable zone contains a range of urban, industrial, recreation, amenity and ecological assets.

The basic approach adopted in the study was to combine information about the physical hazard posed by accelerated sea level rise with data on assets at risk in order to produce a physical/numerical, and then economic estimation of the impacts.

Table 17.4 summarises the costs of protecting the East Anglian hazard zone, and the net benefits value results (i.e. discounted protection costs minus discounted benefits of protection in terms of damage costs avoided) for the 'accommodate' and 'protect' policy options. The former option would involve the continued maintenance of the current defence system, that is a declining standard of protection over time as and when sea level rise takes place. The latter option requires the continual upgrading of defence to keep pace with future sea-level rise.

A protection strategy was found to be the most economically efficient response on a region-wide basis. However, the study area can be divided into 184 individual flood and erosion compartments. At this disaggregated and localised scale, protection is not always the most economic response. Analysis of the 113 flood-hazard compartments indicates that as the sea level rise predictions increase, the number of coastal sections in which retreat (i.e. do nothing) is the most efficient response also increases (Turner et al., 1995a).

17.4 UNCERTAINTIES AND UNKNOWNS

Uncertainty still surrounds the estimation of past rises and falls in sea level and the forecasting of future changes in sea level. Periods of accelerated sea level rise, widely recorded in the recent geological record and from the last interglacial geological record (Streif, 1990) have not been considered in the predictions of sea level rise. It has been demonstrated that rates of sea level change have varied considerably in the past. Sea level change is not uniform around the world already, because of differential heating and therefore thermal expansion, due to ocean circulation and due to geological and tectonic motions. Any future changes to any of these will alter the rates of rise differently around the oceans.

Due to the poor spatial resolution of Global Climate Models (GCMs) it is not possible to quantify regional/local differences between the glo-

Table 17.4: Costs and Benefits of Sea level Rise and Policy Responses in East Anglia (1990-2050)

	Sea-level rise (6% discount rate, £ million)			
	20cm	40cm	60cm	80cm
Protection costs:				
Retreat	-	-	-	-
Accommodate	132	137	151	157
Protect	187	232	292	485
Damage costs avoided (i.e. benefits of defence relative to do nothing 'retreat'):				
Accommodate	1140	1103	1092	1049
Protect	1258	1283	1325	1352
Net benefits:				
Accommodate	1008	966	941	892
Protect	1071	1050	1033	867

Source: Adapted from Turner et al. (1995a)

bal-mean sea level rise and rises affecting the UK coastline. Changes in the frequency track or severity of storms risk potentially significant damage cost risks, because sensitivity studies have been carried out on the surges that would be generated. However, GCMs are not yet capable of resolving such local storm properties thus the degree of likely changes and probabilities cannot be assessed. It is therefore currently not possible to quantify any likely changes in the storm surge regime affecting, for example, the North Sea coast or other UK coasts.

Procedures for the assessment of the socio-economic implications of climate change and sea level rise in the coastal zone require further development, refinement and empirical application. The IPCC's 'Common Methodology' (IPCC 1992b; World Coast Conference, 1994) is data intensive and time consuming; so far it has only been applied in the UK to one limited stretch of East Anglian coastline (Turner et al., 1995a). No national estimates of the likely damage costs from sea level rise, or assessments of the various response options exist for the UK. Within the economic valuation process the meaningful evaluation of a range of non-market assets remains a difficult, and for some, a controversial, procedure.

17.5 PRINCIPAL IMPLICATIONS FOR OTHER SECTORS AND THE UK ECONOMY

A wide variety of different types of valuable assets are potentially at risk in the coastal zone, including environmental functions/services performed by natural systems (Turner, 1988; Barbier, 1989; De Groot, 1992). A number of the coastal assets are non-market goods and services and therefore do not carry easily identifiable appropriate market prices and value. Actual socio-economic empirical studies in the UK coastal zone are relatively few in number and restricted in scope, with the result that many of these assets are ignored in coastal planning (Turner et al., 1995b).

Rising sea level could lead to saline intrusion into coastal aquifers and also increased penetration of salt water along estuaries, posing a risk to low-lying freshwater intakes. Most of the coastal aquifers potentially at risk are along the south and east coasts but yield reductions would be small and easily compensated for. Similarly, studies have shown that impacts on freshwater intakes along estuaries are easily mitigated (see Chapter 7). Offshore oil and gas production and exploration activities, together with energy supply facilities along the coastline, may also be adversely affected by sea level rise and increased storminess (see Chapter 8). Increased protection measures will be site-specific but are unlikely to be economically significant.

Other sectors in which the direct impact of rising sea level (at a rate of 5cm per decade) and increased storm surges is unlikely to be of major or widespread significance are aquaculture, forestry operations, transport, industrial production and financial services (see Chapters 5, 6, 10, 11, 12 and 14).

But while the direct effects may be relatively small, a few indirect effects may be more significant. Thus in the minerals sector, while the large coastline quarries in Scotland will probably be able to adapt to a gradual rise in sea level, they and other quarries may find it less easy to adjust production to meet the national demand for hard rock. This demand may be considerably increased if sea and river estuary defence systems were augmented or maintained by rock amour. The environmental landscape and amenity implications of increased quarrying operations would not be trivial. If domestic production was to be constrained on environmental grounds, the UK's import bill would have to rise.

Changes in the frequency or severity of storm surges would have both direct and indirect health effects. The adverse health impacts would include increased mortality and morbidity, with the latter encompassing stress-related illnesses due to shock, loss of employment and enforced relocation.

The recreational value of the UK's coastal zone is substantial. In some locations recreation benefits may be equal to or greater than the benefits of protecting property. Beach recreation is particularly valuable and dwarfs the other forms of coastal recreation/amenity pursuits such as cliff top walking and promenade walking (Penning-Rowsell et al., 1992). Beaches are also often the frontline coastal defence, protecting cliff sites, coastal landscapes, hard defences and related economic assets at risk.

Climate change and sea level rise will potentially affect coastal tourism and recreation in a dynamic and complex way (see Chapter 16). Improved weather conditions may redress the competitive advantage that southern European resorts currently enjoy over UK seaside resorts.

On the other hand, sea level rise and increased storminess will lead to more erosion and beach loss at certain locations. But some coastal protection works put in place to mitigate erosion and/or to stabilise beaches seem to reduce the enjoyment (and therefore value) of beach recreation for some people. Table 17.5 summarises the results of coastal recreation valuation studies.

The east coast of England from Yorkshire to Essex has a number of traditional beach resorts. Tourism board data suggests that approximately 25 to 26 million visitors (of all types) use these main, coastal resorts during the year. According to the Countryside Commission, in the mid 1980s some 65% of visitors came to use the beach resource and only 35% to use the other coastal recreation assets, e.g. clifftop walks, promenade, etc. If we use these general proportions and assume (very 'conservatively') that each visitor

Table 17.5: Coastal Recreation Valuation Studies

Site Location and Year of Study	Type of Valuation Measure	Value (£) (rounded)
Scarborough (1988)	Value of enjoyment measures by site survey of beach users	5 per visit
Clacton (1988)	"	10 "
Dunwich (1988)	"	7 "
Filey (1988)	"	4 "
Frinton (1988)	"	10 "
Hastings (1988)	"	8 "
Spurn Head (1988)	"	7 "
Bridlington (1989)	"	6 "
Clacton (1989)	"	11 "
Hunstanton (1989)	"	9 "
Morecambe (1989)	"	6 "
Herne Bay (1990)	"	12 "
Herne Bay (1990)	Value of enjoyment: residents survey	4 "
Peacehaven (1988)	Value of enjoyment: cliff top users	3 "
Mablethorpe & Skegness (1991)	Willingness-to-pay for beach recreation	1-2 per visit
Walton on the Naze (1987)	Willingness-to-pay for cliff top recreation	7 per person per annum
Peacehaven (1987)	Willingness-to-pay for cliff top recreation	1-3 per person per annum
Aldeburgh (1988)	Willingness-to-pay for shoreline recreation	6-7.5 per person per annum
Norfolk Broads (1991)	Willingness-to-pay for coastal wetland amenity conservation	67-81 per household per annum
Various Sites (1989)	Mean loss in per adult visit to eroded beach (all visitors)	4 per visit

Quoted in Penning-Rowsell et al., 1992.

only makes one visit per annum to a particular recreation site, then this coastal region attracts around 16.5 million beach visits and nearly 9 million other recreational visits per annum. The economic user value of beach recreation ranges from £33 million to £132 million per annum, depending on whether mean 'value of enjoyment' or mean 'willingness-to-pay' economic value estimates are utilised. Non-beach recreation user value is in the range £27 million to £44.5 million per annum.

Other more unique coastal assets also attract economic user value; the Norfolk Broads, for example, generates an aggregate annual recreation and amenity value of between £5 million and £15.5 million, depending on which visitation data estimates are used. Thus the aggregate zonal recreation user value is approximately £65 million to £192 million per annum. Coastal resources will also probably attract so-called non-use value (or 'existence' value) especially where unique heritage sites are involved. Estimating non-use values is a more difficult task but the studies that have been undertaken suggest that such values are significant.

17.6 POLICY AND RESEARCH ISSUES

17.6.1 Future research effort

- There is a need to develop techniques for monitoring absolute land level changes in order to separate land movement from global sea level rise. A precise global reference frame to which tide gauge benchmarks can be linked has been established. MAFF has funded limited fixing of the UK monitoring network into this system. More funding is required for continued development of the reference frame work.

- Global referencing systems and methodologies based on location and geology are also required for predicting long term land surface changes which can be used for sea defence/coastal protection scheme design and appraisal.

- More localised outputs from GCMs are required to improve prediction and evaluation of the regional effects of changes in sea level and storm frequency, duration, magnitude and direction. These are required both for flood protection scheme design and coastal zone planning purposes.

- Monitoring of and more research into coastline advance and retreat and patterns of sediment redistribution as well as sea level changes, in order to better determine the possible interrelationships is another requirement.

- Research to gain a better understanding of the impact of the incursion of the saline front on sensitive estuaries would also be very useful.

- The UK currently lacks a cost-benefit evaluation on a national basis of different coastal assets at risk and of abatement policy options in response to actual and potential sea level rise. Such a study should be a high priority and should be combined with more detailed regional (coastal cell) studies for vulnerable stretches of the coastline. The evaluation methodology should incorporate economic efficiency, equity and environmental effectiveness criteria. It should further be applied to the full range of abatement options, including the 'managed retreat' option where appropriate.

- An interdisciplinary analysis of the integrated coastal zone management process and its application to the UK situation would provide much useful data and policy guidelines. The large amount of information being collected for Shoreline Management Plan studies provides a useful start for this analysis.

17.6.2 Policy issues

A reasonable policy objective for the UK coastal zone would involve the determination of a 'socially desirable' mix of zonal products and services which can be sustained over time. This social mix can be most efficiently and effectively provided by an integrated approach to coastal zone planning and management. Integrated coastal zone management (ICZM) would be a component of sustainable resource management within the national economic development strategy.

The ground altitudes of some of Britain's coastal lowlands have been and are being further lowered because of drainage activities. Thus any climate-induced rise in sea level will serve only to exacerbate the risk that these lowlands already face. Overall, Britain's coastal zone is being altered by a combination of environmental (geological and climatic) changes and socio-economic changes, plus related feedback mechanisms. The management challenge is therefore to limit the impact of further urbanisation and economic development on adjacent 'natural' areas, while at the same time protecting coastal infrastructure and the social and economic support systems from external (e.g. climate-related) stress and shock.

It seems clear that moves toward the longer-term objective of integrated coastal zone management are required. Such a management approach should be strategic and based on processes rather than end states. In the past in Britain there has been a lack of coordination between coastal planning and coastal defence and the prevailing opinion has been that coastal erosion and flooding were not planning issues (Lee, 1993). More recently MAFF has issued advice on Shoreline Management Planning and the Department of the Environment has commissioned a report on best practice for coastal zone management planning.

The UK coastal zone can provide a range of use benefits but they are not all compatible with each other. Lee (1993) believes that coastal management problems can be focused on a limited number of critical value and use conflicts. These include:

- pressures arising from tourism and recreation on parts of the relatively unspoilt coastline in south and east England and south Wales;
- urban and industrial development pressures on estuaries which also provide natural capital assets of significant conservation value (e.g. Cardiff Bay, the Mersey estuary, the Solent and the Thames);
- marine aggregate extraction activities and possible consequent links to increased erosion on the coastlines of south and east England and Wales;
- residential area protection and conflicting landscape conservation.

During the 1990s there have been a number of changes in the institutional framework underpinning British coastal management. The official view is that more integrated management systems will evolve from a combination of existing statutory and voluntary planning procedures (Department of the Environment, 1993). Coastal groups which comprise all those bodies concerned with coastal defence have been encouraged to produce shoreline management plans. These groups now cover nearly all the coastline of England and Wales. There is informal coordination of these groups through a national coastal defence forum. The coastal groups have a clear role in contributing to local authority development plans in the expectation that they will take account of coastal defence strategies. Central guidelines for the integration of coastal management issues into structure planning were issued in a 1993 Planning Policy Guidance Note (Department of the Environment, 1993).

References

Abegg, B. and Froesch, R. (1994) Climate change and winter tourism. In: Beniston, M. (Ed) Mountain Environments in Changing Climates Routledge, London, 328-340.

Academic Task Force (1995) Preliminary Report of the Academic Task Force on Hurricane Catastrophe Insurance: Restoring Florida's Paradise. The Collins Center for Public Policy, Florida, USA.

Agnew, M. D. and Thornes, J. E. (1995) The weather sensitivity of the UK food retail and distribution industry. Meteorological Applications, 2, 137-147.

Alcamo, J., Bouwman, A., Edmonds, J., Grübler, A., Morita,T. and Sugandhy, A. (1995) An evaluation of the IPCC IS92 emission scenarios In: Houghton, J. T., Meira Filho, L. G., Bruce, J., Lee, H., Callander, B. A., Haites, E., Harris, N. and Maskell, K. (eds.) Climate change 1994: radiative forcing of climate change and an evaluation of the IPCC IS92 emissions scenarios, Cambridge University Press, Cambridge, 247-304.

Alexandratos, N. (1995) The outlook for world food production and agriculture to the year 2010. In: N. Islam (ed.), Population and Food in the Early 21st Century: Meeting Future Food Demand of an Increasing World Population. Occasional Paper, International Food Policy Research Institute, Washington, D.C.

Allen, L.H. (1990) Plant responses to rising carbon dioxide and potential interactions with air pollutants. Journal of Environmental Quality,19, 15-34.

Anon (1990) Cryptosporidium in water supplies. Report of the Group of Experts 1990. HMSO, London.

Armstrong, A. C., Matthews, A. M., Portwood, A. M., Addiscott, T. M., and Leeds-Harrison, P. B. (1994) Modelling the effects of climate change on the hydrology and water quality of structured soils. In: M. D. A. Rounsevell and P. J. Loveland (eds.), Soil Responses to Climate Change. NATO ASI Series 1, Global Environmental Change, 23, 113-136.

Armstrong-Brown, S. A., Rounsevell, M. D. A. and Bullock, P. (1995) Soils and greenhouse gases: management for mitigation. Chemistry and Industry, 16, 647-650.

Arnell, N. W. (1995) Scenarios for hydrological climate change impact studies. In: Oliver, H.R. and Oliver, S. A. (eds.), The Role of Water and the Hydrological Cycle in Global Change. Springer-Verlag: Berlin, 389-407.

Arnell, N. W. & Dubourg, W. R. (1995) Implications for water supply and management. In: Parry, M. & Duncan, R. (eds.), The Economic Implications of Climate Change in Britain. Earthscan: London, 28-45.

Arnell, N.W., Jenkins, A. and George, D.G. (1994) The Implications of Climate Change for the National Rivers Authority. NRA R and D Report 12, HMSO, London.

Arnell, N.W. and N.S. Reynard, (1993). Impacts of Climate Change on River Flow Regimes in the United Kingdom, Report to the Department of the Environment.

Austin, A., Clark, N. A., Greenwood, J. J. D. and Rchfisch, M. M. (1993). An analysis of the occurrence of rare birds in Britain in relation to weather. British Trust for Ornithology, Thetford.

Ball, J. H., Clark, M. J., Collins, M.B., Gao, S., Ingham, A. and Ulph, A. (1991). The Economic Consequences of Sea Level Rise on the Central Southern Coast of England. GeoData Institute Report to MAFF, University of Southampton.

Ball, R. H. and Breed, W. S. (1992) Summary of Likely Impacts of Climate Change on the Energy Sector. Submission to IPCC Working Group II, US Department of Energy, Washington DC.

Ballentine, V. (1994) The use of marketing principles to maximise economic benefits of weather. Meteorological Applications, 1, 165-172.

Barbier, R. B. (1989) The Economic Value of Ecosystems: I - Tropical Wetlands, LEEC Gatekeeper, London Environmental Economics Centre, London, 89-02.

Barkham, J. P., MacGuire, F. and Jones, S. (1992) Sea Level Rise and the UK. Friends of the Earth, Lon-don.

Barrow, E. M. and Hulme, M. (1996) The changing probabilities of daily temperature extremes in the UK related to future global warming and changes in climate variability. Climate Research, 6, 21-31.

Barrow, E. M., Hulme, M. and Jiang, T. (1993) A 1961-90 baseline climatology and future climate change scenarios for Great Britain and Europe. Part I: A 1961-90 Great Britain Baseline Climatology, A Report prepared for the TIGER IV3, a Consortium, Landscape dynamics and climate change. Climatic Research Unit, Norwich.

Batjes, N. H. (1992) Methane. In: Batjes, N. H. and E. M. Bridges (eds.), A Review of Soil Factors and Processes that Control Fluxes of Heat, Moisture and Greenhouse Gases. ISRIC Technical Paper No. 23 (WISE Report 3), 37-66.

Bennet, G (1970) Bristol floods 1968: controlled survey of effects on health of local community disaster. British Medical Journal, 3, 454-458.

Bentham, G. and Langford, I. H. (1994) Climate change and the incidence of food poisoning in England and Wales. GEC CSERGE, University of East Anglia/University College London, 94-15.

Beuker, E. (1994) Long-term effects of temperature on the wood production of *Pinus sylvestris* L. and *Picea abies* (L.) Karst. in old provenance experiments. Scandinavian Journal of Forestry Research 9, 34-45.

Beveridge, M. C. M., Ross, L. G. and Kelly, L. A. (1994) Aquaculture and biodiversity. Ambio, 23, 497-498.

Birtles, A. B., Yates, A. and Bartlett, P. (1994) The BRE Environmental Assessment Method for buildings, BREEAM, Proc. CIBSE National Conference, 1994, Brighton, 2-4 Oct. 1994, Vol. 1, 32-46.

Boardman, B. (1991) Fuel Poverty: From Cold Homes to Affordable Warmth, Belhaven Press, London.

Boardman, J. and Favis-Mortlock, D.T. (1993) Climate change and soil erosion in Britain. The Geographical Journal, 159, 179-183.

Boardman, J., Evans, R., Favis-Mortlock, D.T. and Harris, T. M. (1990) Climate change and soil erosion on agricultural land in England and Wales. Land Degradation and Rehabilitation, 2, 95-106.

Boden, J. B. and Driscoll, R. M. C. (1987) House Foundations - a review of the effect of clay soil volume on design and performance. Municipal Engineering, 4, 181-213.

Bongaarts, J. (1994) Can the growing human population feed itself? Scientific American, January, 36-42.

Boorman, L. A., Goss-Custard, J. D. and McGrorty, S. (1989) Climatic Change, Rising Sea-Level and the British Coast. Institute of Terrestrial Ecology, Research Publication No. 1. HMSO, London.

Bouwman, A. F., Fung, I., Mathew, E. and John, J. (1993) Global analysis of the potential for N_2O production in natural soils. Global Biogeochemical Cycles, 7, 557-597.

Bradbury, N.J. and Powlson, D.S. (1994). The potential impact of global environmental change in nitrogen dynamics or arable systems. In: Rounsevell, M. D. A. and Loveland, P. J. (eds.), Soil Responses to Climate Change. NATO ASI Series Vol. 123, Springer-Verlag, 137-153.

Brignall, A. P. and Rounsevell, M. D. A. (1994). The Effects of Future Climate Change on Crop Potential and Soil Tillage Opportunities in England and Wales. Research Report. Environmental Change Unit, Oxford University, Oxford.

Brignall, A. P., Davies, A., Hossell, J., Parry, M., Pollock, C. (1995) Implications for Agriculture and Land Use. In: Parry, M. and Duncan, R. (eds), The Economic Implications of Climate Change In Britain. Earthscan Publications, 46-63, London.

Brignall, A. P. and Rounsevell, M. D. A., (1995) Land evaluation modelling to assess the effects of climate change on winter wheat potential in England and Wales. Journal of Agricultural Science, Cambridge, 124, 159-172.

Brignall, A. P., Favis-Mortlock, D. T., Hossell, J. E. and Rousevell, M. D. A. (1995) Climate change and crop potential in England and Wales. Journal of the Royal Agricultural Society of England, 156, 140-161.

Brinkman, R., (1990) Soil resilience against climate change? In: Scharpenseel, H.W., Schomaker, M and Ayoub, A. (eds.), Soils on a Warmer Earth. Elsevier Developments in Soil Science, Amsterdam, 51-60.

British Aggregate Construction Industries (BACMI) (1994). BACMI Statistical Year Book 1994.

British Chambers of Commerce (1995a) Comment. Business Briefing, 7 (26), 2.

British Chambers of Commerce (1995b) Comment. Business Briefing, 7 (17), 2.

British Gas plc (1995). Annual Report and Accounts 1994, London.

British Geological Survey [BGS] (1995) United Kingdom Minerals Yearbook, 1995. Keyworth, Nottingham.

British Petroleum Company plc (1995) BP Statistical Review of World Energy, 1995.

British Standards Institute (1972) British Standard Code of Practice CP3: Chapter V Part 2: 1972 Wind Loads, BSI, London.

British Standards Institute (1989), BS 5250:1989 Control of condensation in buildings. BSI, London.

British Standards Institute (1992), BS 8104:1992 Assessing exposure of walls to driven rain. BSI, London.

Building Effects Review Group, (BERG) (1989) The effects of acid deposition on buildings and building materials in the United Kingdom. HMSO, London, ISBN 7114-3404-4.

Building Research Establishment (BRE) (1973) Repair and renovation of flood damaged buildings. Digest No 152, BRE, Garston, Watford.

Building Research Establishment (BRE) (1990) Climate and site development: Part 2 Influence of micro climate, Part 3 Improving micro climate through design. Digest No 350, BRE, Garston.

Building Research Establishment (BRE) (1994) House longhorn beetle: geographical distribution and pest status in the UK. BRE Information Paper 8/94, BRE, Garston.

Buller, P. S. J. (1993a) Wind damage to buildings in the United Kingdom, 1977-1991. Publication N80/93, BRE, Garston.

Buller, P. S. J. (1993b) A thirty year survey of gale damage to buildings in the United Kingdom, Report PD121/93, BRE Garston.

Bullock, A. M. (1988) Solar ultraviolet radiation: a potential environmental hazard In: Muir, J. F. and Robertsthe, R. J. (eds), Cultivation of farmed finfish, Recent Advances in Aquaculture, Vol. 3, Croom Helm, London, 139-224.

Bunch, A. and Smithers, H. (1992) The potential impact of climate change on regional resources In: NWW Ltd. Manuscript. North West Water Limited.

Burd, F. (1995) Managed Retreat: A Practical Guide. English Nature, Peterborough.

Burn, R. (1995) Teaching low energy. Building Services, April, 1995, 19-23.

Burrows, A. M. and House, M. A. (1989) Public's perception of water quality and the use of water for recreation. In: Laikari, H. (ed) River Basin Management - 5 Pergamon, Oxford, 371-379.

Burton, J. F. (1995). Birds and climate change. Christopher Helm Ltd, London. Climate change on winter wheat potential in England and Wales. Journal of Agricultural Science, 124, 159-172.

Burton, R. G. O. and Hodgson, J. M. (1987) Lowland Peat in England and Wales. Soil Survey Special Survey No. 13, Harpenden, England.

Cannell, M. G. R., Grace, J. and Booth, A. (1989) Possible impacts of climatic warming on trees and forests in the United Kingdom: A Review. Forestry 62, 337-364.

Cannell, M. G. R. and Pitcairn, C. (1993) Impacts of the mild winters and hot summers in the United Kingdom in 1988-1990. HMSO, London.

Carnell, R. E., Senior, C. A. and Mitchell, J. F. B. (in press) Techniques for assessing changes in Northern Hemisphere storminess. Climate Dynamics.

Carter, T.R., Parry, M.L., Harasaw, H. and Nishioka, S. (1994). IPCC Technical Guidelines for Assessing Climate Change Impacts and Adaptations, University College London, UK and Centre for Global Environmental Research, Tsukuba, Japan.

Carter, T. R., Parry, M. L., Harasawa, H. and Nishioka, S. (1995) IPCC technical guidelines for assessing climate change impacts and adaptations. UCL/CGER, London/Tsukuba.

Carter R.L. and Falush P. (1995) The London Insurance Market Association of British Insurers, London.

Cavallo, A. J., S. M. Hock and D. R. Smith, (1993) Wind energy: technology and economics. In: T.B. Johansson et al. (eds.), Renewable Energy: Source for Fuels and Energy. Island Press, Washington DC, 121-156.

Central Office of Information (COI) (1995), Britain 1995. An Official Handbook. HMSO, London.

Central Statistical Office (1995a) Economic Trends No. 501 (July 1995). HMSO, London.

Central Statistical Office (1995b) Regional Trends 30. HMSO, London.

Central Statistical Office, (1995c) Annual Abstract of Statistics, No 131, HMSO, London.

Central Statistical Office (1995d) Monthly Digest of Statistics No. 598 (October 1995). HMSO, London.

Central Statistical Office (CSO) (1995e), Monthly Digest of Statistics, No. 592, April 1995, HMSO, London

Central Statistical Office, (1994) Social Trends, HMSO, London.

Chander, K. and Brookes, P. C. (1991) Plant inputs of carbon to metal contaminated and non-contaminated soils and effects on the synthesis of soil microbial biomass. Soil Biology and Biochemistry, 23, 1169-1177.

Chester, P. (1988). The potential for electricity from renewable energy sources. In: National Society for Clean Air, 55th Conference Proceedings. October, Llandudno, 24-27.

Christofides, S., Barlow, C., Michaelides, N., Miranthis, C. (1992) Storm rating in the nineties, General Insurance Study Group, General Insurance Convention of Actuaries, London.

Clark, K. J., Clark, L., Cole, J. A., Slade, S. & Spoel, N. (1992) Effect of Sea Level Rise on Water Resources. WRc plc. National Rivers Authority R and D Note 74.

Climate Change Impacts Review Group [CCIRG] (1991) The Potential Effects of Climate Change in the United Kingdom. Department of the Environment, HMSO, London.

Climatic Research Unit and Environmental Resources Ltd. (1992). Economic Evaluation of Impacts and Adaptive Measures in the EC, London.

Cline, W. R. (1992) The Economics of Global Warming, Institute for International Economics, Washington DC.

Cohen, S. J. (1987) Sensitivity of water resources in the Great Lakes region to changes in temperature, precipitation, humidity and wind speed. In: Solomon, S. I., Beran, M. and Hogg, W. (eds), The Influence of Climate Change and Climate Variability on the Hydrologic Regime and Water Resources. IAHS Publication No. 168.

Coker, A. M., Thompson, P. M., Smith, D. I. and Penning-Rowsell, E. C. (1989) The impact of climate change on coastal zone management in Britain. Conference on Climate and Water, University of Helsinki, Academy of Finland, vol. 2, 148-160.

Cole, J.A. Slade, S., Jones, P.D. & Gregory, J.M. (1991) Reliable yield of reservoirs and possible effects of climate change. Hydrol. Sci. J. 36, 579-598.

Commission of the European Communities (1992) In: Goulding, J.R., Owen Lewis J. and T.C.Steemers J. (eds), Basic climatic data in Energy in Architecture, The European Passive Solar Handbook, Batsford for the CEC, 5-48.

Confederation of British Industry (CBI) (1994), Living with Minerals.

Cook, N. J. (1985) The designers guide to wind loading of building structures, Part 1, Background, damage survey, wind data and structural classification, Building Research Establishment and Butterworths, London.

Cooper, D. M., Wilkinson, W. B. & Arnell, N. W. (1995) The effect of climate change on aquifer storage and river baseflow. Hydrol. Sci. J. 40, 615-631.

Cowden, J. M., Wall, R. G., Adak, G., Evans, H., Le Baigue, S. and Ross, D. (1995) Outbreaks of foodborne infectious intestinal disease in England and Wales: 1992 and 1993. CDR Review, 5 (8), R109-R117.

Craine, N.G., Randolph, S.E. and Nuttall, P. (1995) Seasonal variation in the role of grey squirrels as hosts of Ixodes ricinus, the tick vector of the Lyme disease spirochaete, in a British woodland. Folia Parasitologica, 42, 73-80.

CSERGE (1993) The social cost of fuel cycles, a Report to the UK Department of Trade and Industry, DTI, HMSO, London.

Curwen, M. (1991) Excess winter mortality: a British phenomenon? Health Trends, 22, 169-175.

D'Arifat, J.M. and Warren R.G., (1965) Changes in the amount of organic matter in ley and arable farming systems. Report for Rothamsted Experimental Station for 1958, 42-43.

Data Monitor, (1994) UK Garden Supplies, London.

Davies, A., Shao, J., Bardgett, R.D., Brignall, P., Parry, M.L. and Pollock, C.J. (1994) Specification of climatic sensitivity for UK Farming Systems. Report to MAFF. Institute of Grassland and Environmental Research, Aberystwyth.

Dearnaley, M.P. & Waller, M.N.H. (1993) Impact of Climate Change on Estuarine Water Quality. HR Wallingford. Report to Department of the Environment.

de Freitas, C.R. (1990) Recreation climate assessment. International Journal of Climatology, 10, 89-103.

De Groot, R.S. (1992). Functions of Nature: Evaluation of Nature in Environmental Planning, Management and Decision Making, Wolfers-Noordhott, Groningen.

Department of Energy, (1989). The demand for energy. In: Helm, D., J. Kay and B. Thompson, The Market for Energy, Clarendon Press, Oxford.

Department of the Environment (1988). Possible impacts of climate change on the natural environment in the United Kingdom. Department of the Environment, London.

Department of the Environment (1992) Using Water Wisely. Department of the Environment and Welsh Office. Consultation Paper.

Department of the Environment (1994) Minerals Planning Guidance, Guidelines for Aggregates Provision for England, MPG 6, HMSO, London.

Department of the Environment (1994) Sustainable Development - The UK Strategy. HMSO, London.

Department of the Environment (1994) Advisory Committee on Business and the Environment Seminar: what the City should ask, 16 November (1994), HMSO, London.

Department of the Environment/Welsh Office (1993) Planning Policy Guidance: Coastal Planning. PPG20 HMSO, London.

Department of Trade and Industry (1994) New and Renewable Energy: Future Prospects in the UK, Energy Paper 62. HMSO, London.

Department of Trade and Industry (1994) Regional Trends 29. HMSO, London.

Department of Trade and Industry, (1995a) Digest of UK Energy Statistics 1995. HMSO, London.

Department of Trade and Industry (1995b) Energy Projections for the UK, Energy Paper 65. HMSO, London.

Department of Trade and Industry (1995c) The Energy Report - Volume 1: Competition, Competitiveness and Sustainability. HMSO, London.

Department of Trade and Industry (1995d) The Energy Report - Volume 2: Oil and Gas Resources of the United Kingdom. HMSO, London.

Department of Trade and Industry (DTI) (1993) The Prospects for Coal. Conclusions of the Government's Coal Review, 1993. HMSO, London.

Department of Transport (1989) National Road Traffic Forecast. HMSO, London.

Department of Transport (1994) Transport Statistics Great Britain 1994. HMSO, London.

Department of Transport (1995) Road accidents Great Britain 1994: The casualty report. HMSO, London.

Dewar, R. C. and Watt, A. D. (1992). Predicted changes in the synchrony of larval emergence and budburst under climatic warming. Oecologia 89, 557-559.

Diaz, S., Grime, J. P., Harris, J. and McPherson, E. (1993) Evidence of a feedback mechanism limiting plant response to elevated carbon dioxide. Nature, 364, 616-617.

Dlugolecki A., Clement D., Palutikof J. P., Salthouse R., Toomer C., Turner S., Witt D., (1994) The Impact of Changing Weather Patterns on Property Insurance, Chartered Insurance Institute, London.

Dlugolecki, A., Harrison, P., Leggett, J., Palutikof, J. (1995) Implications for Insurance and Finance. In: Parry, M. and Duncan, R. (eds): The Economic Implications of Climate Change in Britain. Earthscan Publications, London, 83-102.

Doornkamp, J. C. (1993) Clay shrinkage induced subsidence. Geographical Journal, 159, 196-202.

Dukes, M. D. G. and Eden, P. (in press) Climate records and extremes. In: Hulme, M. and Barrow, E.M. (eds.) The Climate of the British Isles: Present, Past and Future. Routledge, London.

Elder, G. H., Hunter, P. R. and Codd, G. A. (1993) Hazardous freshwater cyanobacteria (blue-green algae). Lancet, 341, 1519-1520.

Electricity Council (1988) Annual Report and Accounts 1987/88. Electricity Council, London.

Elmes, G. W. and Free, A. (1994) Climate change and rare species in Britain. HMSO, London.

Emberlin, J., Savage, M. and Jones, S. (1993) Annual variations in grass pollen seasons in London 1961-1990: trends and forecast models. Clin. Exp. Allergy, 23, 911-918.

Emberlin, J. (1994) The effects of patterns in climate and pollen abundance on allergy. Allergy, 49, 15-20.

Eversham, B. C. and Arnold, H. R. (1992) Introductions and their place in British wildlife. In: P. T. Harding (ed.), Biological recording of changes in British wildlife. HMSO, London, 44-59.

Eunson, E. M., (1988) Proof of evidence on system considerations, Hinkley Point 'C' Power Station Inquiry. Central Electricity Generating Board, London.

Evans, A.W. (1995) Major British road accidents: 1946-1994. In Department of Transport (1995) Road accidents Great Britain 1994: The casualty report. HMSO, London.

Evans, J. (1986) A re-assessment of cold hardy Eucalypts in Great Britain. Forestry, 59, 223-242.

Fajer, E. D. and Bazzaz, F. A. (1992). Is carbon dioxide a good greenhouse gas? Effects of increasing carbon dioxide on ecological systems. Global Environmental Change, December, 301-310.

Fankhauser, S. (1995). Protection versus retreat: the economic costs of sea-level rise. Environment and Planning A 27, 299-319.

Fauconnier, C. J., (1992). Fluctuations in barometric pressures as a contributory factor to gas explosions in South African mines. Journal of the South African Institution of Mining and Metallurgy, 92(5), 131-137.

Favis-Mortlock, D. T. (1994) Modelling soil erosion on agricultural land under a changed climate. In: Rounsevell M. D. A. and Loveland P. J. (eds.), Soil Responses to Climate Change. NATO ASI Series Volume 23. Springer Verlag, 211-216.

Firbank, L. G., Watkinson, A. R., Norton, L. R. and Ashenden, T. W. (1995). Plant populations and global environmental change: the effects of different temperature, carbon dioxide and nutrient regimes on density dependence in populations of *Vulpia ciliata*. Functional Ecology, 9, 432-441.

Fitzharris, B. and Garr C. (1995) Climate, Water Resources and Electricity, Greenhouse 94: Proceedings of a conference jointly organised by CSIRO, Australia and the National Institute of Water and Atmospheric Research, New Zealand, CSIRO.

Forestry Commission (1994) Facts and Figures 1993-4. Forestry Commission, Edinburgh.

Foresty Industry Committee of Great Britain (1994) Yearbook 1993-94.

Franz, E. H. (1990) Potential influence of soil organic matter and tropical agroforestry. In: Scharpenseel, H.W., Schomaker, M. and Ayoub, A. (eds), Soils in a Warmer Earth. Elsevier Developments in Soil Science, 20, 109-120.

Gates, W. L., Mitchell, J. F. B., Boer, G. J., Cubasch, U. and Meleshko, V. P. (1992) Climate modelling, climate prediction and model validation. In: Houghton, J. T., Callander, B. A. and Varney, S. K. (eds.) Climate change 1992: the supplementary report to the IPCC scientific assessment. Cambridge Univ. Press, Cambridge, 97-134.

Gilbert, S. and Horner, R. (1984). The Thames Barrier, Thomas Telford, London.

Giles, B. D. and Balafoutis, C. (1990) The Greek heatwaves of 1987 and 1988. International Journal of Climatology, 10, 505-517.

Goldstein, I. F. (1980) Weather patterns and asthma epidemics in New York City and New Orleans, USA. Int. J. Biometeorol., 24, 329-339.

Goossens, D., Offer Z. Y. and Zangvil A. (1993) Wind tunnel experiments and field investigations of eolian dust deposition on photovoltaic solar collectors. Solar Energy, 50 (1), 75-84.

Grant, C. K., Dyer, R. C., and Leggett, I. M. (1995) Development of a new Metocean Design Basis for the N W Shelf of Europe. Paper OTC 7685, Offshore Technology Conference, Houston, May 1995, 415-424.

Grashoff, C., Dijkstra, P., Nonhebel, S., Schapendonk, A. H. C. M. and van de Geijn, S. C. (in press) Effects of climate change on productivity of cereals and legumes: model evaluation of observed year-to-year variability of the CO_2 response. Climate Change Biology.

Graumlich, L. (1991). Subalpine tree growth, climate and increasing CO_2. An assessment of recent growth trends. Ecology, 72, 1-11.

Gregory, J. M. (1993) Sea level changes under increasing atmospheric CO_2 in a transient coupled ocean-atmosphere GCM experiment J. Climate, 6, 2247-2262.

Grime, J. P. (1974) Vegetation classification by reference to strategies. Nature 250, 26-31.

Grime, J. P. (1996) The changing vegetation of Europe: what is the role of elevated carbon dioxide? In: GCTE workshop, August 1994, Leuenberg, Switzerland. (Eds) C. H. Korner and F. Bazzaz. Academic Press.

Grime, J. P., Hodgson, J. G. and Hunt, R. (1988) Comparative plant ecology. Unwin Hyman, London.

Guy, E.C. and Farquhar, R.G. (1991) *Borrelia burgdorferi* in urban parks. Lancet, 338, 253.

Hallegraeff, G.M. (1993) A review of harmful algal blooms and their apparent global increase. Phycologia, 32 (2), 79-99.

Hallett, S.H., Jones, R.J.A. and Keay, C.A., (in press) Environmental information systems developments for planning sustainable land use. International Journal of Geographical Information Systems.

Hallett, S.H., Keay, C.A., Jarvis, M.G. and Jones, R.J.A. (1994) INSURE: Subsidence risk assessment from soil and climate data. AGI 94, 16.2, 1-4.

Hardcastle, R. (1984) The Pattern of Energy Use in the UK. Energy Efficiency Series Number 2, HMSO, London.

Harlfinger, O. 1991 Holiday bioclimatology: a study of Palma de Majorca, Spain. Geojournal, 25 (4), 377-381.

Harrington, R. and Woiwod, I. P. (1995) Insect crop pests and the changing climate. Weather, 50, 200-208.

Harrison, S. J. (1993) Differences in the duration of snow cover on Scottish ski-slopes between mild and cold winters. Scottish Geographical Magazine, 109, 37-44.

Herring, H., R. Hardcastle and R. Phillipson, 1988 Energy Use and Energy Efficiency in UK Commercial and Public Buildings up to the Year 2000. Energy Efficiency Series 6, Energy Efficiency Office, HMSO, London.

Herrington, P. (1995) Climate Change and the Demand for Water. Report to Department of the Environment. University of Leicester.

Hewett, B.A.O., Harries, C.D. and Fenn, C.R. (1993) Water resources planning in the uncertainty of climate change: a water company perspective. In: White, R. (ed.) Engineering for Climatic Change. Thomas Telford: London, 38-54.

Hill, M. O., Wright, S. M., Dring, J. C., Firbank, L. G., Manchester, S. J. and Croft, J. M. (1994) The potential for spread of alien species in England following climatic change. English Nature Research Report, No 90. English Nature, Peterborough.

Hill, S. A., and Walpole, B. J. (1989) National and local spread of barley-yellow mosaic virus in the United Kingdom. OEPP/EPPO Bulletin 19, 555-562.

HM Government (1994) Climate Change: the UK Programme. HMSO, London.

Hodgson, J. M. (1976) Soil Survey Field Handbook. Soil Survey Technical Monograph, No. 5. Harpenden, England.

Hollis, J. M. (1989) A methodology for predicting soil wetness class. SSLRC Research Report for the Ministry of Agriculture, Fisheries and Food, Silsoe.

House of Commons Transport Committee (1994) Transport related air pollution in London. HMSO, London.

Howard, P. J. A. and Howard, D. M. (1994) Modelling effects of land use change and climate change on soil organic carbon stores. In: Howard P. J. A. et al. (eds.), Carbon Sequestration by Soils in the UK. Institute of Terrestrial Ecology report to Department of the Environment. ITE, Merlewood.

Hughes, T. H. (1978) Effect of the environment on processing and handling materials at sea. Trans. R. Soc. London, 290, 161-177.

Hulme, M. (1994) Validation of large-scale precipitation fields in General Circulation Models. In: Desbois, M.and Désalmand, F. (eds.), Global precipitations and climate change. Springer-Verlag, Berlin, 387-405.

Hulme, M., Conway, D., Brown, O. and Barrow, E. M. (1994) A 1961-90 baseline climatology and future climate change scenarios for Great Britain and Europe: Part III, Climate change scenarios for Great Britain and Europe. A Report prepared for the TIGER IV3, a Consortium, Landscape dynamics and climate change, Climatic Research Unit, Norwich.

Hunt, R. (1992) Vegetation and climate: a 35-year study in road verges at Bibury, Gloucestershire, and a 21-year study in chalk grassland at Aston Rowant, Oxfordshire. NERC Unit of Comparative Plant Ecology, University of Sheffield.

Hunt R., Dyer, R. H and Driscoll, R. (1991) Foundation movement and remedial underpinning in low rise buildings. BRE Report.

International Society of Soil Science (1990) Do Soils Matter? Wageningen, The Netherlands.

Institution of Structural Engineers (1994) Subsidence of low rise buildings, London.

Institute of Terrestrial Ecology (1995) Carbon sequestration in vegetation and soils. Report to Department of the Environment Global Atmosphere Division, March 1995.

IPCC (1990) Climate change: the IPCC scientific assessment, Houghton, J. T., Jenkins, G. J. and Ephraums, J. J. (eds). Cambridge University Press, Cambridge.

IPCC (1992a) Climate change 1992: the supplementary report to the IPCC scientific assessment Houghton, J. T., Callander, B. A. and Varney, S. K. (eds.) Cambridge University Press, Cambridge.

IPCC (1992b) Global Climate Change and the Rising Challenge of the Sea, World Meteorological Organisation and United Nations Environment Programme, Geneva.

IPCC, (1994). An evaluation of the IPPC IS92 emission scenarios, in Climate Change 1994, Cambridge University Press, Cambridge.

IPCC (1995) Climate change 1994: radiative forcing of climate change and an evaluation of the IPCC IS92 emissions scenarios (eds.) Houghton,J.T., Meira Filho,L.G., Bruce,J., Lee,H., Callander, B.A., Haites,E., Harris,N. and Maskell,K., Cambridge University Press, Cambridge.

IPCC (1996a) Climate change 1995: the science of climate change, (eds.) Houghton,J.T., Meira Filho,L.G., Callander,B.A., Kattenburg, A. and Maskell,K., Cambridge University Press, Cambridge.

IPCC (1996b) Climate Change 1995: Impacts, adaptations and mitigations: Scientific-technical analysis (eds) Watson, R., Zinyowera, M. and Moss, R., Cambridge University Press, Cambridge.

Jacoby, H.D. (1990) Water quality. in Waggoner, P.E. (ed.) Climate Change and U.S. Water Resources. Wiley: New York, 307-328.

Jarvis, N.J. and Mullins, C.E. (1987) Modelling the effects of drought on the growth of Sitka spruce in Scotland. Forestry 60, 13-30.

Jarvis, P.G. (in press) The role of temperate trees and forests in CO_2 fixation. Vegetatio.

Jeffree, E.P. and Jeffree, C.E. (1994) Temperature and the biogeographical distributions of species. Functional Ecology 8, 640-650.

Jenkins, A., McCartney, M.P., Sefton, C. & Whitehead, P. (1993) Impacts of Climate Change on River Water Quality in the UK. Institute of Hydrology. Report to Department of the Environment.

Jenner, C.F. (1991) Effects of exposure of wheat ears to high temperature on dry matter accumulation and carbohydrate metabolism in the grains of two cultivars. II. Carry-over effects. Australian Journal of Plant Physiology, 18, 179-190.

Jensen, P. H. and F. J. L. Van Hulle (1991) In: Recommendations for an European wind turbine standard load case, E.L. Petersen (ed.). RISO contribution from the Department of Meteorology and Wind Energy to the ECWEC'90 Conference, Madrid, RISO, 33-38.

Joergensen, R.C., Brookes, P.C. and Jenkinson, D.S. (1989) Survival of the soil microbial biomass at elevated temperatures. Soil Biology and Biochemistry 22, 1129-1136.

Johnston, A.E. (1973) The effects of ley and arable cropping systems on the amount of soil organic matter in the Rothamsted and Woburn ley-arable experiments. Report for the Rothamsted Experimental Station for 1972, Part 2, 131-159.

Jones, C. G. and Coleman, J. S. (1991). Plant stress and insect herbivory: towards an integrated perspective. In: H. A. Mooney, W. E. Winner and E. J. Pell, (Eds.) Response of plants to multiple stresses. Academic Press, San Diego, California.

Jones, P. D. (1994) Hemispheric surface air temperature variability - a reanalysis and an update to 1993 J. Climate, 7, 1794-1802

Jones, R.G., Murphy, J. M. and Noguer, M. (in press) Simulation of climate change over Europe using a nested regional climate model Part 1: assessment of control climate, including sensitivity to location of lateral boundaries Quart. J. Royal Meteor. Soc.

Jones, R. J. A. and Thomasson, A. J. (1985) An Agroclimatic Databank for England and Wales. Soil Survey Technical Monograph No. 16, Harpenden, England.

Kalkstein, L. S. (1993) Direct impacts in cities. Lancet, 342, 1397-1399.

Kalkstein, L. S. (1995) Lessons from a very hot summer. Lancet, 346, 857-859.

Kattenburg, A., Giorgi, F., Grassl, H., Meehl, G. A., Mitchell, J. F. B., Stouffer, R. J., Tokioka,T., Weaver, A. J. and Wigley, T. M. L. (1996) Climate models - projections of future climate in: Houghton,J.T., Meira Filho,L.G., Callander,B.A., Kattenburg, A. and Maskell,K. (eds.). Climate change 1995: the science of climate change. Contribution of WGI to the Second Assessment Report of the Intergovernmental Panel on Climate Change. Cambridge University Press, Cambridge.

Kelly, H., (1993) Introduction to photovoltaic technology. In: Renewable Energy: Source for Fuels and Energy, T.B. Johansson et al. (eds.), Island Press, Washington DC, 297-336.

Kelly, P. M. (1991) Global warming: implications for the Thames Barrier and associated defences. in Frassetto, R. (ed.) Impact of Sea Level Rise on Cities and Regions. Marsilio: Venice, 93-98.

Kerstiens, G., Townend, J. A. and Mansfield, T. A. (in press). Effects of water and nutrient availability on physiological responses of woody species to elevated CO_2. Forestry.

Kilbourne, E. (1992) Illness due to thermal extremes. In: Public Health and Preventative Medicine, 13th edition Norwalk, Appleton Lange, 491-501.

Kimball, B. A. (1993) Carbon dioxide and agricultural yield: an assembly and analysis of 430 prior observations. Agronomy Journal, 75, 779-788.

Kramer, K. (1995) Phenotypic plasticity of the phenology of seven European tree species in relation to climatic warming. Plant, Cell and Environment, 18, 93-104.

Langford, I. H. and Bentham, G. (1995) The potential effects of climate change on winter mortality in England and Wales. Int. J. Biometeorol., 38, 141-147.

Lawlor, D. W. and Mitchell, R. A. C. (1991) The effects of increasing CO_2 on crop photosynthesis and productivity: a review of field studies. Plant Cell and Environment, 14, 807-818.

Layton, M., Parise, M. E., Campbell, C. C., Advani, R., Sexton, J. D., Bosler, E. M. and Zucker, J. R. (1995) Mosquito-transmitted malaria in New York City. Lancet, 346, 729-731.

Leatherman, S. P. (1989) Beach response strategies to accelerated sea-level rise. In: Topping, J. C. (ed) Coping with Climate Change. Climate Institute, Washington, DC.

Lee, E. M. (1993). The political ecology of coastal planning and management in England and Wales: policy responses to the implications of sea-level rise. The Geographical Journal, 159, 169-178.

Leeds-Harrison, P. B. and Rounsevell, M. D. A. (1993) The impact of dry years on crop water requirements in Eastern England. Journal of the Institute of Water and Environmental Management, 7, 497-505.

Leggett, J., Pepper, W. J. and Swart, R. J. (1992) Emissions scenarios for the IPCC: an update, In: Houghton, J. T., Callander, B. A. and Varney, S. K. (eds.), Climate change 1992: the supplementary report to the IPCC scientific assessment, Cambridge Univ. Press, Cambridge, 75-95.

Lettenmaier, D. P. and Sheer, D. P. (1991) Climatic sensitivity of California water resources. Journal of Water Resources Planning and Management, 117 (1), 108-125.

Linden, F. (1962) Merchandising weather. The Conference Board Business Record, 19(6), 15-16.

Lines, R. (1987) Choice of seed origins for the main forest species in Britain. Forestry Commission Bulletin 66. HMSO, London.

London Business School (1995) The City Research Project: The competitive position of London's Financial Services. Final Report, March 1995. Corporation of London.

Long, S. P. (1991). Modification of the response of photosynthetic productivity to rising temperature by atmospheric CO_2 concentration: has its importance been underestimated? Plant Cell and Environment, 14, 729-739.

Luxmoore, R. J. (1991). A source-sink framework for coupling water, carbon and nutrient dynamics of vegetation. Tree Physiology, 9, 267-280.

Lynette, R. and Associates (1992) Assessment of Wind Power Station Performance and Reliability. Report EPRI TR-100705, Electric Power Research Institute, Palo Alto.

McBoyle, G. R., Harrison, R., Wall, G., Kinnaird, V. and Quinlan, C. (1986) Recreation and climate change: a Canadian case study. Ontario Geography, 28, 51-68.

McMichael, A. J. (1993a) Global environmental change and human population health: A conceptual and scientific challenge for epidemiology. Int. J. Epidemiol., 22, 1-8.

McMichael, A. J. (1993b). Planetary Overload. Global Environmental Change and the Health of the Human Species. Cambridge University Press, Cambridge.

McMichael, A. J. and Martens, W. J. M. (1995) Assessing health impacts of global environmental change: grappling with scenarios, predictive models, and uncertainty. Ecosystem Health, 1, 15-25.

MacGillivray, C. W. and Grime, J. P. (in press) Genome size predicts frost resistance in British herbaceous plants: implications for rates of vegetation response to global warming. Functional Ecology.

Marsh, T. J., Monkhouse, R. A., Arnell, N. W., Lees, M. L. & Reynard, N. S. (1993) The 1988-92 Drought. Institute of Hydrology and British Geological Survey, Wallingford.

Martens, W. J. M., Niessen, L. W., Rotmans, J., Jetten, T. H. and McMichael, A. J. (1995) Potential impact of global climate change on malaria risk. Environ. Hlth. Pers., 103, 458-464.

Martin, P. and Lefebvre, M. (1995) Malaria and climate: sensitivity of malaria potential transmission to climate. Ambio, 24 (4), 200-207.

Maunder, W. J. (1986) The Uncertainty Business: Risks and Opportunities in Weather and Climate. Methuen, London.

Mikolajewicz, U., Santer, B. D. and Maier-Reimer, E. (1990) Ocean response to greenhouse warming Nature, 345, 589-593.

Milbank, N., (1989) Building design and use: response to climate change. Architects Journal, 96, 59-63.

Miller, B. A. *et al.*, (1992) Impact of Incremental Changes in Meteorology on Thermal Compliance and Power System Operations, Report No. WR28-1-680-109, Tennessee Valley Authority Engineering Laboratory, Norris, Tennessee.

Ministry of Agriculture, Fisheries and Food. (1993a). Strategy for Flood and Coastal Defence. MAFF, London.

Ministry of Agriculture, Fisheries and Food. (1993b). Coastal Defence Works and the Environment: A Guide to Good Practice. MAFF, London.

Ministry of Agriculture, Fisheries and Food, (1993c) Code of Good Agricultural Practice for the Protection of Soil. MAFF, London.

Mintel, (1991) Special Report - Holidays, London.

Mitchell, A. F. (1981) The native and exotic trees in Britain. Arboricultural Research Note 29. Department of the Environment Arboricultural Advisory and Information Service, London.

Mitchell, J. F. B., Johns, T. C., Gregory, J. M. and Tett, S. F. B. (1995) Transient climate response to increasing sulphate aerosols and greenhouse gases. Nature, 376, 501-504.

Mitchell, R. A. C., Lawlor, D. W., Mitchell, V. J., Gibbard, C. L., White, E. M. and Porter, J. R. (1995) Effect of elevated CO_2 concentration and increased temperature on winter wheat: test of ARCWHEAT 1 simulation model. Plant Cell and Environment, 18, 736-748.

Morison, J. I. L. and Butterfield, R. E. (1990) Cereal crop damage by frosts, Spring 1990. Weather, 45, 308-313.

Moutinho, L. (1988) Amusement park visitor behaviour - Scottish attitudes. Tourism Management, 292-300.

Muir, J. F., Smith, A. and Young, J. A. (in press) Aquaculture, economics and development: A European perspective. In: Proceedings, Annual Meeting European Association of Fisheries Economists, Portsmouth, April 1995 (in press).

Mundy, C. (1990) Energy. In: Climate Change: Impacts on New Zealand, New Zealand Ministry of the Environment.

Murphy, J. M. (1995) Transient response of the Hadley Centre coupled ocean-atmosphere model to increasing carbon dioxide. Part I, control climate and flux adjustment J. Climate, 8, 36-56.

Murphy, J. M. and Mitchell, J. F. B. (1995) Transient response of the Hadley Centre coupled ocean-atmosphere model to increasing carbon dioxide. Part II, spatial and temporal structure of the response. J. Climate, 8, 57-80.

Murray, M.B., Smith, R. I., Leith, D., Fowler, D., Lee, H. S. J., Friend, A. D. and Jarvis, P. G. (1994) Effects of elevated CO_2, nutrition and climatic warming on bud phenology in Sitka spruce (*Picea sitchensis*) and their impact on the risk of frost damage. Tree Physiology, 14, 691-706.

Myers, N. (1993) Environmental refugees in a globally warmed world. Bioscience, 43, 752-761.

Myers, N. And Kent, J. (1995) Environmental Exodus: An Emergent Crisis in the Global Arena. Climate Institute, Washington, DC.

National Academy of Sciences, (1992) Policy Implications of Greenhouse Warming, National Academy Press, Washington DC.

National Rivers Authority (1990) Toxic blue-green algae. Water quality series, No 2, September 1990.

National Rivers Authority (1994) Water: Nature's Precious Resource. NRA. March 1994.

National Rivers Authority, (NRA) (1995), Anglian Region. Rebuilding of the Anglian Coastline, 1995.

National Westminster Bank (1994) Nat West Review of Small Business Trends, 4 (2). Small Business Research Trust, London.

NGC Settlements, (1995). Monthly power pool data. Nottingham.

Nie, G. Y., Long, S. P. and Webber, A. N. (1993) The effect of nitrogen supply on down-regulation of photosynthesis in spring wheat grown in an elevated CO_2 environment. Abstract 785, Plant Physiology 102, 5-138. (Now Plant Cell env, 18).s

Nishioka, S., H. Harasawa, H. Hashimoto, T. Okita, K. Masuda, and T. Morita, (1993) The potential effects of climate change in Japan. CGER-I009-'93, Center for Global Environmental Research,

Nuttall, P., Randolph, S. E., Carey, D., Craine, N., Livesley, A. and Gern, L. (1994) The ecology of Lyme Borreliosis in the UK. In: Axford, J.S. and Rees, D.H.E. (eds). Lyme Borreliosis. Plenum Press, New York, 125-129.

Oke, T. R. (1973) City size and urban heat island. Atmos. Environ., 7, 769-779.

Oke, T. R. (1987) Boundary layer climates. Methuen, New York.

Page, J. K. (1992) The impact of climate change in the urban environment of the UK, In: Proc. Cities and Global Change Conference, Toronto, Jun. 1991, Published by the Climatic Institute, Washington, DC, USA 22-42.

Palutikof, J. P. (1983) The impact of weather and climate on industrial production in Great Britain. Journal of Climatology, 3, 65-79.

Parker, D.E., Legg,T. P. and Folland, C.K. (1992) A new daily Central England Temperature series, 1772-1991 Int. J. Climatol., 12, 317-342

Parr, T. and Eatherall, A. (1994) Demonstrating climate change impacts in the UK. The Department of the Environment Core Model Programme. Natural Environment Research Council, Institute of Terrestrial Ecology.

Parry, M.L. and Read, N.J. (1988). The impact of climatic variability on UK industry. Air Report, 1, Atmospheric Impacts Research Group, University of Birmingham, Birmingham.

Parry, M.L. and Rosenzweig, C. (1993) Food supply and the risk of hunger. Lancet, 342, 1345-1347.

Parry, M.L. and Duncan, R. (Eds) (1995) The economic implications of climate change. Earthscan, London.

Parry, M.L., Hossell, J. E., Jones, P. J., Rehman, T., Tranter, R. B., Marsh, J. S., Rosenzweig, C., Fischer, G. and Carson, I. G. (1996). Integrating global and regional analyses of the effects of climate change: a case study of land use in England and Wales. Climatic Change, 32, 185-198.

Pears, N.V. (1967) Present treelines of the Cairngorm Mountains, Scotland. Journal of Ecology 45, 401-439.

Penman, H.L. (1948) Natural evaporation from open water, bare soil and grass. Proc. Roy. Soc. of London (A), 193, 120-146.

Penning-Rowsell, E.C., Green, C.H., Thompson, P.M., Coker, A.M., Tunstall, S.M., Richards, C. and Parker, D.J. (1992). The Economics of Coastal Management: A Manual of Benefit Assessment Techniques. Belhaven Press, London.

Perry, A.H. (1986) A theme for tourism. Geographical Magazine, 58, 2-3.

Perry, A.H. (1987) Why Greece melted. Geographical Magazine, 59, 430-431.

Perry, A.H. (1993) Climate and weather information for the package holiday-maker. Weather, 48, 410-414.

Perry, A.H. and Ashton, S. (1994) Recent developments in the UK's outbound package tourism market. Geography, 79, 313-321.

Perry, A. and Symons, L. (1994) The wind hazard in Great Britain and its effect on road and air transport. Journal of Wind Engineering and Industrial Aerodynamics, 52, 29-41.

Pollock, C. J. and Eagles, C. F. (1988) Low temperature and the growth of plants. In: Long S. P. and Woodward F.I. (eds) Plants and Temperature, Society for Experimental Biology Symposium, 42, 157-180.

Pollock, C., Davies, A., Harrison, J., Turner, L., Gallagher, J. and Parry, M. (in press) Climate change and its implications for grassland agriculture in the United Kingdom: A problem of scale. British Grassland Society Centenary Meeting.

Ponce de Leon, A., Anderson, H.R., Bland, J.M., Strachan, D.P. and Bower, J. (in press). The effects of air pollution on daily hospital admissions for respiratory disease in London: 1987-88 to 1991-92. J. Epidemiol. Comm. Hlth.

Poorter, H. (1993). Interspecific variation in the growth responses of plants to an elevated CO_2 concentration. Vegetatio, 104, 77-97.

Porter, J. R. (1993) AFRCWHEAT2: A model of the growth and development of wheat incorporating responses to water and nitrogen. European Journal of Agronomy, 2, 69-82 .

Powlson, D. S., Brookes, P. C., and Christensen, B. T. (1987) Measurement of soil microbial biomass provides an early indication of changes in total soil organic matter due to straw incorporation. Soil Biology and Biochemistry, 19 (2), 159-164.

Powlson, D. S., Saffinga, P. G. and Kragt-Cottat, M. (1988) Denitrification at sub-optimal temperatures in soils from different climatic zones. Soil Biology and Biochemistry, 20, 719-723.

Public Health Laboratory Service (1995) Communicable Disease Report: Surveillance of waterborne diseases. CDR Weekly, 5 (28), 129.

Pye, K. and French, P. W. (1993). Targets for coastal habitat re-creation. English Nature Science 13, English Nature, Peterborough.

Quine, C. P., Coutts, M. P., Gardiner, B. A. and Pyatt, D. G. (1995) Forests and Wind: Management to minimise damage. Forestry Commission Bulletin 114. HMSO, London.

Radesovich, L. G. and Skinrood, A. C. (1989), The power production operation of Solar One, the 10 MWe solar thermal central receiver pilot plant, Journal of Solar Energy Engineering, 111 (2), 145-151.

Raper, S. C. B., Wigley, T. M. L. and Warrick, R. A. (1996) Global sea-level rise: past and future. In: Milliman, J.D. and Haq, B.U. (eds.), Sea-level and coastal subsidence: causes, consequences and strategies, Kluwer Academic Publishers, Dordrecht, The Netherlands, 11-45.

Robson, J. D. and Thomasson, A. J. (1977) Soil Water Regimes: A Study of Seasonal Waterlogging in English Lowland Soils. Soil Survey Technical Monograph No. 11, Harpenden, England.

Roe, G. M. (1990) Report on the breach to the sea wall in Towyn, North Wales on 26 February 1990. Paper given at the Ministry of Agriculture, Fisheries and Food, Conference of River and Coastal Engineers on 1989-90 Floods, July 1990, Loughborough.

Rosenzweig, C. and Parry, M. L. (1994) Potential impact of climate change on world food supply. Nature, 367, 133-138.

Ross, L. G., Mendoza, E. A. Q. M. and Beveridge, M. C. M. (1993) The application of Geographical Information Systems to site selection for coastal aquaculture: An example based on salmonid cage culture. Aquaculture, 113, 165-178.

Rounsevell, M. D. A. (1993) A review of soil workability models and their limitations in temperate regions. Soil Use and Management, 9, 15-21.

Rounsevell, M. D. A. and Jones, R. J. A., (1993) A soil and agroclimatic model for estimating machinery workdays: the basic model and climatic sensitivity. Soil and Tillage Research, 26, 179-191.

Rounsevell, M. D. A. and Brignall, A. P. (1994) The potential effects of climate change on autumn soil tillage opportunities in England and Wales. Soil and Tillage Research, 32, 275-289.

Rounsevell, M. D. A., Jones, R. J. A, and Brignall, A. P. (1994) Climate change effects on autumn soil tillage opportunities and crop potential in England and Wales. Proceedings 13th International Soil Tillage Research Organisation Conference, Aalborg, Denmark, 1175-1180.

Royal Commission on Environmental Pollution (1994) 18th report: Transport and the environment. CM2674, HMSO, London.

Sakai, S. (1988) The impact of climate variation on secondary and tertiary industry in Japan. Meteorological Research Note 180, 163-173.

Santer, B. D., Wigley, T. M. L., Barnett, T. P. and Anyamba, E. (1996) Detection of climate change and attribution of causes. In: Houghton, J. T., Meira Filho, L. G., Bruce, J., Lee, H., Callander, B. A., Haites, E., Harris, N. and Maskell, K. (eds.). The second IPCC scientific assessment of climate change. Contribution of WGI to the second assessment report of the Intergovernmental Panel on Climate Change, Cambridge University Press, Cambridge, 407-443.

Seiler, W. and Conrad, R. (1987) Contribution of tropical ecosystems to the global budgets of trace gases, especially CH_4, H_2, CO and N_2O. In: R.E. Dickinson (ed.), The Geophysiology of Amazonia. Vegetation and Climate Interactions, Wiley and Sons, New York, 133-160.

Seligman, N. G. and Sinclair, T. R. (1995) Climate change, interannual weather differences and conflicting responses among crop characteristics: the case of forage quality. Global Change Biology, 1, 157-160.

Semenov, M. A. and Porter, J. R. (1995) Climatic variability and the modelling of crop yields. Agricultural and Forest Meteorology, 73, 265-283.

Shand, J. (1995) Braer found adrift in stormy waters. Page 15, Post Magazine, 2 November 1995.

Shand, J. (1995) Tillinghast Insurance Pocket Book, 1995 NTC Publications, London.

Shennan, I. (1989) Holocene crustal movements and sea-level changes in Great Britain J. Quaternary Research, 4, 77-90.

Shope, R. (1991) Global climate change and infectious diseases. Environ. Hlth. Persp., 96, 171-174.

Singh, B., (1987) Impacts of CO_2-induced climate change on hydro-electric generation potential in the James Bay Territory of Quebec. In: The Influence of Climate Change and Climate Variability on the Hydrologic Regime and Water Resources, S.I.

Skea, J. (1992) Physical impacts of climate change. Energy Policy, 20 (3), 269-272.

Skea, J. (1995) 'Energy' In: Parry M. and Duncan R. (eds.), Economic Implications of Climate Change in Britain. Earthscan, London.

Smith, J. B. and D. A. Tirpak (eds.) (1989) The Potential Effects of Global Climate Change on the United States, EPA-230-05-89, Office of Policy, Planning and Evaluation, US Environmental Protection Agency, Washington DC.

Smith, K. (1990) Tourism and climate change. Land Use Policy April, 176-180.

Smith, K. (1993) The influence of weather and climate on recreation and tourism. Weather 48, 398-404.

Solomon, M. Beran and Hogg W. (eds.), IAHS Publication No. 168.

Steele, A.T. (1951) Weather's effect on the sales of a department store. Journal of Marketing, 15, 436-443.

Sterba, H. (1995) Forest decline and increasing increments: a simulation study. Forestry, 68, 153-167.

Stitt, M. (1991). Rising CO_2 levels and their potential significance for carbon flow in photosynthetic cells. Plant, Cell and Environment 14, 741-762.

Streif, H. (1990). Quaternary sea-level changes in the North Sea: an analysis of amplitudes and velocities. In: Brosche, P. and Sundermann, J. (eds.), Earth's Rotation from Eons to Days. Springer-Verlag, Berlin, 201-214.

Sutherst, R.W. (1993) Arthropods as disease vectors in a changing environment. In: Lake, J. V. (ed) Environmental Change and Human Health. Ciba Foundation Symposium. John Wiley and Sons, New York, 124-145.

Taylor, K. E. and Penner, J. E. (1994) Response of the climate system to atmospheric aerosols and greenhouse gases. Nature, 369, 734-737.

Tester, R. F., Morrison, W. R., Ellis, R. H., Piggott, J. R., Batts, G. R., Wheeler, T. R., Morison, J. I. L., Hadleigh, P. and Ledward, D. A. (1995) Effects of elevated growth temperature and carbon dioxide levels on some physiochemical properties of wheat starch. Journal of Cereal Science, 22, 63-71.

Thomas, H. and Norris, I. B. (1977) The growth responses of Lolium perenne to the weather during winter and spring at various altitudes in mid-Wales. Journal of Applied Ecology, 14, 949-964.

Thomas, J. A., Moss, D. and Pollard, E. (1994) Increased fluctuations of butterfly populations towards the northern edges of species' ranges. Ecography, 17, 215-220.

Thomasson, A. J. (1979) Assessment of Soil Droughtiness. In: M.G. Jarvis and D. Mackney (eds.): Soil Survey Applications. Soil Survey Technical Monograph No. 13, Harpenden, England, 43-50.

Thomasson, A. J. and Jones, R. J. A. (1992) An empirical approach to crop modelling and the assessment of land productivity. Agricultural Systems, 37, 351-367.

Thornes, J. E. (1983) The effect of weather on attendance at sports events. In: Bale, J. and Jenkins, C. (Eds) Geographical Perspectives on Sport, Workshop Paper, University of Birmingham, 201-209.

Thornes, J. (1991) Applied climatology: severe weather and the insurance industry. Progress in Physical Geography, 15, 173-181.

Tinker, P. B. and Ineson, P. (1990) Soil organic matter and biology in relation to climate change. In: Scharpenseel, H.W., Schomaker, M. and Ayoub, A (eds.), Soils on a Warmer Earth. Elsevier Developments in Soil Science 20, 71-88.

Titus, J. G. (1992) The costs of climate change to the United States. In: S.K. Majumdar, L.S. Kalkstein, B. Yarnal, E.W. Miller, and L.M. Rosenfeld (eds) Global Climate Change: Implications, challenges and Mitigation measures, Pennsylvania Academy of Science.

Tooley, M. J. and Turner, R. K. (1995) The effects of sea level rise. In. Parry, M. and Duncan, R. (eds) The Economic Implications of Climate Change in Britain, Earthscan, London, 8-27.

Treharne, K. (1989) The implications of the 'greenhouse effect' for fertilizers and agrochemicals. In: R.M. Bennet (ed.) The greenhouse effect and UK agriculture. Centre for Agricultural Strategy, Reading, 67-78.

Turner, R. K. (1988) Wetland conservation: economics and ethics. In. Collard, D., Pearce, D. W. and Ulph, D. (eds.) Economics, Growth and Sustainable Environments. Macmillan, London, 121-159.

Turner, R.K., Adger, N. and Doktor, P. (1995a) Assessing the economic costs of sea level rise. Environment and Planning A 27, forthcoming.

Turner, R.K., Subak, S. and Adger, N. (1995b) Pressures, trends and impacts in the coastal zone. Environmental Management, forthcoming.

Tyler, A.L., Macmillan, D.C. and Dutch, J. (in press) Models to predict the general yield class of Douglas fir, Japanese larch and Scots pine on better quality land in Scotland. Forestry.

UK Renewables Energy Advisory Group (1992) Report to the President of the Board of Trade, Department of Trade and Industry. HMSO, London.

US Department of Energy (1989) Controlling summer heat islands, Proc. of the Workshop on saving energy and reducing atmospheric pollution by controlling summer heat islands. Berkeley, California, Feb. 23-24, 1989., U.S. Department of Energy and Financial Assistance, Washington, D.C.

US Environmental Protection Agency (in press) Preliminary Assessment of the Benefits to the US of Avoiding or Adapting to Climate Change. US Environmental Protection Agency, Climate Division, Washington, DC.

Viner, D. and Hulme, M. (1994) The Climate Impacts LINK Project: providing climate change scenarios for impacts assessment in the UK DoE/CRU, Norwich.

Wall, G. (1992) Tourism alternatives in an era of global climate change. In: Smith, V. L. and Eadington, V.R. (eds), Tourism Alternatives, J. Wiley, London.

Wall, G. and Badka, G. (1994) Tourism and climate change: an international perspective. Journal of Sustainable Tourism, 2, 193-203.

Wardlaw, I. F., Dawson, I. A., Munibi, P. and Fewster, R. (1989) The tolerance of wheat to high temperature during reproductive growth. Australian Journal of Agricultural Research, 40, 1-13.

Warrick, R. A., Barrow, E. and Wigley, T. M. L. Climate and sea level change: observations, projections and implications Cambridge University Press, Cambridge.

Webb, B.W. (1992) Climate Change and the Thermal Regime of Rivers. Report to Department of the Environment, University of Exeter.

Weller, S. and Prior, J. (1993) The weather sensitivity analysis in manufacturing and retailing. Proc. 1st European Conference on Applications of Meteorology, Oxford, 27 September-1 October, 1993.

Whittle, I. R. (1990) Lands at risk from sea-level rise in the UK. In: The Greenhouse Effect and Rising Sea Level in the UK (Ed. J.C. Doornkamp), M1 Press, Long Eaton, 85-94.

Wigley, T. M. L. (1994) Outlook becoming hazier. Nature, 369, 709-710.

Wigley, T. M. L. (1995) Global mean temperature and sea level consequences of greenhouse gas concentration stabilization Geophys. Res. Letts., 22, 45-48.

Wigley, T. M. L. and Raper, S. C. B. (1987) Thermal expansion of sea water associated with global warming. Nature, 330, 127-131.

Wigley, T. M. L. and Raper, S. C. B. (1992) Implications of revised IPCC emissions scenarios. Nature, 357, 293-300.

Wilby, R. L. (1994) Stochastic weather type simulation for regional climate change impact assessment Water Resources Research, 30, 3395-3403

Williams, M., Shewry, P. R., Lawlor, D. and Harwood, J. L. (1995) The effects of elevated temperature and atmospheric carbon dioxide concentration on the quality of grain lipids in wheat (*Triticum aestivum* L.) grown at two levels of nitrogen application. Plant Cell and Environment, 18, 999-1009.

Williamson, R.B. and Beveridge, M.C.M. (1994) Fisheries and aquaculture. In: The Freshwater Resources of Scotland: A National Resource of National Significance (Eds. P.S. Maitland, P.J. Boon and D.S. McLusky), John Wiley and Sons Ltd., Chichester, 317-332.

Willis, A. J., Dunnett, N. P., Hunt, R. and Grime, J. P. (in press) Does Gulf Stream position affect vegetation dynamics in western Europe? Oikos, 70.

World Coast Conference (1994). Preparing to Meet the Coastal Challenges of the 21st Century. Conference Report, World Coast Conference 1993, November, Noordwijk, Netherlands.

World Meteorological Organisation (1984) Urban climatology and its applications with special regard to tropical areas. WMO Publication No 652, WMO, Geneva, Switzerland.

Worrell, R. and Malcolm, D.C. (1990) Productivity of Sitka spruce in Northern Britain. 1. The effects of elevation and climate. Forestry, 63, 105-118.

Wurr, D. C. E. (1994) Global warming: impact on crops. Grower, 122, 29-31.

Zeisel, H. (1950) How temperature affects sales. Printers Ink, 223, 40-42.

Annex I

Glossary

Abstractions
Withdrawals of water from rivers or aquifers

Acclimation
Species performance in particular environmental conditions

Airport malaria
When the transmission of malaria occurs around international airports due to persons being bitten by an infected mosquito from an incoming aircraft

Algal blooms
A reproductive explosion of algae in a lake, river or ocean

Biodiversity
Variation of living organisms at the genetic, species and population levels

Biogeochemical processes
Processes which alter chemical composition of a substance (such as water)

Biological oxygen demand (BOD)
The consumption of oxygen by biogeochemical processes

CIBSE Guide
The standard design handbook containing climatological data used by engineers in the design of engineering services for construction

Climate sensitivity
The eventual surface warming of the Earth following an instantaneous doubling of atmospheric CO_2 concentration

Condensation
In construction refers to the deposition of water on building surfaces either internally or within construction elements from the surrounding moist air

Day-degrees
The number of degrees per day by which temperature exceeds a stated threshold, usually 5.6 °C

Denitrification
The conversion of nitrate to nitrogen gas or to other gases such as nitrous oxide which are unavailable to plants and lost from the soil

De-oxidation
Removal of oxygen

Desorption
Soils contain organic matter and clay minerals which have the properties to adsorb nutrients, would-be pollutants and gases. Under certain conditions, this process can be reversed and components that were formerly held may be desorbed i.e. released

Determinate (crops)
Crops in which each successive growth stage is finite

Driving rain
In the construction industry refers to rainfall driven by the wind onto the external surfaces of construction

Ecophysiology
Study of physiological responses to environmental variables

Endemic
Occurrence of an organism in a specified region

Eutrophication
Enrichment with mineral nutrients, especially with reference to freshwaters

Exotic species
Species introduced to areas in which they do not naturally occur

Greenhouse gas forcing
The radiative perturbation to the atmosphere resulting from changed concentrations of greenhouse gases

Groundwater recharge
Recharge of groundwater resources from rainfall

Incidence
The number of cases of illness commencing, or of persons falling ill, during a given period of time within a specified population

Interdeterminate (crops)
Crops where growth stages are not finite and continue as long as conditions permit

Ion exchange
The transfer of electrically charged atoms or groups of atoms between the soil solution and soil particle surface or *vice versa*

Irradiance
The quantity of light energy reaching a surface

Just-in-time
The procedure by which business units do not hold large stocks of goods (raw materials for manufacturing, or retail commodities in the retail sector) but rely on the supplier to provide the required amounts rapidly on request

Leaching
The removal of substances in solution or suspension by water percolating through the soil

Migration
The spread of species and their colonisation of new areas

Mineralisation
Release of mineral components of organic matter during its decomposition

Nitrogen deposition
Wet and dry deposits of gaseous and aerosol nitrogenous compounds from the atmosphere

Off-line reservoir
Reservoir filled by pumping from a river

On-line reservoir
Reservoir filled directly from a river

Oxidation-reduction
Conditions in the soils alternating between aerobic (oxidising) and anaerobic (reducing) which influence the chemical nature of several soil processes

Photoperiod
Length of daylight within a 24 hour period

Phenology
The timing of recurring natural events, such as flowering and budburst

Phytotoxicity
Toxic to plants

Potential evaporation
The amount of evaporation demanded by the atmosphere and which would be satisfied if enough water were available

Prevalence
the proportion of persons within a population who are currently affected by a particular disease

Process-based models
Mathematical expression of the physiological responses of organisms to environmental stimuli

Provenance
Location from which seed is collected

Rainfall-runoff model
Computer model to simulate the translation of rainfall into streamflow

Re-aeration
Addition of oxygen

Return period
The average interval between the occurrence of events of a specified magnitude

Rhizosphere
The zone around plant roots in the soil in which the microbial population lives

Root exudates
Roots exude a wide range of substances including amino acids, organic acids and sugars which provide a source of nutrients for the large number of bacteria which occupy the root zone

Rotation
The period from establishment to harvesting of a tree stand

Roundwood
Harvested tree stems before manufacturing

Saline intrusion
Intrusion of salt water into fresh water

Salinisation
The enrichment of soils with salts

Salmonellosis
Intestinal disease caused by salmonella bacteria.

Smectite
A group of clay minerals which have the particular capacity to shrink when dry and to swell when re-wetted

Soil moisture deficit
A soil moisture content less than the maximum water holding capacity of the soil, a situation which occurs when evapotranspiration exceeds rainfall

Soil moisture balance
The balance between water inputs into the soil (precipitation, irrigation) and water losses (evapotranspiration, run off, drainage)

Specification
A widely used design guide used in the support of design by architects

Water distribution network
The network of service reservoirs and mains pipes that distributes water from source to consumer

Windthrow
Uprooting of trees by wind forces

Vector
An organism which acts as an essential intermediate or definite host for a human pathogen and plays an active role in its transmission

Annex 2

Acronyms

ΔSL	change in sea level	CSO	Central Statistical Office
ΔT	change in temperature	DANI	Department of Agriculture, Northern Ireland
BBSRC	Biotechnological and Biological Sciences Research Council	DTI	Department of Trade and Industry
BCC	British Coal Corporation	EC	European Community
BERG	Building Effects Review Group	ECHAM	European Centre/Hamburg climate model (Germany)
BGS	British Geological Survey	FACE	Free Air Carbon dioxide Enrichment
BMRC	Bureau of Meteorology Research Centre (Australia)	FICGB	Forestry Industry Committee of Great Britain
BP	British Petroleum Company plc	FPSO	Floating Production Storage and Offloading
BRE	Building Research Establishment	GCM	Global climate model
BREEAM	Building Research Establishment Environmental Assessment Method	GDDs	Growing degree days, defined here as accumulated mean temperature above a base of 5.5°C
BSI	Building Standards Institute		
C	Carbon	GDP	Gross Domestic Product: the value of the total economic activity taking place in the UK territory
CBI	Confederation of British Industry		
CCC	Canadian Climate Centre		
CCGT	Combined cycle gas turbine	GFDL	Geophysical Fluid Dynamics Laboratory (USA)
CCIRG	Climate Change Impacts Review Group	GHG	Greenhouse gas
CDDs	Cooling degree days, defined here as accumulated mean temperature above a base of 18°C	GIS	Geographical Information Systems
		GISS	Goddard Institute for Space Studies (USA)
CEC	Commission of the European Communities	HCFCs	Hydrocloroflourocarbons
CFCs	Chloroflourocarbons	HDDs	Heating degree days, defined here as accumulated mean temperature below a base of 15.5°C
CH_4	Methane		
CHP	Combined heat and power	hPa	Hectopascals, a measurement unit for vapour or atmospheric pressure
CIBSE	Chartered Institution of Building Services Engineers		
		HRLAM	High resolution limited area model
CO_2	Carbon dioxide	IPCC	Intergovernmenal Panel on Climate Change
COI	Central Office of Information		
CSERGE	Centre for Social and Economic Research on the Global Environment	IS92a	International emissions scenario a, defined by the IPCC in 1992
		K	Potassium
CSIRO	Commonwealth Scientific and Industrial Research Organisation (Australia)	KT/sq.km	Kilotonnes/square kilometre

LLNL	Lawrence Livermore National Laboratory (USA)	**SOAFD**	Scottish Office Agriculture and Fisheries Department
MAFF	Ministry of Agriculture, Fisheries and Food	**SPECTRE**	Spatial and Point Estimates of Climate change due to TRansient Emissions. A statistical model developed by the Climatic Research Unit, University of East Anglia, in 1994.
MAI	Mean annual increment i.e. tree volume production divided by age - a measure of productivity		
MAGICC	Model for the Assessment of Greenhouse gas Induced Climate Change	**SSSI**	Sites of Special Scientific Interest
MNC	Multi-national corporation - a business with headquarters in one country but with offices/production plants/distribution outlets, etc. in a number of other countries and with products and services generally sold internationally.	**Tmin**	Minimum temperature
		UKCS	United Kingdom Continental Shelf
		UKOOA	United Kingdom Offshore Operators Association Limited
		UKHI	UK Meteorological Office high resolution equilibrium experiment
MVA	Millivolt-amps	**UKLO**	UK Meteorological Office low resolution equilibrium experiment
ms^{-1}	Metres per second (windspeed)		
m^{3}ha^{-1}yr^{-1}	Cubic metres per hectare per year	**UKTR**	UK Meteorological Office high resolution transient experiment
NERC	Natural Environment Research Council		
		UVB	Ultra-violet radiation
NFFO	Non-fossil fuel obligation	**WMO**	World Meteorological Organisation
NGC	National Grid Company	**Wm^{-2}**	Water per meter square
NGL	Natural Gas Liquids		
NNR	National Nature Reserves		
NOx	Oxides of Nitrogen		
NRA	National Rivers Authority		
N$_2$O	Nitrous oxide		
OFWAT	Office of Water Services		
OSU	Oregon State University (USA)		
P	Phosphorous		
PE	Potential evapotranspiration		
ppmv	Parts per million by volume, a measurement unit for gas concentrations		
PV	Photovoltaics		
SME	Small and medium sized enterprises variously defined by number of employees (DTI definition) or by annual (financial) turnover (banking definition). A typical SME employs under 200 people and has an annual turnover of less than £5 million		
SO$_2$	Sulphur dioxide		

Annex 3

Membership of Climate Change Impacts Review Group

Professor M.L. Parry *(Chairman)*
Jackson Environment Institute
University College London
5 Gower Street
London WC1E 6BT

Dr. N. Arnell
Department of Geography
University of Southampton
Highfield
Southampton SO17 1BJ

Professor P. Bullock
Soil Survey and Land Research Centre
Cranfield University
Silsoe
Bedford
MK45 4DT

Dr. M.G.R. Cannell
Institute of Terrestrial Ecology
Bush Estate
Penicuik
Midlothian EH26 0QB

Dr. A.F. Dlugolecki
General Accident Fire and Life Assurance
Corporation, plc
Pithealvis
Perth PH2 0NH

Dr. M. Hulme
Climatic Research Unit
University of East Anglia
Norwich NR4 7TJ

Professor A.J. McMichael
London School of Hygiene and Tropical Medicine
University of London
Keppel Street
London WC1E 7HT

Dr. D. Malcolm
Institute of Ecology and Resource Management
Schools of Forestry and Ecological Sciences
University of Edinburgh
Darwin Building, King's Buildings
Mayfield Road
Edinburgh EH9 3JU

Professor J. Page
Emeritus Professor
University of Sheffield

Dr. J. Palutikof
Climatic Research Unit
University of East Anglia
Norwich NR4 7TJ

Dr. A. Perry
Department of Geography
University of Wales Swansea
Singleton Park
Swansea SA2 8PP

Professor C. Pollock
Institute of Grassland and Environmental Research
Plas Gogerddan
Aberystwyth
Dyfed SY23 3EB

Professor D. Potts
Department of Mineral Resources Engineering
University of Nottingham
University Park
Nottingham NG7 2RD

Professor J.F. Skea
Science Policy Research Unit
Mantell Building
University of Sussex
Falmer
Brighton BN1 9RF

Professor K. Smith
School of Natural Sciences
University of Stirling
Stirling FK9 4LA

Ms. H. Thompson
Environmental Management

National Westminster Bank plc
1st Floor
41 Lothbury
London EC2P 2BP

Dr. M. Tight
Institute for Transport Studies
University of Leeds
Leeds LS2 9JT

Professor K. Turner
CSERGE
University of East Anglia
Norwich NR4 7TJ

Dr. P. Bramwell *(Executive Secretary)*
Department of the Environment
Romney House
43 Marsham Street
London SW1P 3PY

Dr. S.M. Cayless *(Executive Secretary)*
Building Research Establishment
Garston
Watford WD2 7JR

Mrs. C.J. Parry *(Technical Secretary)*
Jackson Environment Institute
University College London
5 Gower Street
London WC1E 6BT

Annex 4

List of persons consulted

Chapter 2 (Changing Climate and Sea Level) Lead Author: M. Hulme.
Persons consulted: B. Atkinson; E. Barrow; M. Beran; O. Brown; J. Gregory; J. Harrison; P. Jones; M. Kelly; A. Olecka; S. Raper; D. Viner; P. Whetton; T. Wigley; R. Wilby.

Chapter 3 (Soils) Lead Author: P. Bullock.
Persons consulted: M. Beran; E. Bridges; P.C. Brookes; J. Gould; K. Goulding; P. Gregory; N. Konijn; D.S. Powlson; D. Rimmer; M.D.A. Rounsevell; K. Smith.

Chapter 4 (Flora, Fauna and Landscape) Lead Author: M.G.R. Cannell.
Persons consulted: M. Beran; J. Grime; M. Hill; H. Kruuk; I. Newton; S. Woodin.

Chapter 5 (Agriculture, Horticulture and Aquaculture) Lead Author: C. Pollock.
Persons consulted: R. Baxter; M. Beran; M. Beveridge; P. Brignall; P. Costigan; J. Farrar; M. Haines; D. Lawlor; S. Long; T. Mansfield; J. Morison; M. Parker; S. van de Geijn; D. Wilkins; M. Williams.

Chapter 6 (Forestry) Lead Author: D. Malcolm.
Persons consulted: M. Beran; J. Evans; P.G. Jarvis; A. Petty; A. Rook; P. Savill.

Chapter 7 (Water Resources) Lead Author: N. Arnell.
Persons consulted: M. Beran; C. Binnie; A. Brook; S. Cohen; G. Davies; C. Fenn; M. Knowles; M. Sitton; G. Spragge.

Chapter 8 (Energy) Lead Author: J.F. Skea.
Persons consulted: N. Adger; M. Beran; N. Burdett; J.P. Finnegan; C. Grezo; D. Hall; I. McPherson; G. Manners; D. Western.

Chapter 9 (Minerals Extraction) Lead Author: D. Potts.
Persons consulted: A. Bann; M. Beran; H. Cohen; A. Goode; C.K. Grant; E. Hassall; D.E. Highley; M. Jones; H. Lunt; J. McLaughlin; G. Manners; M.J. Richards; W. Rowell; J.K.W. Stevenson; B.G.S. Taylor; M.A. Tuck.

Chapter 10 (Manufacturing, Retailing and Service Industries) Lead Author: J. Palutikof.
Persons consulted: M. Beran; Q. Chiotti; D. Etkin; G. Manners; S. Weller; B. Whittaker.

Chapter 11 (Construction Industry) Lead Author: J. Page.
Persons consulted: M. Beran; M. Bury; J. Clarke; M.J. Hotchkiss; G. Henderson; G.K. Jackson; P. Jones; N. Lawrence; J. Miller; H. Nahapiet; P. O'Sullivan; L.J. Parrot; C.H. Saunders; H. Thomas.

Chapter 12 (Transport) Lead Author: M. Tight.
Persons consulted: R. Allsopp; M. Beran; A. Bristow; P. Goodwin; F. Hodgson; M. Page; A. Perry; H. Pyne; Y. Ling Siu; J. Thornes.

Chapter 13 (Insurance Industry) Lead Author: A.F. Dlugolecki.
Persons consulted: M. Beran; C. Brett; S. Christofides; D. Clement; R. Keeling; J. Lynch; J. Parker; S. Reilly; C. Toomer.

Chapter 14 (Financial Sector) Lead Author: H. Thompson.
Persons consulted: M.Beran; M. Kelly; M. Pummell; T. Tennant.

Chapter 15 (Health) Lead Author: A.J. McMichael.
Persons consulted: G. Bentham; M. Beran; D. Bradley; C. de Freitas; P. Epstein; A. Haines; R.S. Kovats; J. Powles; M. Rowland; D. Warhurst.

Chapter 16 (Recreation and Tourism) Lead Authors: A. Perry and K. Smith.
Persons consulted: J. Adams; M. Beran; E. Brogan; T. Mackie; V. Middleton; R. Morrison-Smith; T. Whittome; S.C. Woodward.

Chapter 17 (Coastal Regions) Lead Author: K. Turner.
Persons consulted: M. Beran; L. Bijlsma; E. Bird; B. McCartney; R. Nicholls; T. O'Riordan; D.W. Pearce; E. Penning-Rowsell; D. Pugh; R. Purnell; A. Stebbing; S. Subak; R. Willows.